HYDRAULIC ENGINEERING

By GARDNER D. HISCOX, M.E.

www.KnowledgePublications.com

HYDRAULIC ENGINEERING

A TREATISE ON THE PROPERTIES, POWER AND RESOURCES OF WATER FOR ALL PURPOSES

INCLUDING THE MEASUREMENT OF STREAMS; THE FLOW OF WATER IN PIPES OR CONDUITS; THE HORSE-POWER OF FALLING WATER; TURBINE AND IMPACT WATER-WHEELS; WAVE-MOTORS; CENTRIFUGAL, RECIPROCATING, AND AIR-LIFT PUMPS

BY

GARDNER D. HISCOX, M.E.

AUTHOR OF "MECHANICAL MOVEMENTS," "GAS, GASOLINE, AND OIL ENGINES," "COMPRESSED AIR AND ITS USES," "MODERN STEAM ENGINEERING," ETC.

300 ILLUSTRATIONS WITH 36 PRACTICAL TABLES

NEW YORK

THE NORMAN W. HENLEY PUBLISHING COMPANY

132 NASSAU STREET

1908

ISBN:978-1-60322-019-4

COMPOSITION, ELECTROTYPING, AND PRINTING
BY THE TROW PRESS, NEW YORK, U. S. A.

www.KnowledgePublications.com

PREFACE

THE need of a general yet compact treatise on hydraulic engineering has long been recognized by students and engineers. In the writer's endeavor to supply such a volume he now submits after presenting a brief technical sketch of the development of hydraulic engineering from the earliest times, a systematic and progressive statement of the mechanics of water and fluids in general, including hydrostatics or the equilibrium of fluids, hydrodynamics which treats of the laws of liquids in motion, and hydraulics in which the motion of water in pipes and canals is considered. It has been the writer's constant purpose to make every detail perfectly clear and to supply the necessary formulas in their simplest expression, which are further explained by figured examples, that the reader seeking information may have his search fully rewarded. A farmer, for example, wishing to improve his property by installing suitable machinery for domestic water-supply, for fire protection, or the selection and installation of a water-wheel for operating labor-saving machines, will find sufficient information to enable him to proceed intelligently. The mechanic will be informed as to the construction and operation of hydraulic machines; the mechanics of water and of fluids in general will be sufficiently covered for his needs; the general design as well as the minor details of accumulators, pumps, presses, and machine tools are clearly illustrated and described.

The student desiring an easily comprehended statement of the mathematical theory of the motion of fluids proceeding to the consideration of the motion of water in pipes and canals, and finally to the practical application of fluid pressures in combination with suitable trains of mechanism adapted to any given problem, will doubtless find this book useful as a preliminary guide to a complete understanding of all the practical questions involved.

To the engineer this book will be a ready reference in the construction of dams and storage reservoirs for irrigation, city and domestic water-supply, or for driving water-wheels for manufacturing purposes. Illustrations are given showing the construction of the simpler form of log and timber dams, followed by illustrated descriptions of the more permanent ones of masonry and concrete. The illustrations are accompanied by the formulas for stability, completely worked out. The preliminary data necessary to the location and construction of water-works for town and city supply are outlined, such as the sources of supply, size of reservoirs, settling basins, and filtration systems, and the consumption of water by families in cities. The flow of water in pipes for water-supply, the friction and slope elements of canals, ditches, and pipe-lines, for irrigation supply and its distribution, of growing importance to agricultural industry, is presented in form for engineers and will be easily understood by the general reader.

The essential features of Artesian wells and the geology of the Artesian areas of the United States are fully discussed and illustrated. So also the work of the United States Reclamation Service in the irrigation of the arid districts of the Western States has been described and illustrated.

The air-lift method of raising water has been given an entire chapter for its complete presentation; especial prominence being given the Pohlé Air-Lift, which is fully described, including single and multi-stage applications with illustrations showing the arrangement of air- and water-pipes; rules for calculating the volume of air required for raising water are also given.

The various water-wheels for the utilization of water-power are illustrated and described, with rules for calculating the horse-power for given conditions. Impact water-wheels, now extensively used in the far West, are illustrated together with a description of the methods employed for determining the power of a jet, and formulas for determining its dynamic force. A table giving the power of small motors is added. Turbine water-wheels are described and illustrations of high-powered wheels at Niagara Falls.

Centrifugal, Rotary, and Screw pumps are sufficiently explained that the general reader may easily understand their operation, and formulas are given for measuring the work which may be expected of Centrifugal pumps. Reciprocating pumps, of which several varieties are illustrated, are accompanied by formulas for horse-power required for raising a given quantity of water to any given height; so also the relative areas for steam- and water-cylinders, concluding with a sectional elevation of the Allis Pumping Engine and a synopsis of duty trial of the St. Louis Triple Expansion Pumping Engine, which yielded the remarkable duty of more than 181 million foot-pounds.

A number of tables, some thirty-six in all, will be found useful for reference, including, as they do: the properties of water, coefficients for hydraulic grades, discharge of water from orifices and nozles, pressure lost by friction in fire hose, velocity discharge and horse-power of nozles, flow of water over weirs, loss of head by friction of water in pipes, etc.

GARDNER D. HISCOX.

April, 1908.

CONTENTS

CHAPTER XII

CHAPTER XIII

CHAPTER XIV

CHAPTER XV

CHAPTER XVI

CHAPTER XVII

CHAPTER XVIII

CHAPTER XIX

CHAPTER XX

LIST OF ILLUSTRATIONS

CHAPTER I

HISTORICAL—INTRODUCTION

Hydraulics, its theory and practice, like other of the arts and their science, has an interesting history, bearing upon the development of civilization and its culmination at different periods in the progress of the human race from its earliest stages. But although we deplore the want of exact information relating to details of the arts in general at the several periods of ancient civilization, it is probable that few of their crude devices for raising water and for irrigation have been wholly lost.

The early periods of civilization seem to have culminated in regal pomp and show, and the people existed only in the frugal and simple life. The only arts that appear to have been progressive to a point of perfection were architecture and sculpture; while inventive progress in mechanics and especially in the methods of water-supply seems to have been in abeyance for many centuries and only came into prominence through architectural and engineering methods for supplying the great cities of those times.

The power that might have been developed from the falls and barrages of their streams seems not to have invaded their mental capacity, and so their working implements remained the same simple models as those of their ancestors, and are still the same over a large portion of our habitable globe.

The implements of husbandry, modes of irrigation, and devices for raising water are similar to those in use when Ninus and Nebuchadnezzar, Sesostris, Solomon, and Cyrus flourished.

The most ancient implements for conveying water and used for a drinking-cup was of Nature's origin, and is still the model of beauty in our vases and other art vessels; the gourd was long in use, followed by pottery, glass, and metal utensils modelled after its lines, of which the remains of this ancient type are still abundant.

Of this beautiful form of household water service, illustrated in

Fig. 1, the three at the left are from Thebes in Egypt; the fourth, Etruscan, the people of Italy; fifth, an ancient Chinese model, and at the right the pitcher model of the Egyptians.

These models prevailed in the common and precious metals throughout the Greek and Roman periods and to the present day.

Egyptian. Etruscan. Chinese. Egyptian

FIG. 1.—Ancient Water Implements.

In the earlier ages water was dipped from the brooks and large streams and drawn from wells of even great depth by cords and in vessels of various forms and carried upon the head or in pairs on yokes. Many of the wells had circular steps on the inside of the walls by which the water-carriers descended to the water surface.

FIG. 2.—Water-carrier.

FIG. 3.—At the well.

In ancient Alexandria, where the arts were cultivated and science flourished to an extent perhaps unequalled in any older city, water was drawn up from the cisterns, with which every house was provided, with the simple cord and bucket. This city was supplied with water from the Nile; it was admitted into vaulted reservoirs or cisterns, which were constructed at the time the foundations of the city were laid by Alexander. They were sufficiently capacious to contain water for a whole year, being filled only at the annual inundation of the river through a canal made for the purpose. Apertures or well openings, through which the water was raised from these reservoirs, are still to be seen. Whole lines of ancient streets are traceable by the wells, recurring every six or seven yards, by which the contiguous houses, long since crumbled away, drew water from the vast cisterns with which the whole city was undermined. The pavement of the old city is from ten to thirty feet below the surface of the modern streets, and excavations are frequently made by the Pasha's workmen for the stones of the old pavement and of the buildings. In this manner the marble mouths of the vaulted reservoirs or cisterns are frequently brought to light, and they invariably exhibit traces of the ropes used for raising the water.

The agricultural pursuits of man must at a very early period have convinced him of the value of water in increasing the fruitfulness of the soil; he could not but observe the fertilizing effects of rain, and the rich vegetation consequent on the periodical inundation of rivers. Hence Nature taught man the art of irrigating land, and confirmed him in the practice of it by the benefits it invariably produced.

Babylonia, which was chiefly watered by artificial irrigation, was the most fruitful of all the countries visited by Herodotus.

Fig. 4 represents the practice of irrigation at an early age in Egypt, taken from the sculptures at Beni Hassan.

Fig. 5 represents the method of water-supply in India from the earliest ages.

The advantages of artificial irrigation have not only been known from the earliest ages, but some of the most stupendous works which the intellect of man ever called into existence were designed for that purpose: works so ancient as to perplex our chronologists, and so vast as to incline some historians to class them among natural formations. Ancient writers unite in asserting that Lake Mœris was "the work of men's hands," and constructed by a king of that name; its prodigious extent, however, has led some modern authors to question its alleged origin, although artificial works, still extant, equal it in the amount of labor required, as the Wall of China, the Pyramids, and other works of ancient Egypt. Sir William Chambers, when comparing the works of the remote ancients with those of Greece, observes that the city of Babylon would have covered all Attica; that a greater number of men were employed in building it than there were inhabitants of Greece; that more materials were consumed in a single Egyptian Pyramid than in all the public structures of Athens; and that Lake Mareotis could have deluged the Peloponnesus, and ruined all Greece. But incredible as the accounts of Lakes Mœris and Mareotis may appear, these works did not surpass, if they equalled, another example of Egyptian engineering, which had previously been executed. This was the removal of the Nile itself! In the reign of

Fig. 4.—Irrigation, Egypt.

Fig. 5.—Water-supply, India.

Menes (the first, or one of the first sovereigns) it swept along the Libyan chain of mountains, that is, on one side of the valley that constitutes Egypt; and in order to render it equally beneficial to both sides, a new channel was formed through the centre of the valley, into which it was directed: an undertaking which indicates a high degree of scientific knowledge at that early period.

Before the lakes and canals of Egypt or China could have been undertaken, the inhabitants must have been long under a regular government, and one which could command the resources of a settled people, and of a people, too, who from experience could appreciate the value of such works for the purpose of irrigation, as well as the inefficiency of previous devices for the same object; that is, of machines for raising the water—for if it be supposed the construction of canals to convey, of reservoirs to contain, and of locks and sluices to distribute water, preceded the use of machines for raising it, it would be admitting that men in ignorant times had the ability to conceive and the skill to execute the most extensive and perfect works that civil engineering ever produced—to have formed lakes like oceans and conveyed rivers through deserts ere they well knew how to raise water in a bucket or transmit it through a pipe or a gutter. The fact is, ages must necessarily have elapsed before such works could have been dreamed of, and more before they could have been accomplished. Individuals would naturally have recourse to rivers in their immediate vicinity, from which (the Nile, for example) they must long have toiled in raising water before they would ever think of procuring it from other parts of the same stream at distances vary-

FIG. 6.—Hand-irrigation. The Méntal.

ing from ten to a hundred miles, or consent to labor for its conveyance over such extensive spaces.

How extremely ancient, then, must hydraulic machinery be in Egypt, when such works as we have named were executed in times that transpired long before the commencement of history—times that have been considered as extending back to the infancy of the world!

However, from the information derived from sculptured monuments and legends, which fail to furnish evidence of the engineering methods of construction of the greater works for irrigation and water-supply in the early ages of those countries, we can only imagine that their civilization must have been far more advanced than is shown in the legends of history. Their pictured work in its primitive details is at hand, and we must be content to imagine their methods in their greater structures.

A small trench is dug on the edge of the river, on the borders of which two men stand opposite each other. They hold in each hand a cord, the ends of which are attached to a basket of palm leaves covered with leather. After launching it into the water, they lean backward so as to be half seated on small mounds of earth raised for the purpose, by which the weight of the body assists in raising the load, as it is swung toward the gutter or basin formed on the bank to receive it. The movements of the men are regulated by chanting, a custom of great antiquity, and adopted in all kinds of manual labor where more than one person was engaged.

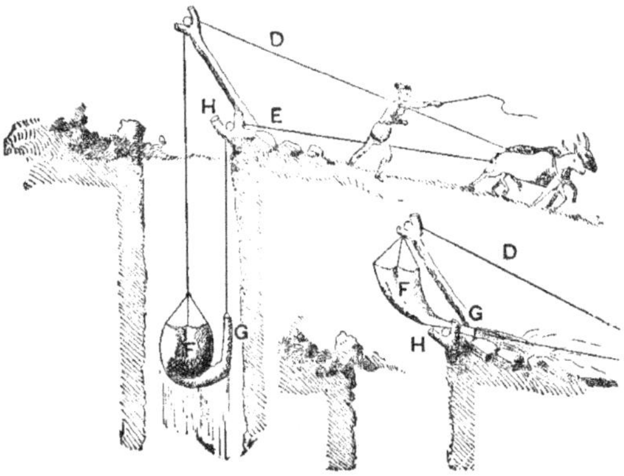

FIG. 7.—Oriental irrigation works.

The raising of water from deep wells for irrigation, as practised in Egypt, Assyria, Babylonia, Persia, and India, is illustrated in Fig. 7, and yet in use, was a most ingenious device for the age in which it originated. It consisted of an ox-hide bucket and spout drawn by oxen with lines D and E so arranged that the spout line was stopped at H and the bucket raised to automatically empty the water into the conveying trough.

FIG. 8.—Theban swape.

Of machines for raising water, the swape has been more extensively used in all ages, and by all nations, than any other. In Fig. 8 is a sketch of a swape which in Egypt is named the shadoof, from the sculptures at Thebes, as used in the time of Moses, 1550 B.C., a period extending beyond the Exodus. A pole is used for connecting the top of the swape by a rope and the bale of the bucket for the purpose of more readily sinking the bucket.

This ancient device may still be seen operated in its primitive form in the United States.

On the upper Nile where the banks are high the swapes are stepped with intervening tanks, so that irrigation is possible on the higher plateaus.

Of the swape, it may be remarked that the most ancient portraiture extant of any hydraulic machine is a sculptured representa-

FIG. 9.—High-lift swapes on the Nile.

tion of it, between three and four thousand years old, and even at that remote period it was in all probability a very old affair, and in common use. These sculptures, moreover, prove that it has remained in Egypt unaltered in its form, dimensions, mode, and material of construction and methods of using it during at least thirty-four centuries! and this, notwithstanding the political convulsions to which that country has ever been subject since its conquest by Cambyses; its inhabitants having been successively under the Persian, Grecian, Roman, Saracenic, and Turkish yoke, thus literally fulfilling a prophecy of Ezekiel that "there shall be no longer a prince of the land of Egypt"—a descendant of its ancient kings; yet through all these mighty revolutions that have swept over it like the fatal simoom, and destroyed every vital principle of its ancient grandeur, this simple machine has passed through them all unchanged, and is still applied by the inhabitants to the same purposes, and in precisely the same way for which it was used by their more enlightened progenitors.

We have seen it used by the Greeks and Romans, and we find it still in the possession of their descendants, wherever they dwell, as well as among those of more ancient people, the Hindoos, Arabs, and Chinese. And although we may be unable to keep it constantly in view in Europe in those ages which immediately followed the fall of the Roman power, when the ferocious tyranny of the Saracens established a despotism over the mind as well as the body, and by the characteristic zeal of Omar entailed ignorance on the future by consuming the very sources of knowledge under the baths of Alexandria; yet when in the fifteenth century the human intellect began to shake off the lethargy which during the long night of the dark ages had paralyzed its energies, and printing was introduced—that mighty art which is ordained to sway the destinies of our race forever—among the earliest of printed books, with illustrations, this interesting implement may be found portrayed in vignettes, in views of cities and of rural life, tangible proofs of its universal use throughout Europe at that time, as well as during the preceding ages.

Although allusions to machines for raising water have been found in the works of several ancient authors, it does not appear that any general account or comprehensive treatise of them was ever written by the ancients. If such a work was executed, it has perished in the general wreck of ancient records.

About the beginning of the Christian era, a Roman architect and engineer, Vitruvius, published a brief description of some hydraulic engines of the water-lifting type, viz., the Tympanum, Noria, Chain of Pots, the Screw of Archimedes, and the Machine of Ctesibius or pump. He wrote at a period the most favorable for acquiring and transmitting to posterity a knowledge of the mechanic arts of the ancient civilized nations; for he flourished during the last scenes of the mighty drama when Rome had become the arbitress of the world, and the enlightened nations of the East—their wealth, learning, arts, and artisans—were prostrate at her feet; so that if the previous intercourse of the Romans with Asia Minor, Egypt, Carthage, and Greece had not made them familiar with the arts of those countries, nothing could have prevented them from possessing such knowledge when they became Roman provinces.

The water-raising wheels described by Vitruvius are circular revolving frames driven by the current impinging upon a set of buckets and blades as shown, by which the Tympanum (Fig. 10) discharges

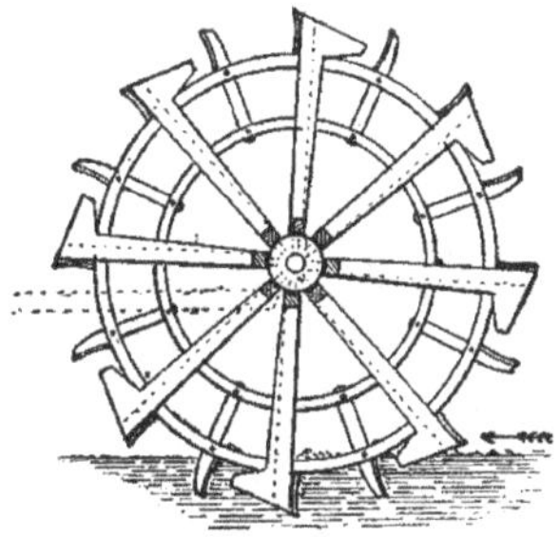

FIG. 10.—Tympanum.

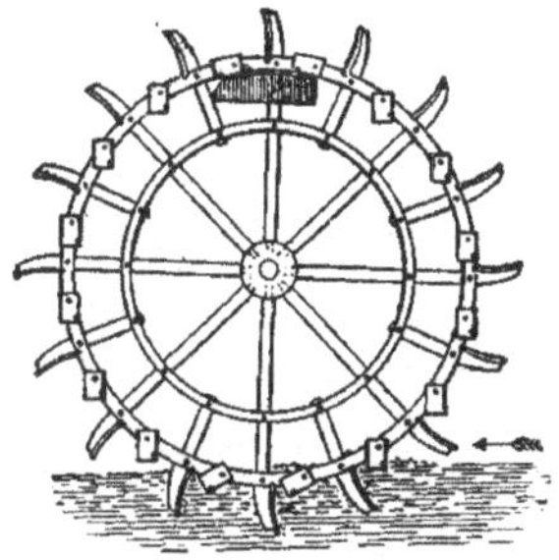

FIG. 11.—Noria.

the water through a side outlet in the hollow arms to a trough just below the axle. The Noria, of the same general construction, is provided with swinging buckets on the side of the wheel that are tipped by contact with a trough discharging the water near the top of the wheel.

The Persian wheel has long rectangular buckets across the rim of the wheel with side orifices that discharge into a trough near the top

FIG. 12.—Persian wheel.

FIG. 13.—Screw lift.

of the wheel. The screw lift of Archimedes was driven by a current wheel; but was also operated in still water by hand and animal power.

The chain of pots, operated by a sprocket-wheel of crude form, was in use at an early period by all the nations of antiquity, and is still employed to a great extent in Egypt, Asia, and Europe, and in the form of the chain-pump, with buckets enclosed in a tube, is now in use in all countries.

The old chain-and-bucket-pump as depicted by Vitruvius, and its use credited by Heron in his "Spiritalia," is illustrated in Fig. 14, and a later model used by the Moors, derived from the earlier forms of Egypt and Asia Minor, is illustrated in Fig. 15. The ox and the ass had been advanced to the continuous work of irrigation.

The piston-pump described by Vitruvius, and its use credited by Heron in his "Spiritalia" as used for extinguishing fires, does not establish its use for water-supply, either domestic or for irrigation; it rested in its crude form throughout the centuries of the dark ages, when at the dawn of the modern epoch the science of mechanics advanced its construction for all purposes.

FIG. 14.—Roman chain-pump.

The fountain of Heron is the oldest pressure-engine known, and in it a volume of air is used as a substitute for a piston. It is not certain that it was invented by him, for it may have been an old device in his time, and one which he thought worthy of preservation, or of being made more extensively known, and therefore inserted an account of it in his book. The two vessels A B, of any shape, are made air-tight. The top of the upper one is formed into a dish or basin, in the centre of which the jet pipe is inserted, its lower end extending to near the bottom of A; a pipe, C, whose upper orifice is soldered to the basin extends down to near the bottom of the lower vessel, either passing through the top of B, as in the figure, or inserted at the side. Another pipe, D, is connected to the top of B, and continued to the upper part of A. This pipe conducts the air from B to A. Now suppose the vessel A filled with water, through an aperture made for the purpose, and which is then closed; the object is to make this water ascend through the jet, and it is accomplished thus: water is poured into the basin, and of course it runs down the pipe C into B; and as it rises in the latter, the air within is necessarily compressed, and having no way to escape but up the pipe D, it ascends into the upper part of A, where, being pressed on the surface of the water, the latter is compelled to ascend through the jet pipe, as shown in the cut.

A pressure-engine (Fig. 17) on the principle of Heron's fountain was erected for draining the mines at Chemnitz in Hungary about the

FIG. 15.—Moorish chain-pump.

middle of the eighteenth century. From a spring at 140 feet above the mouth of the mine shaft, which was 104 feet in depth, a difference of about 16 pounds pressure was obtained as a power for the drainage of the mine. The chamber B is submerged that the water of the pump may enter the chamber by gravity through the cock or valve at 4; the pressure being off by opening 2 and closing 1; 3 having been opened for discharging the water in A; with 2, 3, and 4 closed and 1 open, the apparatus represents Heron's fountain and the water in the lower chamber is discharged at E. In this apparatus 18 cubic feet of water are raised 104 feet and discharged at each operation.

During all the centuries of the dark ages of war and devastation and up to the beginning of the eighteenth century, we find no recorded efforts to derive power from running or falling water to operate industrial machinery, and not until the mechanical idea enlightened the minds of Bacon, Worcester, Papin, and others did the mechanical arts seem to have taken a permanent footing in the industrial world.

The early progress in architectural construction seems to have led to the aqueduct system of water-supply of the great cities of ancient

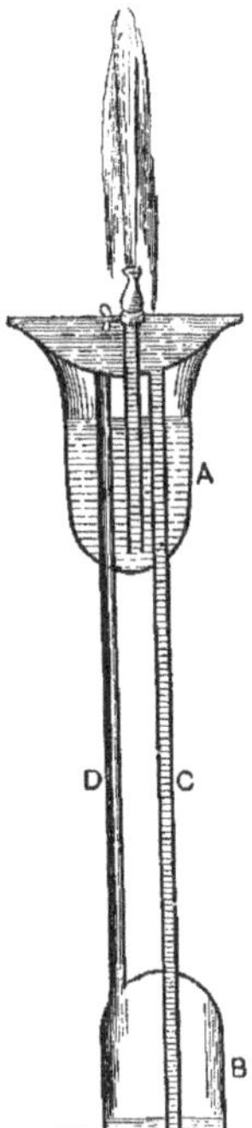

FIG. 16.—Heron's fountain.

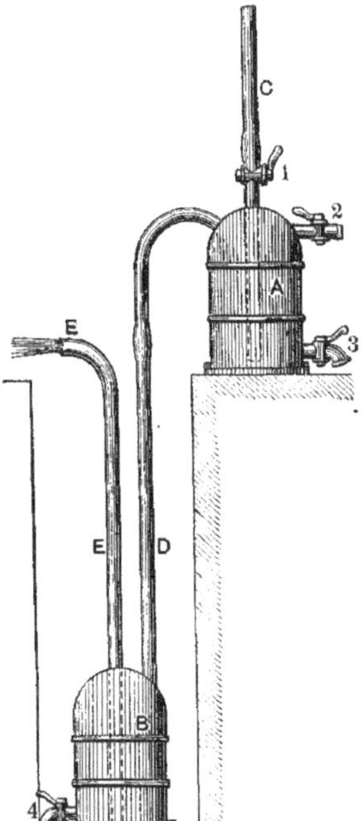

FIG. 17.—Pressure-engine, Chemnitz.

times. If history furnishes but few and scattered accounts of the art of supplying cities with water, we find in the national traditions and monuments the proof that the science of hydraulics dates back to the most remote antiquity.

Persia possesses canals for irrigation, the origin of which is lost in the "Night of Ages," and which continue to the present day to perform their functions.

In the province of Anachosia, ruins of canals attest the former existence of a vast system of irrigation. At Memphis is the ancient aqueduct of Sesostris, and Babylon had its aqueduct.

Jerusalem was supplied with water from the aqueduct of Solomon. Among the ruins of the ancient aqueducts of Greece, that of Patara in Lycia is most interesting. It consists of a siphon constructed of cut stone supported by an arch in cyclopean masonry.

The current wheel lost none of its earlier prestige in comparatively modern times, and is still in use with the improvement of wing dams for increasing the velocity of the current. In Fig. 18 is illustrated the current wheel and beam operating pumps of the London, England, water-works in the seventeenth and eighteenth centuries A.D. The arches of the old London Bridge at one time contained

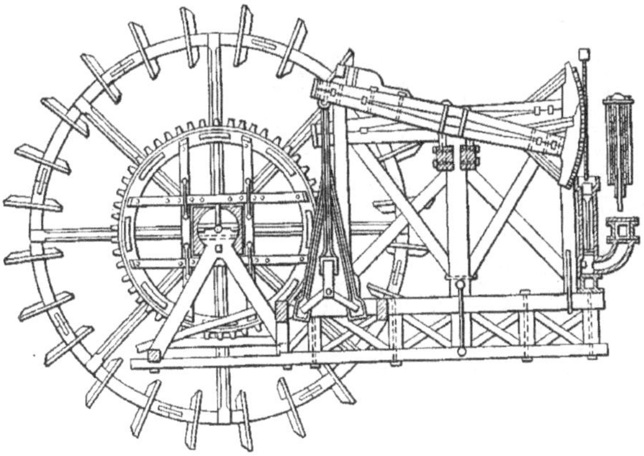

FIG. 18.—Current wheel, 1731, London Bridge.

three of these great wheels, operating 68 pumps, and furnishing over 2,000 gallons per minute to reservoirs 120 feet above the river.

"The tide and current wheel, erected first in the vicinity of the north end of London Bridge, and subsequently under its northern arch, was erected by Peter Morice, a Dutchman, in 1582, and operated force-pumps which supplied a part of London with water. The stand-pipe from the pump was 120 feet high, and conducted the water to a cistern at that height, where it was distributed to the dwelling-houses in the vicinity, and by four lead pipes to cisterns at Bishopsgate, Aldgate, the Bridge, and Wall-brook. The amount raised was about 216 gallons per minute. The wheel worked 16 pumps, each 7 inches in diameter, and having a stroke of 30 inches. Several other similar machines were erected at other points, and were similarly driven.

"The axle of the trundle was prolonged at each end, and had quadruple cranks which connected by rods to the ends of four walking-beams 24 feet long, whose other ends worked the piston-rods of the pumps. The axis of oscillation of the lever supporting the wheel, and by which it was adjusted to the state of the tide, was coincident with the axle of the trundle, so that the latter engaged with the 8-feet cog-wheel in any condition of vertical adjustment. Each end of the walking-beam was made effective.

"During the seventeenth and eighteenth centuries the works were extended from time to time, and occupied one after another of the arches."

FIG. 19.—Aqueduct of Antioch.

But it is, above all, the Romans who particularly attract our attention by the architectural development of their hydraulic construction. Among the earlier examples of their water-supply aqueducts, we illustrate a section of the aqueduct of Antioch with its massive cyclopean walls pierced by irregular arches.

Its length is about 700 feet and its height 200 feet; though solidly built, it is the rudest example of Roman aqueduct construction, and contrasts strangely with their later work.

The beauty of the aqueduct architecture of the Romans is well illustrated in the Pont du Gard aqueduct at Nîsmes, France, which spans the Valley of the Gardon, at a height of 157 feet above the water. It was one of the earliest constructed by the Romans out of Italy, about the time of Augustus. The top course of arches is about 1,000 feet in length and the entire aqueduct over 25 miles; the dimensions of the water-way is 4 feet wide by 4 feet 9 inches high;

the fall throughout its entire length is 2.112 inches per mile and it supplied from 14 to 18 millions of gallons of water per day. This noted aqueduct has a roadway at its side on the first tier of arches

Fig. 20.—Pont du Gard Aqueduct.

as shown in Fig. 21 in which the section at the left is the water-way at the top, with a section of the aqueduct and roadway at the right.

The grandest of the aqueducts of Rome was the triple tier of conduits conveying waters of different qualities. The water of the purest kind, for drinking use, was conveyed through the upper channel, the Aqua Julia, and the water for the baths through the middle channel, the Aqua Tepula, while the lower channel, Aqua Martia, water was used for washing and irrigation. The upper tiers were built at successive periods.

The city of Rome had no less than nine aqueducts in the time of the Emperor Nerva, with a combined capacity of 252,000,000

Fig. 21.—Details of the Pont du Gard.

gallons per day, at which time there was a population of about 1,000,000. The distribution within the city was much hampered by their system of low pressure in lead and earthen pipes—fed through bronze gates of regulated dimensions.

Constantinople, the Rome of the East, was also celebrated for its water-supply brought to the city by the Emperors Valens and Justinian aqueducts. The magnificent Carthaginian aqueduct carried its supply of water through mountains and over valleys for 70 miles, and at the town of Undena it was supported by an arcade of more than a thousand arches.

It is evident from what has been said in regard to water-supply that modern cities have not advanced beyond ancient Rome; indeed, in regard to abundance, no city has ever yet even contemplated supplying to its inhabitants such a large quantity. There was an abundant supply for every purpose; that which was used for drinking purposes was brought from a great distance, and its freshness was retained by bringing it through conduits of stone and keeping in covered reservoirs, where it was exposed to the action of the air, and at the same time protected from the rays of the sun. Great care was taken to prevent any pollution. As to the abuses in ancient Rome of the public water, it is not the place of the writer, nor his object, to remark upon them, or to make any comparison with those of modern times. Many of the larger cities of both the Old and New World have recently imitated ancient Rome or are now contemplating so doing, by procuring their supply of water from long distances, from localities removed from all causes of pollution, and bringing it to the inhabitants by means of gravity through aqueducts.

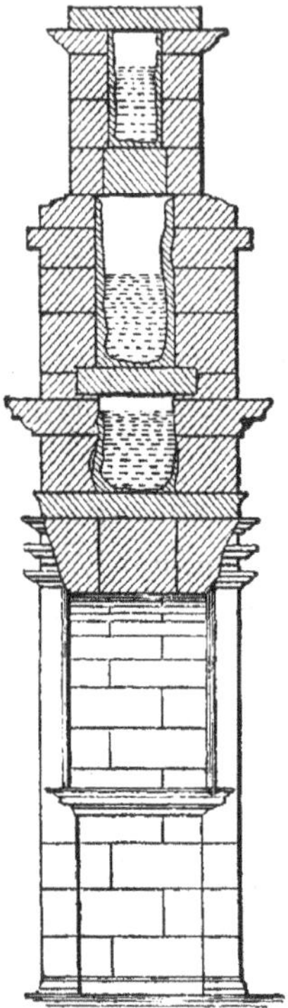

FIG. 22.—Triple aqueduct—Julia, Tepula, Martia.

Through the long period of a thousand years from the fall of the Roman Empire to the dawn of the "Age of Iron" in the eighteenth century of the Christian era, the distribution of water in cities and towns made little or no progress. The manufacture of cast-iron and steel pipe has led to the modern and more perfect methods in city distribution and made a most economical substitute for lofty aqueducts in the crossing of valleys. The tedious drawing of water from the deep wells of old has drifted by the progress in the iron industry into the rapid method of the power-pump and the air-lift.

CHAPTER II

HYDRAULICS—PROPERTIES OF WATER

THE term hydraulics is a general name covering the scientific and engineering properties of fluids and in its special consideration in this work is its relation to water, its use and power. In its practical treatment there are two subdivisions, viz., Hydrostatics, which treats of fluids or water at rest and their pressure; Hydrodynamics, which treats of fluids or water in motion; the running stream or falling water.

THE PROPERTIES OF WATER

Water, the active principle in hydraulic effect, is composed principally of two elastic gases: oxygen 8 parts, hydrogen 1 part, chemically united by combustion, resulting in the production of water; which may be reconverted into its gaseous elements by chemical and electrical effect.

The standard weight of commercial purity has been accepted at 62.37 pounds per cubic foot, at 62° F. and at 29.92 inches, the mean barometric pressure. Its density changes slightly with changes of temperature, with its greatest density 62.424 pounds per cubic foot at 39.3° F. as shown in the following table:

TABLE I.—RELATIVE DENSITY AND WEIGHT PER CUBIC FOOT OF WATER AT VARIOUS TEMPERATURES

Temp. ° F.	Relative density.	Pounds per cubic ft.	Temp. ° F.	Relative density.	Pounds per cubic ft.	Temp. ° F.	Relative density.	Pounds per cubic ft.
32	.99987	62.416	60	.99907	62.366	140	.98338	61.386
35	.99996	62.421	70	.99802	62.300	150	.98043	61.203
39.3	1.00000	62.424	80	.99669	62.217	160	.97729	61.006
40	.99999	62.423	90	.99510	62.118	170	.97397	60.799
43	.99997	62.422	100	.99318	61.998	180	.97056	60.586
45	.99992	62.419	110	.99105	61.865	190	.96701	60.365
50	.99975	62.408	120	.98870	61.719	200	.96333	60.135
55	.99946	62.390	130	.98608	61.555	212	.95865	59.843

Water in ordinary use is considered as incompressible and therefore in this respect is treated practically as a solid. Its elasticity is very slight and is given at $\frac{1}{50000000}$ of its volume at a pressure of one atmosphere and temperature of 39.3° F.; increasing very slightly to $\frac{1}{45000000}$ at 80° F.

Taking 0.00005 as a mean value of linear compression at ordinary temperatures per atmosphere, the co-efficient of elasticity of water is $E = \frac{14.7}{0.00005} = 294{,}000$ pounds per square inch, which is far less than the co-efficient of elasticity of most solid bodies and the metals.

A column of water hence increases in density from the surface downward; applying this value to the density of the ocean at great depths, say 27,000 feet in the Pacific, off the coast of Japan, where the pressure may be $27{,}000 \times .436 = 11{,}762$ pounds per square inch, 800 atmospheres, the weight of a cubic foot of sea water would be increased from 64 pounds at the surface to 66.56 pounds at that great depth.

HYDROSTATICS

In hydrostatics are involved the principle of static pressure in fluids, as water; which is found to be of equal value in all directions at any given point; the same downward, upward, and lateral, receiving the direction of its pressure from the mobile condition of free fluidity and its power value from its gravity or weight, which may be considered as potential power, *stored power*, when in the static state of pressure. This condition of fluid pressure constitutes the essential difference between fluids and solids, solids pressing only downward or in the direction of gravity.

The upper surface of a gravitating fluid is always horizontal, but not a plane. Its form is spherical, corresponding to a radius of the earth's semidiameter.

The line of sight in levelling is always a tangent to the spherical surface and must be taken into account in laying out extended waterways and canals.

The pressure of water on every portion of the surface of the vessel containing it and in contact with it, is equal to the weight of a column of water whose base is equal to its unit area and whose height is equal to its depth below the surface to the vertical centre of the unit, as shown in Fig. 22a.

The level due to hydrostatic pressure in a series of areas connected by pipes and at various distances from a common reservoir, is the same irrespective of the form or the length of the connecting passages.

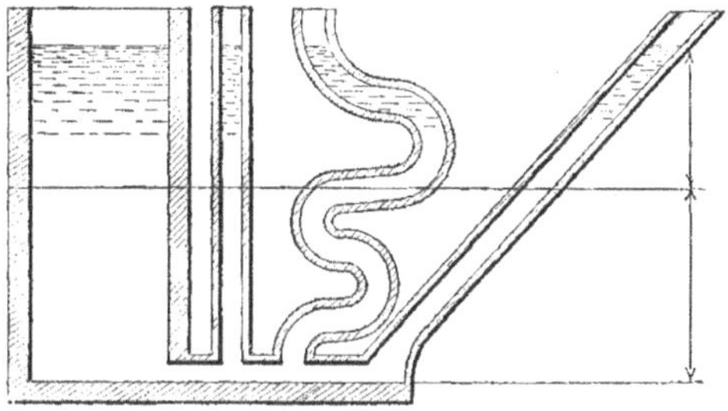

FIG. 22a.—Hydrostatic level.

The pressures beneath the surface are also the same for any given vertical depth throughout the system, as shown in the illustration, Fig. 23.

To illustrate the pressure due to gravity for any head or natural level upon a unit of surface, say one square foot, whether vertical, inclined, or horizontal, we have for an open head, the weight of the water per cubic foot of the column from the surface, vertically to the centre of the unit area of say one square foot, is equal to the pressure per square foot at the central point of the unit area.

For example, a square foot of area at a depth of its centre from

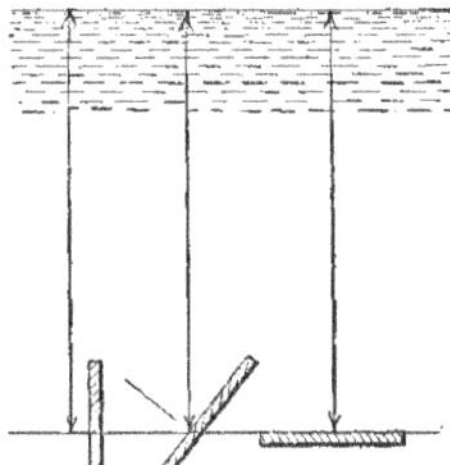

FIG. 23.—Centre of pressure.

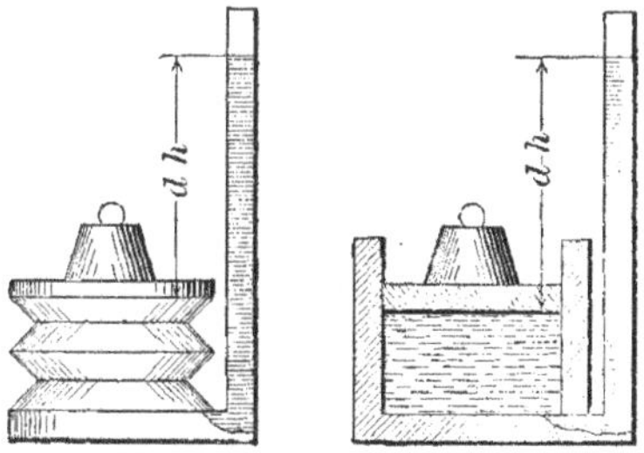

FIG. 24.—Differential pressure.

the surface of 10 feet will receive a pressure against the sides of a vessel, tank, or retaining wall of

$$62.424 \times 10 = 624.24 \text{ pounds per square foot, or}$$

$$\frac{624.24}{144} = 4.33 \text{ pounds per square inch, and}$$

$$\frac{4.33}{10} = .433 \text{ pounds per square inch per foot of depth.}$$

For the pressure under differential heads and areas, we have the differential head multiplied by the differential areas in corresponding units, as illustrated in the hydrostatic bellows and piston, Fig. 24.

The units of area must be the same in name, viz., 1 square foot or 1 square inch, without reference to the actual area of the head, which is often misunderstood by the novice.

For example: With square inches as the units of areas, and with the height dh = 10 feet; then $10 \times .433 = 4.33$ pounds will be the uplifting pressure per square inch on the head of the bellows and on the piston; then if the areas of the bellows head or piston are 100 square inches, the total lift will be $100 \times 4.33 = 433$ pounds, exclusive of friction.

ATMOSPHERIC PRESSURE AND ITS EFFECT

The pressure of the atmosphere upon the surface of water at sea-level is always due to its barometric condition, the mean of which is 14.7 pounds per square inch, with the barometer at 29.92 inches. It will sustain a column of water, by a vacuum in a closed tube, of $\frac{14.7}{.433} = 34.39$ feet in height; but, as the evaporation of water under a vacuum is equal to .556 of an inch height in barometric pressure, it reduces the atmospheric counter-pressure to 14.42 pounds per square inch, and $\frac{14.42}{.433} = 33.3$ feet, the highest water head possible in practice under a vacuum at 62° F.—varying inversely as the temperatures.

POTENTIAL ENERGY OF WATER

Water when stored in reservoirs or by dams and barrages may be said to possess a potential or stored energy in foot-pounds, equal to the available weight multiplied by the height in feet through which its dynamic effect can be obtained.

The pressure of a stored body of water against the sides of a reservoir, a dam, or barrage, has a horizontal direction, tending to push forward the mass, equal to its whole vertical area in units of surface, multiplied by the weight and depth of its centre of gravity in the same units of measurement and weight.

Thus on a vertical wall or face exposed to water pressure, the pressure at any point of depth is found to be equal to the depth in feet, multiplied by the weight of water per cubic foot, 62.424 for a square foot, or .4335 pound per square inch, for a square or circular area; but if the area be triangular, as at A, C, B, Fig. 25, the centre of pressure will be found at the centre of gravity of the given area of whatever form it may be.

In a triangle, as in Fig. 25, it is at one-third the total height, G, F.

The mean effective pressure for a unit width for the whole depth collectively is found by adding the total pressure at each increment

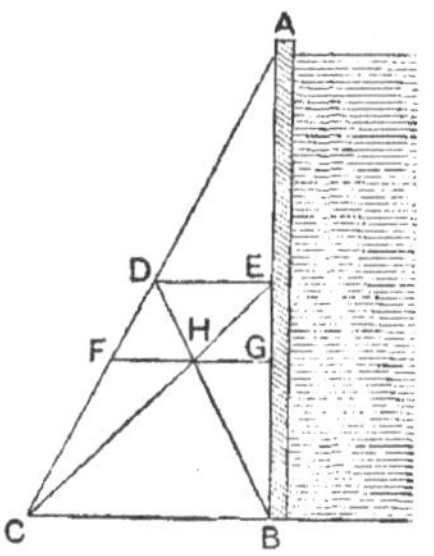

FIG. 25.—Centre of pressure.

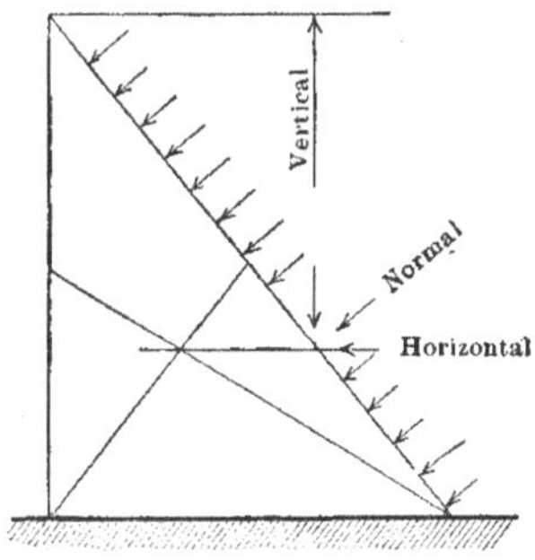

FIG. 26.—Direction of pressure.

of depth together and dividing the sum by 2, which gives the mean pressure of the whole column at a point at two-thirds of the total height from the surface.

This is proved by example in that the sum of the unit pressures of the upper two-thirds of the total height or head is equal to the sum of the unit pressures of the lower one-third of the total height or head.

The direction of pressure on the face of a retaining wall is normal to its surface at right angles, as shown in Fig. 26; with the mean centre of pressure at the centre of gravity of the rectangle; or the total vertical pressure may equal its unit pressure P $\times$ secant of the angle of the wall from the vertical $\times \frac{1}{2}$ the total height.

In the construction of walls for resisting only the hydrostatic pressure of water, as that pressure is in proportion to the depth, the strength of the wall should be in the same proportion. If strength were not given to the lower layers by superincumbent pressure, the inclination of the slope should be 45°; but in consequence of this pressure it may be less, varying with the materials and their manner of being put together. In the construction of dams or barrages the varying circumstances of cases allow of the display of a good deal of engineering skill. A barrage suitable for restraining a body of water which is never strongly moved in a lateral direction against it, as at the outlet of a canal or a reservoir fed by an insignificant stream, would not be adapted to a mountain torrent, where the surface of the reservoir can scarcely ever be large enough to prevent, by the inertia offered by a large mass of water, the walls from being subjected to a strong lateral force from the action of the current. Under such circumstances it is usual to give a curved surface to the facings, in a vertical as well as in a horizontal direction; the curves in both directions being calculated from the following elements: 1, the ascertained hydrostatic pressure; 2, the nature of the materials, such as the weight of stone and tenacity of the hydraulic cement used; and 3, an estimate of the maximum force of flowing water which may at any time be brought against the structure during a freshet. This force, it will readily be seen, will have a different direction and a different point of application in different cases, depending upon the depth and extent of the reservoir. The top of the dam is therefore given a greater horizontal section than would be called for if hydrostatic pressure alone had to be opposed. The hydrostatic pressure at any point against the surface of a containing vessel is the resultant of all the forces collected at that point, and is therefore at right angles to that surface. In a cylindrical or spherical vessel these resultants are in the direction of the radii, and in the sphere vary in direction at every point.

FLOTATION AND THE STABILITY OF FLOATING BODIES

When a body floats upon water it is sustained by an upward pressure of the water equal to its own weight, and this pressure is the same as the weight of the volume of water displaced by the body.

But in many cases, when a body is only partially immersed, the centre of gravity may be above that of buoyancy, and yet the action of turning cannot take place, so that a condition of stable equilibrium will be attained under these circumstances. If a flat body, such as a light wooden plank, is placed in water, it will float, and a portion will be above the surface, as shown in Fig. 27, and therefore, if the centre of gravity is not below the centre of volume, it will be above the centre of buoyancy, and yet the body will be in a state of stable equilibrium. For if it be tipped as represented in Fig. 28, the centre of buoyancy will be brought to the position B′, on the depressed side of the vertical passing through the centre of gravity, and this will cause the body to return to its former position. But if the body has such a shape that when it is displaced the centre of buoyancy is brought to that side of the vertical passing through the centre of gravity, which is elevated as represented in Fig. 29, then the body will turn over. When the body is in the new position, a vertical drawn

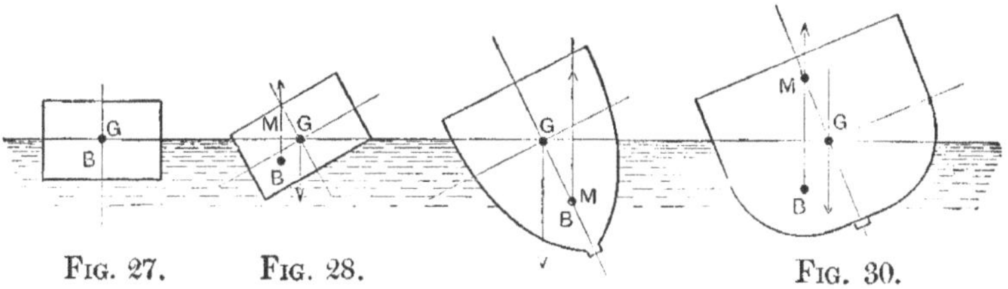

FIG. 27. FIG. 28. FIG. 29. FIG. 30.

The metacentre of flotation.

through the changed position of the centre of buoyancy will intersect the line which in the first position passed vertically through the centre of gravity, and this point of intersection is called the metacentre, represented at M in Figs. 29 and 30. When the metacentre is above the centre of gravity, as in Fig. 30, the body will tend, by the action of the centre of buoyancy, to return to its former position; but when it is below, as in Fig. 29, the action of the centre of buoyancy, being upward on the elevated side, will tend to turn the body over. Its proper place, therefore, as its name would indicate, is above the centre of gravity, but it cannot be a fixed point. For example, in Fig. 30, as long as increase of inclination of the vessel carried the centre of buoyancy B to the left, the point M might remain at nearly the same distance from G, because it would also move to the left. But if the inclination of the vessel in the same direction carried the centre of buoyancy to the right, the height of the metacentre M would diminish until it would be in G, when the equilibrium would be indifferent, and at last below G, when the ship would turn over. It is desirable to have the metacentre as far as possible above the centre of gravity; and this condition is secured by bringing the centre of gravity to the lowest practicable point, by loading the ship with the heaviest part of the cargo nearest the keel, or by employing ballast, and in yachts by deep loaded keel.

DIFFERENTIAL PRESSURES FROM UNEQUAL HEIGHT OF WATER ON EACH SIDE OF A GATE OR WALL

As before stated the mean pressure against a barrier is found to be at a point two-thirds of the total height from the water surface, so that an opposing water pressure at a less height has also its mean pressure at a point two-thirds of its height from the surface. The gross pressure of the high water side is $P_1 = w \times h_1 \times \frac{1}{2}h_1 = \frac{1}{2}wh_1^2$, which shows that the gross pressure varies as the square of the height, also that the gross pressure of the lower level is also $\frac{1}{2}\ wh_2^2$; the centre of whose pressures is at one-third of their height respectively above the bottom, or sill if it is a gate.

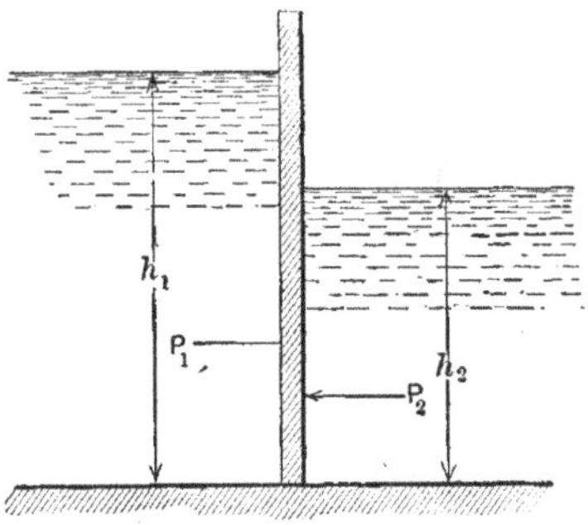

FIG. 31.—Differential pressures.

DIFFERENTIAL PRESSURES FROM UNEQUAL HEIGHT OF WATER ON THE SIDES OF A GATE OR BARRAGE

As before stated, the mean centre of pressure against a barrier is found to be at two-thirds of the total height from the water surface; so that an opposing water pressure from a less height has also its mean centre of pressure at a point two-thirds of the total height from its surface.

For example, the resultant pressures from the unequal levels of the head and tail water of a floodgate or barrier are horizontal in direction, but not on the same level.

Each of the mean centres of pressure are at two-thirds of their depth from their respective surfaces, as shown in Fig. 32 and Fig. 33.

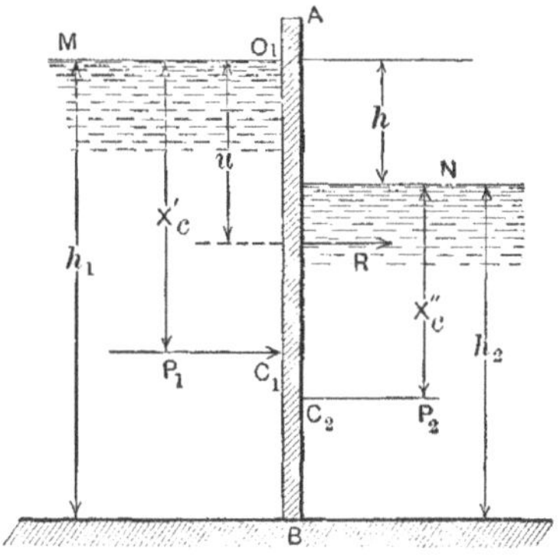

FIG. 32.—Differential pressures.

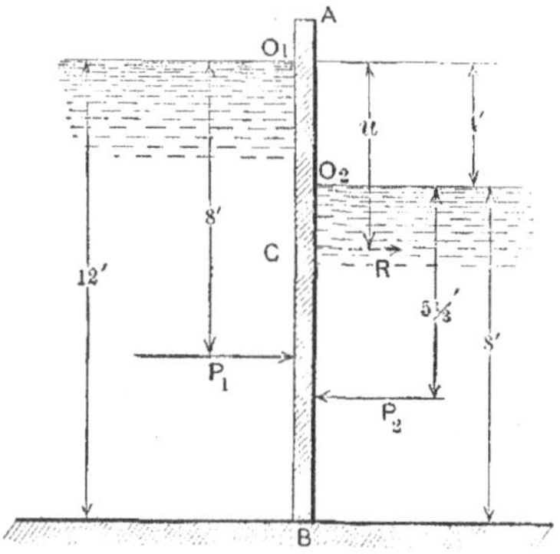

FIG. 33.—Differential dimensions.

The resultant centres of opposite pressures are shown at C_1 and C_2 and equal to P_1 and P_2. Their resultant pressure, $R = P_1 x'_c - P_2 (x''_c + h)$ and the distance of u from the surface at O, is

$$u = \frac{[P_1 x'_c - P_2 (x''_c + h)]}{P_1 + P_2}$$

Solving by the figures in diagram Fig. 33, the resultants $R = (P_1 - P_2)\,w$, the weight of a cubic foot, are for one foot in width of a vertical section as follows:

$$R = (12 \times 1 \times 6 - 8 \times 1 \times 4) \times 62.424 = 2{,}497 \text{ pounds, and}$$

$$u = \frac{(12 \times 1 \times 6 \times 8 - 8 \times 1 \times 4 \times 9\tfrac{1}{3})\ 62.424}{2{,}497} = 6.93 \text{ feet, the distance}$$

from the higher surface O_1 at which the centre of effect C is found.

THE HYDROSTATIC ACCUMULATOR

The accumulator is an arrangement of a cylinder with a plunger loaded by weights for the purpose of delivering a larger volume of water in a given time than the uniform action of the pump makes it possible to meet the intermittent and quick action of a hydraulic crane, elevator, press, or for testing pipes, when a large body of water at high pressure may be obtained quickly without a sudden increase in the action of the supply-pump.

One of the advantages of this type of accumulator is its uniform pressure throughout its range of action.

The accumulator is shown in Fig. 34; it consists of the large cast-iron cylinder *a*, fitted with the plunger *b*, which works water-tight by means of the gland *c*, and packing. To this plunger is attached, by means of the bolts *f*, and strong cast-iron cross-head *e*, the loaded weight-case *d*. Thus a pressure is obtained upon the water in the cylinder equal to a column of water 3,000 feet

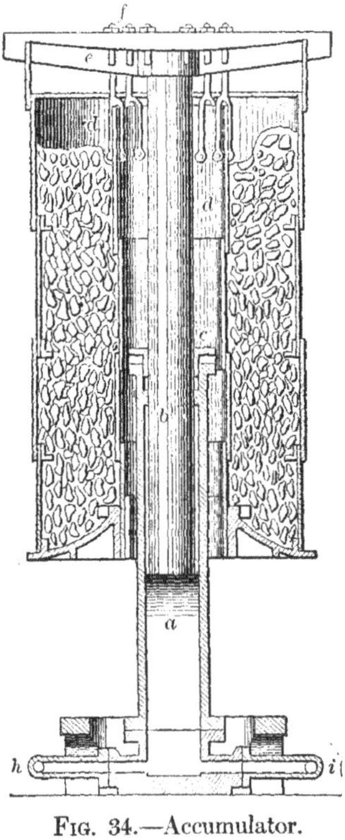

Fig. 34.—Accumulator.

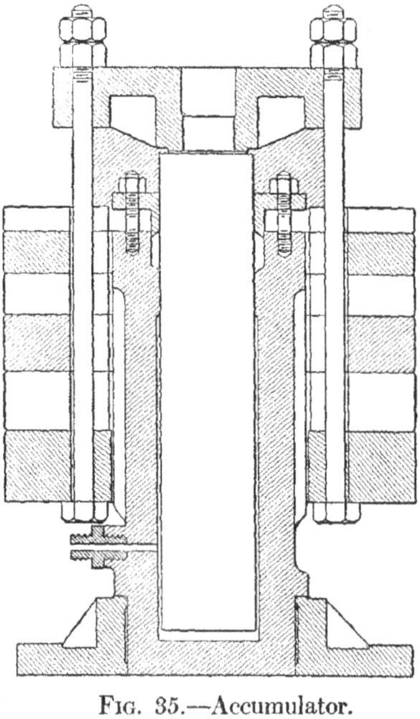

Fig. 35.—Accumulator.

high, or 1,300 pounds per square inch. As the water is pumped into the cylinder by the pumping engines through the pipe *h*, the piston, with the weighted case, rises, being guided by a strong framework, and is made to regulate the amount of water pumped in, by actuating a throttle-valve in the steam-pipe of the pumping engine, which it closes after having reached a certain height. When the crane, press, or elevator is in operation, the water passes from this cylinder through the pipe *i*, to those actuating the motion of the crane, press, etc., and the weighted plunger descends, always keeping up a constant pressure upon the water; in descending, the same causes the throttle-valve to open again, and the water is again pumped in. Another design weighted with iron blocks is shown in Fig. 35.

Another type of accumulator is much in use, consisting of a closed high-pressure tank, containing air, into which water is pumped, compressing the air to the desired pressure; it does not give a constant pressure during its action. An open tank placed at an elevation suited to the desired pressure is much in use in the elevator service with the pumping plant in the basement. It gives a constant pressure.

Hydrostatic pressure obtained from a natural fall of water, or a force pump, has a great range of usefulness in the many methods of its application for mechanical effect.

The hydrostatic press, by which enormous pressures may be obtained by multiplying a small power by differential areas has a large range of applications in the industrial arts. Among these may be mentioned the hydraulic push and pulling jacks, crank-pin presses, wheel-presses, shears, riveters, lifts, and elevators.

The principles involved in all these devices for accumulating power, are the relative areas of the pump plunger, the main plunger of any device, and the pressure upon the pump plunger, all in uniform units of area and pressure.

For example: As most of these devices are operated by hand, a person may easily exert a force of 60 pounds upon the lever handle and with the lever pivoted at a point 10 to 1, then with a pump piston of one-half square inch, the pressure will be $60 \times 10 \times 2 = 1{,}200$ pounds per square inch, and if the forcing piston be 3 inches in diameter as in many jacks, say 7 square inches area, then $7 \times 1{,}200 = 8{,}400$ pounds, the lifting capacity of the jack. In the case of a hydraulic punch the total pressure of the punch upon the metal to be punched will be $\frac{8.400}{\text{area of punch}}$ which for one-half inch diameter = .196 and $\frac{8.400}{.196} = 44.382$ pounds per square inch, sufficient to punch one-half-inch holes in iron one-half inch thick.

The detailed working parts of the hydraulic punch are shown in Fig. 36, in which a small plunger in the cylinder D, with inlet valve E and discharge valve below, is operated by the lever B, to give great pressure to the ram H, to which is attached the punch. On opening the by-pass valve K, the ram is lifted by the lever L and revolving wedge M, pushing the oil back into the reservoir A.

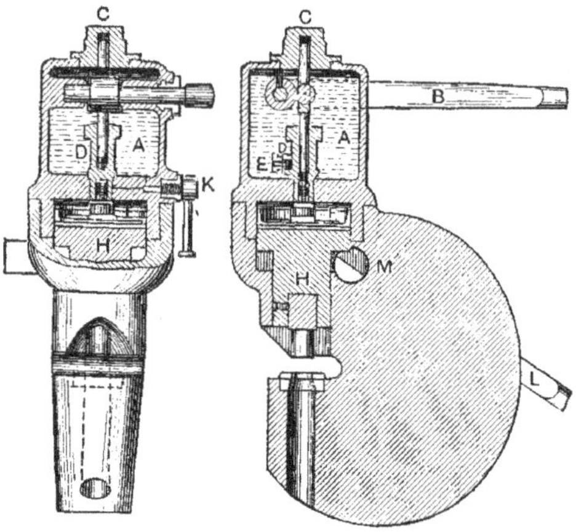

FIG. 36.—Hydraulic punch.

The hydraulic lifting jack is made in many designs; the leading principles of operation and action are shown in Fig. 37. The lifting power as above computed is derived from the formula $P = pD^2 \div d^2$ in which p = pressure per square inch, D = diameter of the piston in large cylinder, P = the total lifting pressure. As in the example above, $\frac{600 \times 9}{.64} = 8{,}437$ pounds, the total lifting power of the jack.

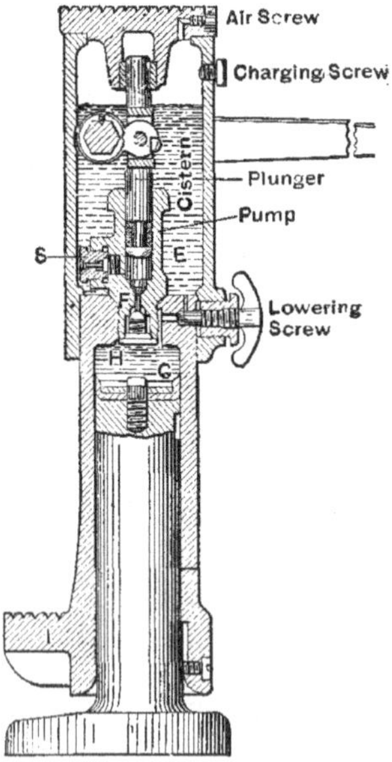

FIG. 37.—Hydraulic jack.

The shell of the jack extends over and nearly to the foot of the ram to enable a lift at both head and foot of the jack. The small plunger and valves are operated by the lever and arm D, for lifting, and the by-pass valve serves for lowering a load or closing the jack by its own weight, which will send the oil back to the cistern. S, suction valve; F, discharge valve; G, leather cup packing.

The fundamental formulas for the conditions of the relations of the force, p; diameters of the pump piston d, ram D, and the total pressure P, are expressed as follows:

$$p = \frac{P\,d^2}{D^2};\quad D = \sqrt{\frac{P\,d^2}{p}};\quad d = \sqrt{\frac{p\,D^2}{P}}.$$

The hydraulic rail punch and rail bender and straightener are important implements in the line of their usefulness and are operated in the manner before described as to their power. Their usual designs are illustrated in Fig. 38 and Fig. 39; the loops are for suspension and easy handling.

The hydraulic press is made in many designs to meet the require-

FIG. 38.—Rail punch.

FIG. 39.—Rail bender.

ment of its work, while its power is derived from the hydrostatic principles before described and illustrated.

The model of press much in use in factories for pressing goods, embossing, and stamping is illustrated in Fig. 40 and one for compressing bale goods in Fig. 41.

To facilitate speed in embossing, the upper platen is provided with a large screw, which is quickly run down to make contact, when

FIG. 40.—Embossing press.

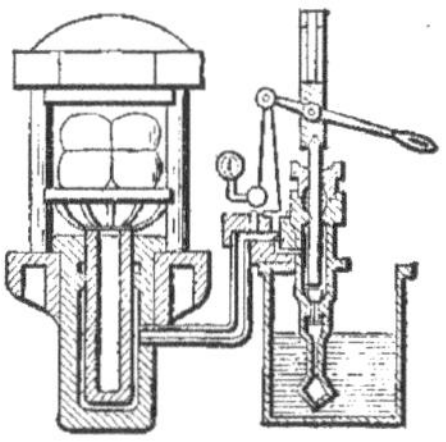

FIG. 41.—Bale press.

a few strokes of the pump brings up the desired pressure, which has a uniform tally by the pressure gauge.

The cotton and other presses used for large platform areas require very high initial pressure from the pump and large areas in the ram.

Pressures up to 5 tons per square inch are in use for special work with steel cylinders; while 3 tons per square inch is about the limit with cast-iron cylinders with a safe factor.

The gauges used for indicating the pressure are of the Bourdon type with dials figured up to 6,000 pounds, or to any required maximum pressure per square inch. For these high pressures the bent spring is made of strong steel tubing flattened so that the pressure tends to compress the inner wall and cause a tension on the outer wall and thus tend to straighten it and by its free end attachment through a rack and pinion, moves the indicator over the dial. Among the many hydraulic or rather hydrostatic devices in use we may further name the pulling jack, suspension weighing scale, bolt ram, cranes for foundries and for loading at docks, bar and beam straighteners, plate benders, punches, and shears. A novel device for forcing close-fitted bolts that are rusted in, such as the bolts holding the blades of propellers, and which allows a blow to be given to the ram to start the bolt is shown in Fig. 43. Its novelty is in the loose plunger and head within the ram

FIG. 42.—Hydraulic gauge.

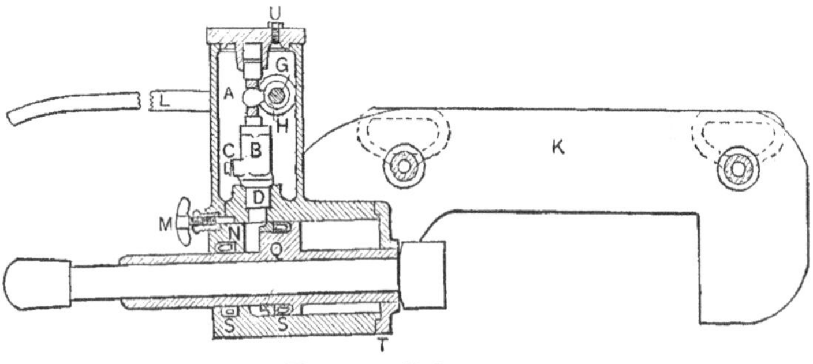

FIG. 43.—Bolt ram.

which allows a blow to be given to start the bolt. The hook claw is double to allow the bolt head to clear.

The pipe-bending ram shown in Fig. 44 is one of the conveniences of every establishment where pipe bent work is made. A set of curved dies is used to fit each size of iron pipe so that the bends can be made without flattening the pipe, a most necessary condition for inserting electrical conductors. A rack and pinion operates the ram for closing upon the pipe, when the hydraulic plunger finishes the work with a total pressure of from 30 to 40 tons.

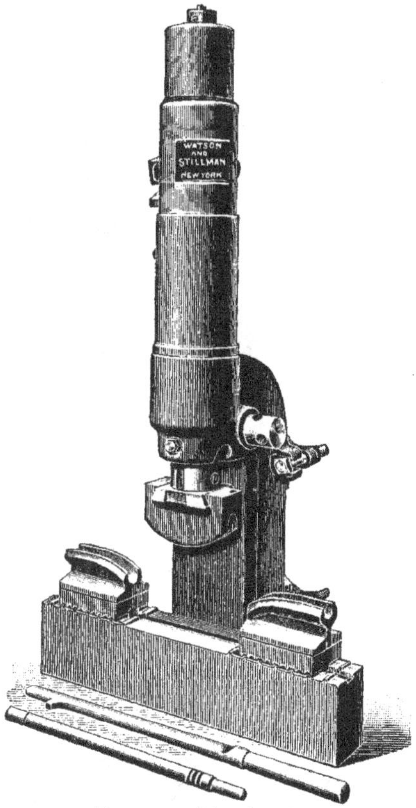

FIG. 44.—Pipe bender.

The hydraulic testing pump, Fig. 45, is a most convenient apparatus for high-pressure testing and is made for pressures from 1,000 to 10,000 pounds per square inch. The piston has two areas: the larger one with the hand lever works up to 2,000 pounds and the small one carries the pressure up to 10,000 pounds per square inch. The pump is mounted upon an iron tank for its water- or oil-supply and with wheels for portable use.

The hydraulic two-plunger belt-pump, Fig. 46, is a special design for continuous work as for supplying accumulators, and is suitable for all pressures up to 6,000 pounds per square inch. These pumps are also supplied with a start-and-stop device operated by the weight platform of the accumulator.

The general arrangement of the moving parts of a hydraulic press is shown in Fig. 47, consisting of the force pump operated by a lever, its valves, the safety valve, and the lifting ram.

The vital parts of such devices for the various kinds of work to which they are assigned are their design and construction, so as to resist the required hydrostatic pressure with a safe factor for the tensile strength of the material surrounding the points of greatest stress.

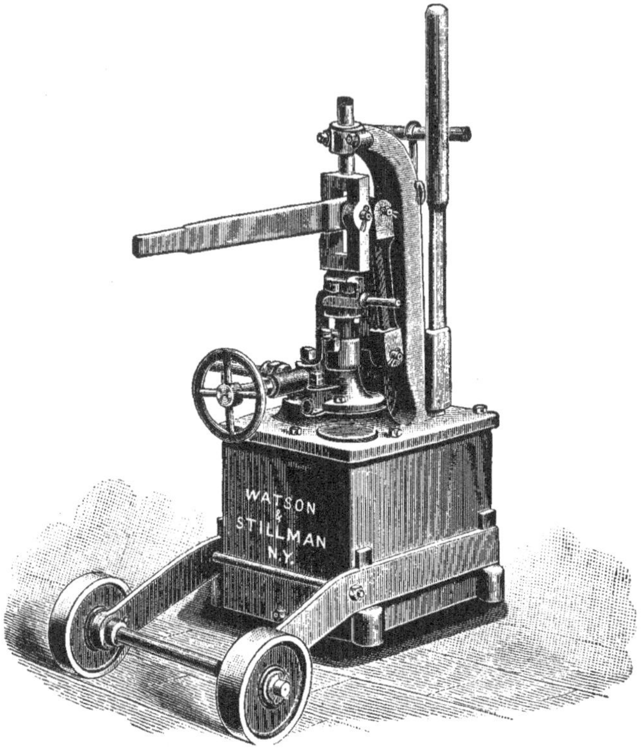

FIG. 45.—Hydraulic testing pump.

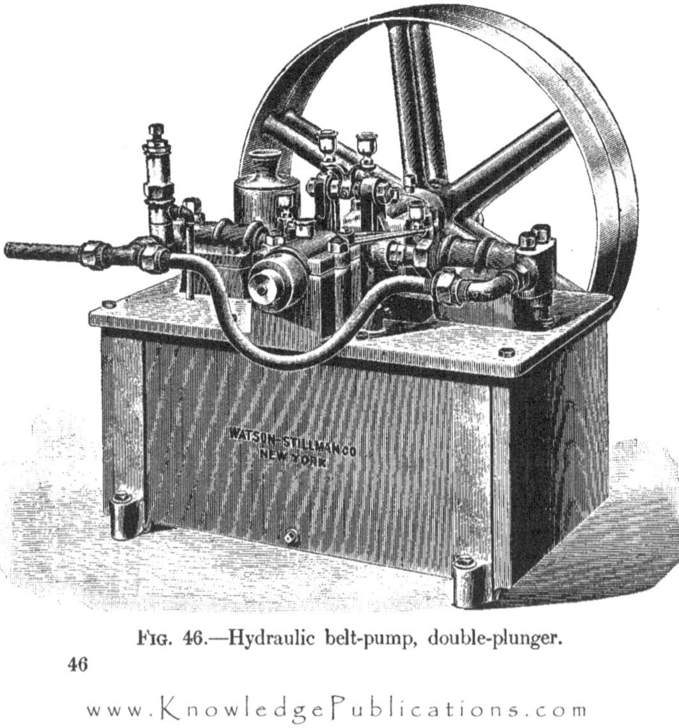

FIG. 46.—Hydraulic belt-pump, double-plunger.

For the thickness of the cylinder walls of a hydraulic ram, we have the equation—

$\frac{P\ r}{.7854\ (CD^2)\ -\ P}$ = thickness in inches; in which P = the total pressure on the ram, r = radius of the ram cylinder, C = tensile strength of cast-iron or other metal cylinder, D^2 = square of the diameter of the ram, or the area may be substituted for .7854 D^2.

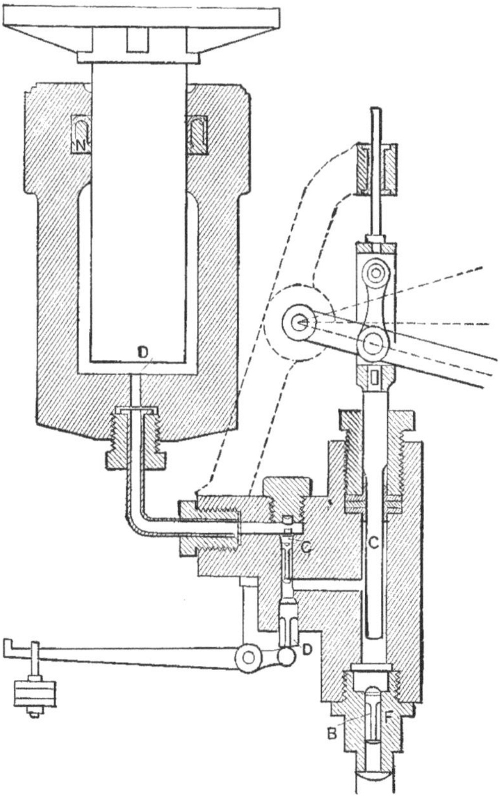

FIG. 47.—Hydraulic press.

For example, for the ultimate tensile strength of a cast-iron cylinder for a pressure of 2,000 pounds per square inch, with a ram 5 inches in diameter and 17,000 pounds per square inch, as the tensile strength of cast iron, we have the total load or pressure P = 19.635 × 2,000 = 39,270 pounds. Then,

$\frac{39,270 \times 2\frac{1}{2}}{333,795 - 39,270} = \frac{98,175}{294,525}$ = .333 inch, and if a factor of 3 be used, makes the thickness 1 inch, to which may be added a small percentage for defective casting or finish. By dividing the tensile strength by the factor of safety and multiplying by the area in the place of .7854 CD^2 in the equation the result in full will be obtained.

The assumed friction of a ram is quite small but should be added to the total load. It has been estimated at 4 per cent. of the load divided by the diameter of the ram in inches.

The following table represents the theoretical practice for thickness very nearly:

TABLE II.—OF THE TESTING OR HIGHEST PRESSURE ALLOWABLE FOR CAST-IRON CYLINDERS OF GOOD METAL AND EVEN THICKNESS AND OF THE INSIDE DIAMETER AND THICKNESS IN THE VERTICAL COLUMNS, FOR A SAFE FACTOR OF ONE-THIRD THE TENSILE STRENGTH OF GOOD CAST IRON.

Inside diam. Inches.	LBS. PER SQUARE INCH.				TONS PER SQUARE INCH.								
	800	1,000	1,200	1,500	1	1 1/4	1 1/2	1 3/4	2	2 1/4	2 1/2	2 3/4	3
3	3/8	3/8	1/2	1/2	5/8	3/4	7/8	1 1/8	1 1/4	1 1/2	1 5/8	1 7/8	2
3 1/2	7/16	1/2	1/2	1/2	3/4	7/8	1	1 1/4	1 3/8	1 5/8	1 7/8	2 1/8	2 1/4
4	1/2	1/2	1/2	9/16	7/8	1	1 1/8	1 3/8	1 1/2	1 7/8	2 1/8	2 3/8	2 5/8
5	1/2	9/16	5/8	3/4	1	1 1/4	1 3/8	1 5/8	1 7/8	2 1/4	2 1/2	2 7/8	3 1/8
6	1/2	5/8	5/8	3/4	1 1/8	1 3/8	1 5/8	2	2 1/4	2 5/8	3	3 3/8	3 3/4
7	9/16	5/8	3/4	7/8	1 1/4	1 1/2	1 7/8	2 1/4	2 1/2	3	3 1/2	3 7/8	4 1/4
8	9/16	3/4	7/8	1	1 3/8	1 3/4	2	2 1/2	2 7/8	3 3/8	3 7/8	4 3/8	4 7/8
9	5/8	3/4	7/8	1 1/8	1 1/2	1 7/8	2 1/4	2 3/4	3 1/4	3 7/8	4 3/8	4 7/8	5 3/8
10	3/4	7/8	1	1 1/8	1 5/8	2 1/8	2 1/2	3	3 1/2	4 1/4	4 3/4	5 3/8	6
11	3/4	7/8	1	1 1/4	1 3/4	2 1/4	2 3/4	3 3/8	3 7/8	4 5/8	5 1/4	6	6 5/8
12	7/8	1	1 1/8	1 3/8	2	2 1/2	3	3 5/8	4 1/4	5	5 3/4	6 1/2	7 1/8
13	7/8	1	1 1/4	1 1/2	2 1/8	2 3/4	3 1/4	3 7/8	4 1/2	5 3/8	6 1/8	7	7 3/4
14	7/8	1 1/8	1 1/4	1 1/2	2 1/4	2 7/8	3 3/8	4 1/8	4 7/8	5 3/4	6 5/8	7 1/2	8 1/4
15	1	1 1/8	1 3/8	1 5/8	2 3/8	3	3 5/8	4 1/2	5 1/4	6 1/8	7 1/8	8	8 7/8
16	1	1 1/4	1 1/2	1 3/4	2 1/2	3 1/4	3 7/8	4 3/4	5 1/2	6 1/2	7 1/2	8 1/2	9 1/2
17	1 1/8	1 1/4	1 1/2	1 7/8	2 5/8	3 3/8	4 1/8	5	5 7/8	7	8	9	10
18	1 1/8	1 3/8	1 5/8	2	2 3/4	3 5/8	4 3/8	5 3/8	6 1/4	7 3/8	8 3/8	9 1/2	10 1/2
20	1 1/4	1 1/2	1 3/4	2 1/8	3	3 7/8	4 3/4	5 7/8	6 7/8	8 1/8	9 3/8	10 5/8	11 3/4
22	1 3/8	1 5/8	1 7/8	2 3/8	3 3/8	4 3/8	5 1/4	6 3/8	7 1/2	8 7/8	10 1/4	11 5/8	12 7/8
24	1 3/8	1 5/8	2 1/8	2 1/2	3 5/8	4 3/4	5 3/4	7	8 1/4	9 3/4	11 1/8	12 5/8	14
26	1 1/2	1 7/8	2 1/4	2 3/4	3 7/8	5 1/8	6 1/4	7 5/8	8 7/8	10 1/2	12	13 5/8	15 1/8
28	1 5/8	2	2 3/8	2 7/8	4 1/4	5 1/2	6 5/8	8 1/8	9 1/2	11 1/4	13	14 5/8	16 1/4
30	1 3/4	2 1/8	2 5/8	3 1/8	4 1/2	5 7/8	7 1/8	8 3/4	10 1/4	12 1/8	13 7/8	15 5/8	17 3/8

A formula for the tensile strain per square inch on the walls of a hydraulic press cylinder, due to any required load, is

$$ts = \frac{P\ (R^2 + r^2)}{R^2 - r^2}.$$

P = the total pressure or area × pressure in pounds per square inch; R = radius of outside of cylinder; r = radius of inside of cylinder or of ram if it is a close fit.

For example, for total load of 25,000 pounds, $12\frac{1}{2}$ tons, with a cylinder 6 inches diameter and a 4-inch plunger we have:

$\frac{25{,}000 \times 13}{5} = \frac{65{,}000}{\text{area } 12\frac{1}{2}} = 5{,}200$ pounds per square inch strain upon the metal, about one-third its tensile strength.

The more active operations of hydraulic power in the use of lifts, cranes, and hammers will be treated in chapters farther on.

CHAPTER III

HYDRAULICS

MEASUREMENT OF THE FLOW OF STREAMS

WATER power will be sought, utilized, and economized in all countries wherever fuel becomes expensive or scarce, and is now the only recourse for power in many of the mining regions of the United States, Mexico, and South America.

The utilization of the water of the arid districts of all countries for the purposes of irrigation and power has long been a necessity in the older countries, and will soon become, if not now, of the utmost importance as the tide of population covers the great plains of our Western hemisphere, and to this end the necessities of human existence require the utmost economy in utilizing this great element in nature that is so essential to our being.

The flow of water in streams and the measurement of the power that may be derived from its mechanical transmission into work, in a form that may be understood by persons not versed in mathematics or engineering expedients, is a much-desired want with many people who have water power to develop and are unable to obtain the service of experts for the solution of a preliminary or uncertain problem.

For such we will endeavor to make the problems appertaining to the flow and power of water of easy solution by any one with ordinary arithmetical tact.

The numerical value of the power of water as a natural mechanical agent is derived from three elements, namely, gravity or weight, height or pressure due to height, and volume of flow, against which, form of ajutage or orifice, friction and leakage or spillage are the coefficients that make a reduction in the realization of its full value.

Descending fluids are actuated by the same laws as falling bodies for all the practical purposes of computation, and by virtue of their vertical fall descend 16.08 feet in the first second and attain a velocity at the end of the first second of 32.16 feet per second.

As one second is assigned as the unit of time in hydraulic computations, gravity or the velocity attained by a falling body during one second of time becomes a symbolic expression of which the letter g is always the exponent, and which represents the value 32.16.

The measurement of the mechanical effect of the flow of water in streams or over dams may be obtained from a few observations, beginning with the measurement of a quiet section of a stream in a general way, for obtaining the approximate volume of flow.

For this purpose select a place in the stream a few yards or rods in length, according to its size, where the channel is nearly straight.

Measure the depth at regular intervals of space from shore to shore, making a section of from 6 to 12 or more divisions, according to the width of the stream, as shown in Fig. 48.

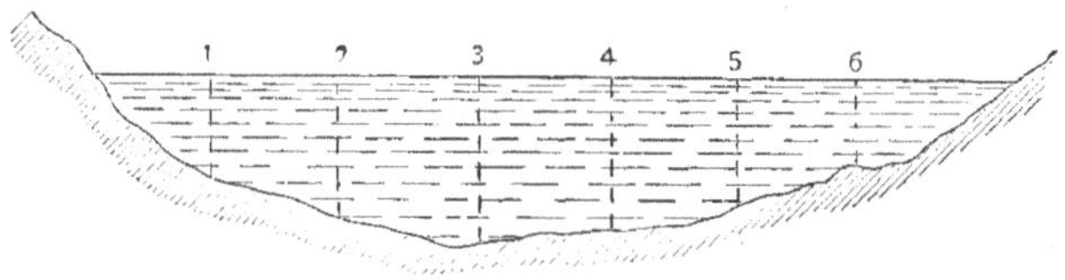

FIG. 48.—Measured depths.

Adding these measurements together and dividing their sum by the number of measurements, will give the mean depth, which multiplied by the width gives the mean area.

If the channel is not uniform for two or three rods and the best results are required, two or more sectional measurements may be made, their means added and the sum divided by the number of sections for the mean of the range.

All the measurements should be made in feet and tenths of a foot for convenience of computation.

If measured in inches, the change is readily made by multiplying the measured inches by 10 and dividing by 12.

The velocity of the stream may then be taken with a float made of a small block of wood three or four inches square, with nails driven into the under side, and a nail in the centre of the upper side for observation, as in Fig. 49, and to which a string attached to a rod will make a convenience in handling. A fishing pole with bob and sinker answers every purpose for small streams.

A measured distance of from 10 to 50 feet on the centre line of the stream should be made by driving stakes on each side of the stream, or setting a very thin stake near the centre, if accessible.

The float or bob being dropped into mid-current just above the upper line of sight or stake, the moment of passing the line is noted to the nearest second, and the time of passing the second line or stake also noted to the nearest second. This may be repeated, as the mean of two or more observations makes the work more reliable.

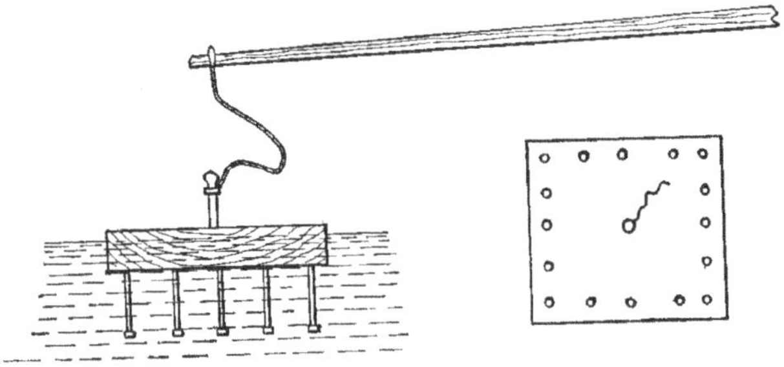

FIG. 49.—Float block.

The double float, one submerged for subsurface velocities, is used in deep and rapid rivers, where current meters are not available for anchorage.

The distance in feet between the two marks or stakes divided by the number of seconds gives the number of feet and decimals of a foot as the velocity of the centre line of the surface of the stream.

From this should be deducted 10 per cent. for friction of the bottom and sides of the stream or channel when of a depth equal to one-third of their width, if the bottom is smooth, 15 per cent. for rough, stony bottoms, and 20 per cent. for very shallow streams with rough bottom.

The corrected velocity of the stream in feet per second, multiplied by the corrected sectional area, gives the volume of water in cubic feet per second.

In a deep stream, its velocity in different parts of its sectional area varies to a considerable amount, being greatest near the centre of gravity of its sectional area. In the measurement of ordinary streams with a rough bottom a deduction of 10 per cent. from the surface velocity at the centre of the stream may be made for the mean velocity, while in canals, ditches, and flumes, 8 to 5 per cent. may be a fair deduction according to the smoothness of their walls and form of area.

The velocity close to the bottom or next to the wet perimeter is very much less than at the surface at centre of stream, and varies with the condition of roughness and form of the bottom; 20 to 30 per cent. difference has been observed, but the conditions of form and roughness are so uncertain, that the above rule will be found practical for all ordinary purposes.

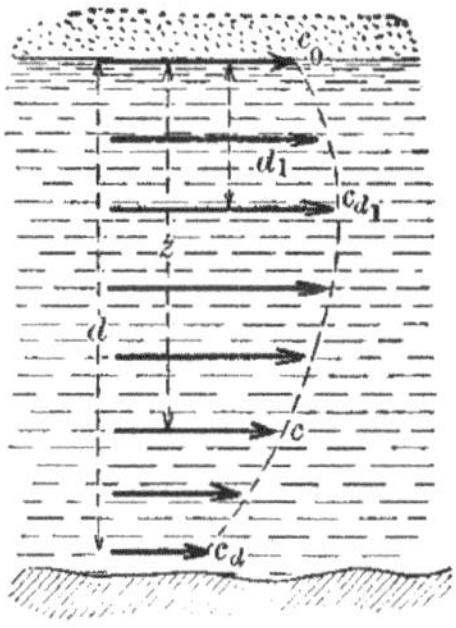

FIG. 50.—Curve of velocity.

The velocity curve, Fig. 50, vertically has a parabolic form in which the force of the wind either up or down stream makes a slight difference in the mean velocity. A horizontal velocity section also shows a lagging at the sides of the stream, which produces a vortical motion of the surface toward the centre causing floating *débris* to gather in the centre of a river and also of the slight elevation of the centre above the water along the banks, equal to $\frac{V^2 - V_1^2}{2g}$.

For the more accurate velocity measurement of large streams, propeller or wheel meters are in use. The current meter, Fig. 51,

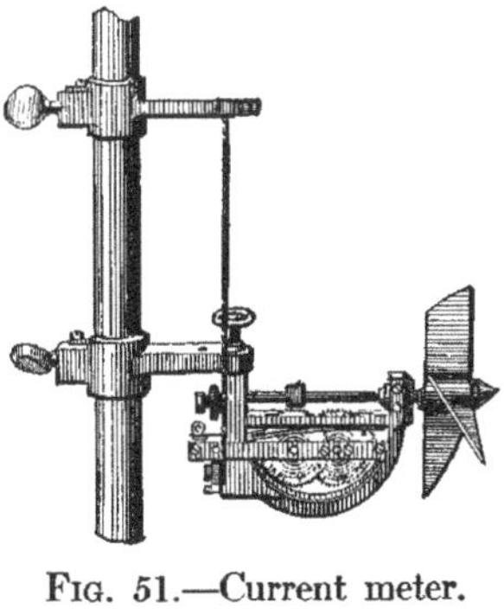

FIG. 51.—Current meter.

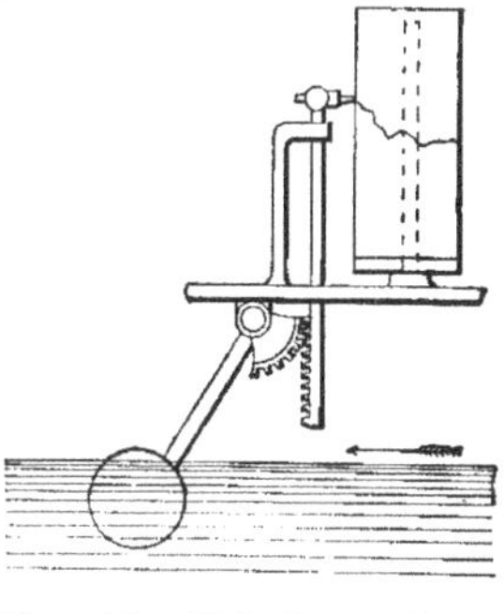

FIG. 52.—Velocity register.

is a propeller on a spindle with a worm operating geared register wheels graduated to the number of revolutions. Started and stopped for time by a string or rod and spring pawl.

The velocity register, Fig. 52, is to show the variation in velocity of streams by the angular position of a float, sector and rack, with a pencil to mark the variations of velocity on a revolving cylinder.

THE FLOW OF WATER IN OPEN STREAMS AND DITCHES

The flow of water in streams, canals, and ditches and their hydraulic grades or slope is a somewhat complex problem, arising from the friction of the bottom and sides varying with the kind of material with which they are lined, as well as the varying velocity due to varying areas; the friction being inversely as their areas.

For irrigation ditches, where the lining is of fine material, as sand, loam, or clay, the flow will have the least friction; but curves and angular bends have an influence in retarding the flow due to a given slope or hydraulic grade.

A few definitions may here make plain the terms of computation for velocity and slope of streams, canals, and ditches, under various conditions.

The area of a stream, canal, or ditch is the width multiplied by the mean depth as described in the following factors of measurement.

THE WET PERIMETER

The wet perimeter is the measure of a stream, canal, or ditch following the irregularities of the bottom and sides, as shown in Fig. 53.

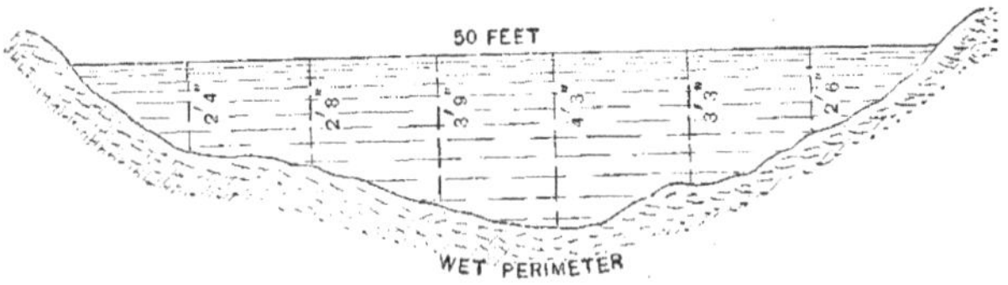

FIG. 53.—The wet perimeter.

As the wet perimeter is an important factor in the computation of the slope of a stream, canal, or ditch, the means for approximately obtaining it from the set of measurements for area is desirable for natural streams, while for canals and ditches of known dimensions, the wet perimeter can be measured on a sectional drawing.

For natural streams, to the sum of the outside measures for mean depth add the difference between these and the next measure successively, and also one-half the sum of the outside measures as follows, from Fig. 53.

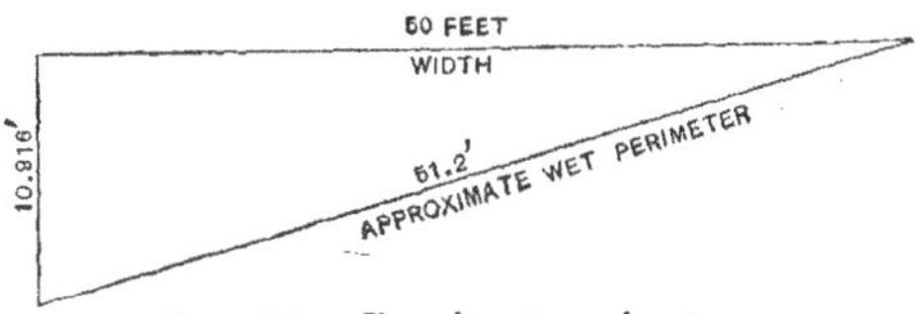

FIG. 54.—Resultant perimeter.

Make this sum and the width of the stream the two legs of a triangle, Fig. 54, and to the square of the width add the square of the sum of the depths, and from this sum take the square root for the angular wet perimeter:

Sum of outside measures .2′ 4″+2′ 6″=	4′ 10″
Differences . . . . .4″−13″−6″−12″−9″=	3′ 8″
One-half the sum of outside measures	2′ 5″
	10′ 11″
Or decimally	10.916

Then—

$$50^2 = 2,500$$
$$10.916^2 = \quad 119.14$$
$$\sqrt{2,619.14} = 51.2'$$

The angular wet perimeter being found to be 51.2′, add, for roughness and inequalities of the bottom between the measured divisions, 5 per cent. to the angular amount, making a wet perimeter of 51.2 + 5 per cent. = 53.76′, a very near approximation to the measured wet perimeter of such a stream.

For canals and ditches with flat bottoms and straight sloping sides, the operation becomes an easy one, as, for example, the following form, Fig. 55:

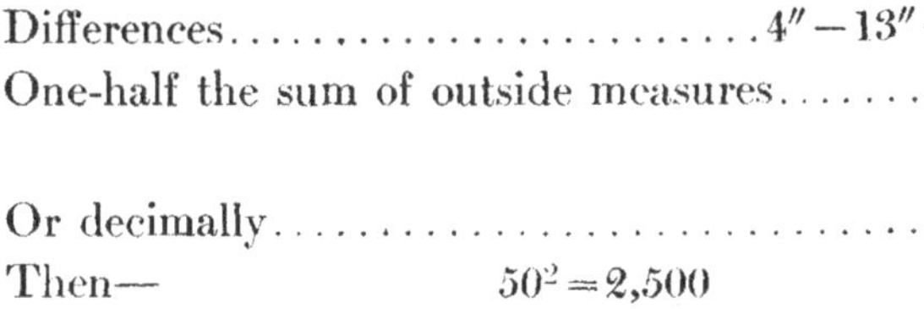

FIG. 55.—Canal or ditch perimeter.

Here the slopes of the sides make equal and similar triangles, and as both triangles make a rectangle 4′ by 3′, the solution for area becomes a very simple one, viz., $4' + 4' \times 3' = 24$ square feet.

The wet perimeter is then found by obtaining the length of the slope; which is equal to the square root of the sum of the squares of the horizontal and vertical sides of the triangle, viz.:

$4^2 + 3^2 = 25$, and the square root of $\sqrt{25} = 5$, the slope. Then $5' + 5' + 4' = 14$ feet, for the wet perimeter.

THE HYDRAULIC RADIUS

The hydraulic radius, another factor in the problem of the slope or hydraulic gradient and its consequent current, is the area of the transverse section of a stream, canal, or ditch, in square feet, divided by the wet perimeter, in lineal feet.

For example, as in the first case, Fig. 53, the sum of the measures for depth amount to 18′ 9″, or 18.75 feet, which, divided by 6, the number of measures, $=3.225 \times 50$ feet $=161.25$ square feet, the area of the stream, and $\frac{161.25}{53.76} = 2.99$, the hydraulic radius sought.

For the second case, Fig. 55, for a ditch:

$$\frac{24'}{14'} = 1.714,$$

the hydraulic radius sought.

The trapezoidal form of canals and ditches, as above described and proportioned, approximates very nearly to the best form for irrigation ditches in gravel, sand, and loam, as the slopes of the sides will hold with mean velocities below 2 feet per second.

The formula for the velocity of flow of earth-lined channels and streams, by Kutter, requires a coefficient varying with the roughness and square root of the hydraulic radius, as expressed, is, velocity $=$ $C\sqrt{r} \times \sqrt{s}$ in which $s=$ the fall in feet divided by the distance in feet.

TABLE III.—REPRESENTS THE COEFFICIENTS FOR VARIOUS HYDRAULIC GRADES AND FOR TWO CONDITIONS OF ROUGHNESS; THE FIRST SECTION FOR SMOOTH CANALS AND DITCHES AND THE SECOND SECTION FOR ORDINARY STREAMS WITH ROUGH BOTTOMS.

Slope 1 in.	$\sqrt{r}$ in feet—Smooth.					$\sqrt{r}$ in feet—Rough.				
	0.4	1.0	1.8	2.5	4.0	0.4	1.0	1.8	2.5	4.0
	C	C	C	C	C	C	C	C	C	C
1.000	35.7	62.5	80.3	89.2	99.9	19.7	37.6	51.6	59.3	69.2
1.250	35.5	62.3	80.3	89.3	100.2	19.6	37.6	51.6	59.4	69.4
1.667	35.2	62.1	80.3	89.5	100.6	19.4	37.4	51.6	59.5	69.8
2.500	34.6	61.7	80.3	89.8	101.4	19.1	37.1	51.6	59.7	70.4
3.333	34.0	61.2	80.3	90.1	102.2	18.8	36.9	51.6	59.9	71.0
5.000	33.0	60.5	80.3	90.7	103.7	18.3	36.4	51.6	60.4	72.2
7.500	31.6	59.4	80.3	91.5	106.0	17.6	35.8	51.6	60.9	73.9
10.000	30.5	58.5	80.3	92.3	107.0	17.1	35.3	51.6	61.5	75.4
15.840	28.5	56.7	80.2	93.9	112.2	16.2	34.3	51.6	62.5	78.6
20.000	27.4	55.7	80.2	94.8	115.0	15.6	33.8	51.5	63.1	80.6

Interpolations may be made for intermediate values of slope or $\sqrt{r}$ in computation. For example, for a slope of 2 feet per mile, $\frac{5280}{2}=1$ in 2,640 and with a hydraulic radius of 2.99 we find by the table the approximate coefficient 94.9; then velocity $=94.9\sqrt{2.99}\times\sqrt{\frac{1}{2640}}=1.73\times.01946=.0336658\times94.9=3.19$ feet per second.

THE SLOPE OR HYDRAULIC GRADIENT

The slope of a ditch or stream is its fall in feet per mile or 1 divided by any measured length, as the case may be named in the formula.

For the velocity of flow in rivers having a uniform area, we have the following simple formula for velocity from a given slope, and, by inversion, the slope for a given velocity:

1.2 $\sqrt{\text{hydraulic radius}\times\text{slope in feet per mile}}=$ velocity in feet per second.

Thus, for example, as for Fig. 53, with a hydraulic radius of 2.99 and an assumed slope of 2 feet per mile:

1.2 $\sqrt{2.99\times2}=\sqrt{5.98}\times1.2=2.93$ feet velocity per second.

For the velocity of flow in canals and ditches of uniform area and smooth bottom, we have the formula:

2d. $\sqrt{\text{hydraulic radius}\times2\times\text{slope in feet per mile}}=$ velocity in feet per second.

Assuming the hydraulic radius of the canal or ditch as in Fig. 55, with a slope or hydraulic grade of 1 foot per mile,

Then $\sqrt{1.714\times2\times1}=1.85$ feet velocity per second.

The widening of such a ditch to 24 feet, with the sides of the same slope and depth, makes a marked increase in the velocity of the current with the same hydraulic grade, or will allow of a less grade for the same current, as an example will show.

Width at top 24 feet, bottom 16 feet, sides as in Fig. 55: Then the area will be $20\times3=60$ square feet, and the wet perimeter will be $5+5+16=26$ feet, and $\frac{60}{26}=2.307$, the hydraulic radius.

Then $\sqrt{2.307\times2\times1}=\sqrt{4.614}=2.15$ feet per second for velocity of the current. As this is too swift a current for canals or ditches with a lining of sand or loam, some other grade may be adopted, say three-quarters of a foot to a mile, which with the above formula will give a velocity of 1.86 feet per second, or an inversion of the formula may be used to adapt the grade to a required current, as shown farther on.

For ascertaining the required slope or hydraulic grade for any given velocity of current per second with an assigned hydraulic radius for a canal or ditch, the formulas Nos. 1 and 2 may be inverted as follows:

3d. $$\frac{\text{velocity in feet per second}^2}{\text{Hydraulic radius} \times 2} = \text{gradient in feet per mile.}$$

Taking the second example, Fig. 55, the computation becomes

$$1.85^2 = \frac{3.428}{1.714 \times 2} = 1 \text{ foot per mile slope or hydraulic gradient.}$$

With these data for an irrigation ditch twice the width with a required velocity of 2 feet per second, the equation

3d $$\frac{2^2}{2.307 \times 2} = 0.866 \text{ foot per mile for the slope or hydraulic gradient.}$$

The ditch, as in Fig. 55, with a current velocity of 1.85 feet per second and a hydraulic grade of 1 foot per mile, will deliver 24 square feet × 1.85 = 44.4 cubic feet of water per second, or over 3,800,000 cubic feet per day of 24 hours.

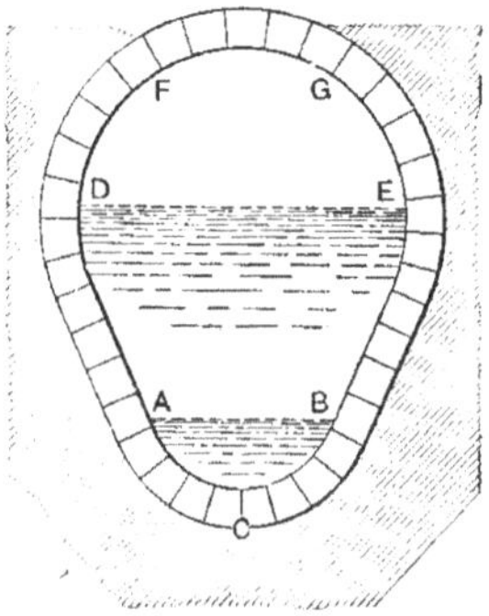

FIG. 56.—Sewer.

A good form of the section of a sewer, as shown in Fig. 56, is essential to the uniformity of effective flow at different depths and for the purpose of preventing the lodgment of silt at the lowest stage of flow.

The lower section A, B, has a larger hydraulic radius than the high stage section D, E, and thus compensates for the greater velocity in the larger area.

The computation for velocity and volume of flow is the same as for the trapezoid or open channel, Fig. 55.

The slope of the old Croton aqueduct is 1.1 feet per mile. The new aqueduct has a less slope, due to its larger area, being 8½ inches to a mile.

These low-grade slopes were made to meet the requirement of a nearly perfect wet perimeter.

In practice, the ditches for irrigation purposes, where the length is less than five miles, should be made level, with depth enough to balance the hydraulic grade, so that waste weirs will not be required, that the water catch may all be saved during a time of drought.

Reservoirs wherever practicable along the line of main ditches are of great importance, in arid districts, especially where the source of supply is a flood stream that may become dry during a drought. At such times a single irrigation from a stored supply at the opportune time will often save a crop.

In regard to the quantity of water required to save a crop on a given size farm, no definite figure can be given under the varied conditions of climate and rainfall, yet it has been estimated that, for a larger part of the arid lands of the United States, a depth of 3 inches on the whole acreage during the dry season will save a crop.

A depth of 6 inches will be sufficient in the more arid districts of Arizona and New Mexico. (See chapter on Irrigation.)

Taking for example a farm with 100 acres under special cultivation, the quantity required for a seasonal depth of 3 inches will be about 1,089,000 cubic feet, and for 6 inches in depth 2,178,000 cubic feet.

These figures look large, but a ditch of 5 square feet water area, or say $4\frac{1}{2}$ feet wide by $1\frac{1}{2}$ feet in depth, at a velocity of 1 foot per second, will meet the supply of 3 inches in depth over 100 acres in 14 days, and of 6 inches in depth in one month.

This will be more than enough for a continuous drought of two months.

The possibilities of storage by artificial ponds or lakes is of the utmost importance where the source of supply is limited; and although not always feasible by dams in the streams, a series of reservoirs along the line of ditch may often be made available for the storage of one or two million cubic feet of water, and in this way save the cost of large ditches and make available the smallest streams by the cumulative process through the rainy season.

Where pumping is required we refer the reader to a subsequent chapter.

If a dam or any obstruction is to be made in the stream for the purpose of obtaining a fall, a carpenter's level, or, if not at hand, a water-level may be improvised and set up at the proposed site and height of the dam, so ranged that the fall may be measured at any desired distance downstream and the backwater line of the proposed dam observed and staked.

The water-level, Fig. 57, may be made of a board 4 inches wide by 2 feet in length, planed straight upon the edges and slightly concave in the centre and oiled. It should be laid upon another flat board, which may rest upon a pile of stones, or on stakes, so arranged that it may be leveled by wedging up the under board. A little water in the slightly hollowed part of the upper board showing an approximate level suitable for mill engineering.

This arrangement will allow the level to be swept around, up, down, or across the stream for observation.

A pole with its foot set at the water's edge at any point downstream may be observed by the eye along the upper edge of the water-

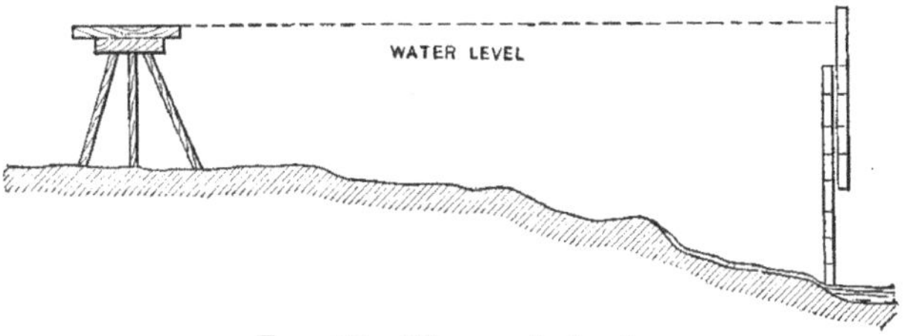

FIG. 57.—The grade level.

level, and the level height marked by a second person by shoving a stick up to the point of sight, and thus obtain a measure reliable for all practical purposes.

The cubic feet of water flowing per second multiplied by its weight per cubic foot (62.5 pounds) and this product multiplied by the observed height in feet gives the number of pounds falling 1 foot per second, which, divided by 550, gives the horse-power of the stream.

To make this plainer to the mind of the amateur, the following example is given:

2′ 4″
2′ 8″
3′ 9″
4′ 3″
3′ 3″
2′ 6″
———
18′ 9″

Let the measurements be taken as described for Fig. 53 at

$$18'\ 9'' = \frac{18.75'}{6} = 3.225 \text{ mean depth.}$$

Let the stream be 50 feet wide, then $50 \times 3.225 = 161.25$ square feet area.

Let the surface velocity, as found by several measurements, be 40 feet in 20 seconds, or 2 feet per second, with a fairly smooth bottom, for which we deduct 10 per cent., making the mean velocity 1.8 feet per second. Then $161.25 \times 1.8 = 290.25$ cubic feet. Let the observed fall that can be had equal 10 feet.

Then $\frac{290.25 \times 62.5 \times 10}{550} = 329.8$ horse-power as the value of the stream.

From this 25 per cent. should be deducted for loss of head by the flume, leakage, and the coefficient of the best water-motors, which leaves for work 247 horse-power. If an ordinary water-wheel is to be used, not less than 40 per cent. should be deducted from the gross power of the stream.

Allowance should also be made for the water stage of a stream at the time of measurement, so that a computation for the average of a season may be made.

The backwater of a dam is of some value, if it can be made large enough to be made available on small streams, when the power required is nearly equal to the flow. The proportions for increase of backwater service in large streams where the full power is utilized become enormous, as a few figures will show.

For the above stream the constant flow for 10 hours is $290.25 \times 36{,}000$ seconds $= 10{,}449{,}000$ cubic feet for a gross power of 329.8 horse-power.

To increase this power 10 per cent. by impounding the night flow will require an area to hold 1,044,900 cubic feet of water, or the raising of the dam 1 foot, and with a pond area 200 feet wide by 5,224 feet or nearly one mile long. Hence the utility of storage of water for power purposes is a useless expense, except where the natural conditions greatly favor it.

The raising of the water-head or dam 10 per cent. is expensive, in proportion to the gain; and, finally, the most effective way to do more work with water power is to extend the running time to 11 or 12 hours, or making 2 shifts of 8 or 10 hours each per 24 hours, as is so much in practice among our great mills using water or steam power.

THE SCOURING ACTION OF WATER

As undue velocity has an abrading effect upon the material of the bottom and sides of channels, various experiments have given us a gauge for guidance in laying out the hydraulic slope of canals and ditches. Its effect has been found to be for a mean velocity of—

0.4 of a foot per second, moves fine clay.
0.5 of a foot per second, moves loam and fine sand.
0.8 of a foot per second, moves fine gravel.
1.6 feet per second, moves coarse gravel size of beans.
3.5 feet per second, moves shingle 1 inch in diameter.
4.5 feet per second, moves stones $1\frac{1}{2}$ inches diameter.
6.0 feet per second, moves stones 3 to 6 inches diameter.
10.0 feet per second, moves bowlders and rocks.

DISCHARGE FROM FLOODED WEIRS AND SUBMERGED ORIFICES

When a dam, weir, or orifice has a backwater obstructing their flow, as shown in Figs. 58 and 59, the formula for an obstructed dam or weir for its flow is,

$$C\ l\ \sqrt{2g\ h}\ (d+\tfrac{2}{3}\ h) = \text{volume in cubic feet per second.}$$

In which C is a coefficient of from 0 to 1, increasing with the difference in levels of head and tail water.

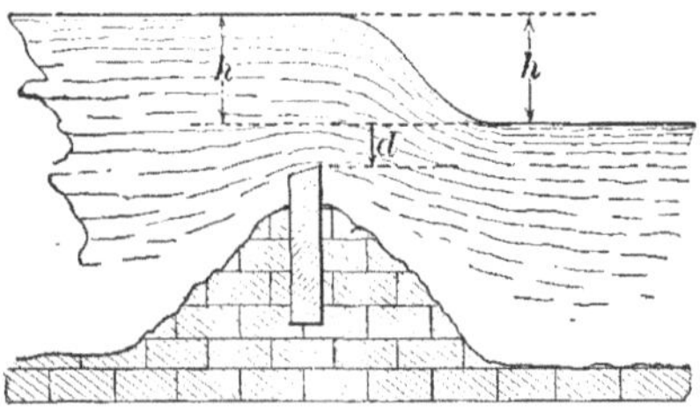

FIG. 58.—Submerged dam.

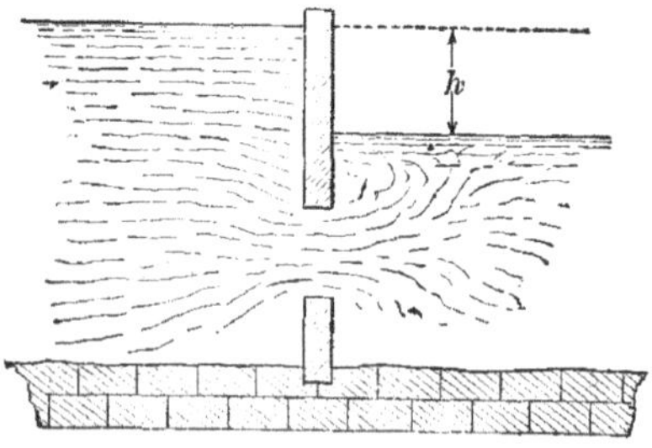

FIG. 59.—Submerged orifice.

l = length of dam or weir; h = difference in height of head and tail water in feet; d = difference in height between the edge of dam or weir and the tail water in feet. Then for example, with a submerged dam and overflow of 4 feet with h=3 feet; d=1 foot; l=5 feet; the figured expression will be with a coefficient of .6 as follows:

$$.6\times5\times\sqrt{64.12\times1}\ (1+2)=72.18 \text{ cubic feet per second.}$$

For a submerged orifice with a backwater overflow, as shown in Fig. 59, the formulas are $C^1\sqrt{2g\ h}$ = the velocity in the orifice in feet per second, and $C^1\ a\ \sqrt{2g,\ h}$ = the volume in cubic feet per second when a = square feet of orifice.

For example, with a differential height h=4 feet, and C=.6, the velocity in the orifice will be $.6\times8.02\times2=9.624$ feet per second, and if the area is 2 square feet, the volume will be 19.248 cubic feet per second.

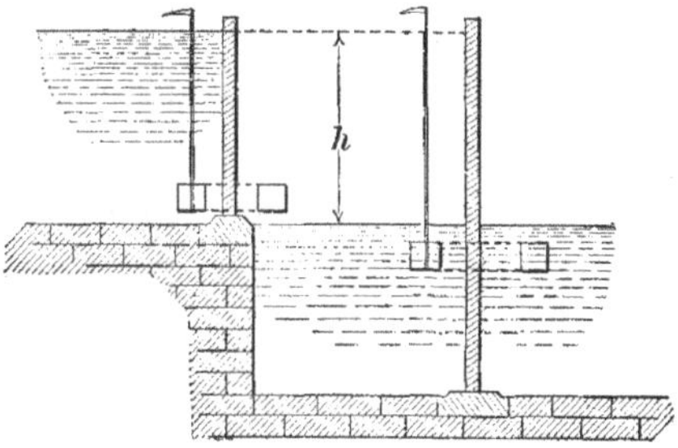

FIG. 60.—Time of discharge.

A formula applicable for the time for discharging a tank or a canal lock, requires the discharge gate to be submerged and the height (*h*) to be equal to the difference in water-levels at the commencement of discharge.

A = the area or contents in cubic feet, and a = area of the gate or gates, C=.6.

Then $\dfrac{2\ A\ \sqrt{h}}{C\ a\ \sqrt{2\ g}}$ = time in seconds.

For example, a lock 80′×20′×16′ fall with a discharge gate of 8 square feet $\dfrac{2\times1{,}600\times4}{.6\times8\sqrt{64.32}}=332$ seconds or $5\frac{1}{2}$ minutes for its discharge.

CHAPTER IV

FLOW FROM SUBSURFACE ORIFICES AND NOZLES

THE velocity of a fluid issuing from a hole or aperture of the best form in a reservoir is theoretically the same as it would be if it fell from the height of the surface of the water in the reservoir to the center of the hole or aperture. It varies as the square root of the height and is found to be equal to the square root of the sum of twice gravity (64.32) and the height in feet, the formula of which is:

$$\sqrt{2\ g \times h} \text{ or } 8.02 \times \sqrt{\text{height}}.$$

Thus for a fall of 1 foot the velocity is $\sqrt{64.32 \times 1} = 8.02$ feet per second, and for a fall of 16 feet it is $\sqrt{64.32 \times 16} = 32.08$ feet per second.

Practically, friction and the forms of orifices of issue make a reduction from the theoretical values, except in some peculiar forms in which the velocity exceeds the theoretical value.

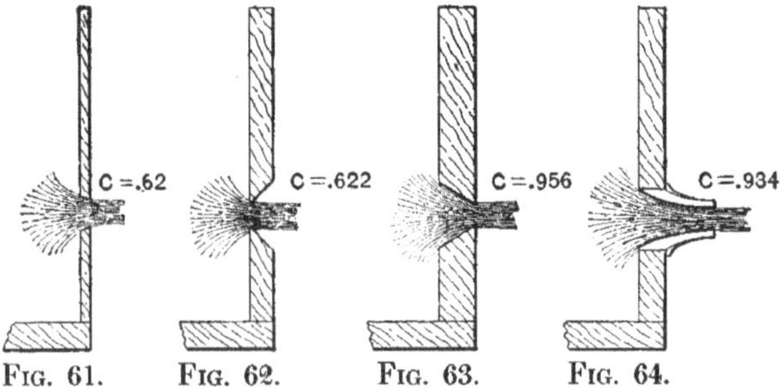

FIG. 61. FIG. 62. FIG. 63. FIG. 64.

Forms of nozles.

The coefficient of discharge may be briefly stated for all ordinary tests for volume of discharge to be from an orifice in a thin plate, or a hole in a plank with a bevel on the outside brought to a sharp edge up on the inner side or a nozle not longer than three times the diameter of the smallest part, as shown in Figs. 61, 62, 63, and 64, in which the coefficients are marked for each form of orifice.

The measure of any orifice, opening, or nozle should always be its smallest diameter.

For the quantity delivered from these forms of orifice, the rule for which is, the velocity in feet per second multiplied by the area in square feet or decimals of a foot gives the volume in cubic feet per second.

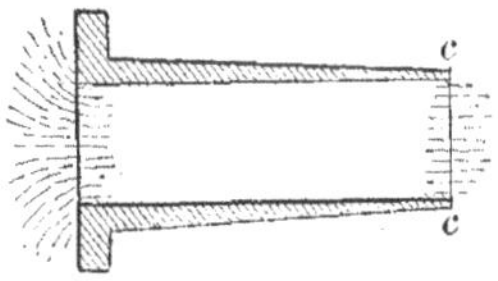

FIG. 65.—Straight nozle.

For obtaining the area, multiply the square of the diameter in inches by 0.7854 and divide the product by 144: which will give the square feet or decimals of a square foot as a multiplier in the above rule for cubic feet per second for round orifices.

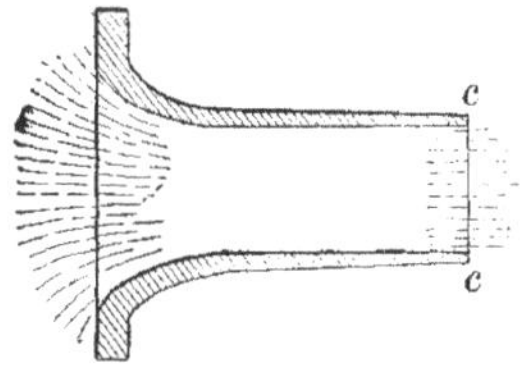

FIG. 66.—Curved entrance.

As an example say for a 3-inch orifice or nozle the area will be

$$3'' \times 3'' = 9'' \times 0.7854 = \frac{7.0686''}{144''} = 0.04909$$

of a square foot.

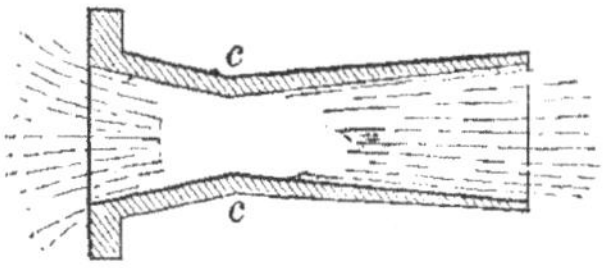

FIG. 67.—Double cone.

The velocity for a spouting jet will be, theoretically for 16 feet head, equal to $\sqrt{64.32 \times 16'} = 32.08$ feet per second, and $32.08 \times 0.04909 = 1.5748$ cubic feet per second.

The coefficient of discharge as found by experiment varies in the different forms of tubular nozles as follows:

A straight nozle $1\frac{1}{2}$ to $2\frac{1}{2}$ times its internal diameter in length at *c*, *c*, is .823; and for the same tube rounded to eliminate the vein of contraction, is .934.

A double conical tube $3\frac{1}{2}$ times its smallest diameter in length of diverging end is for *c*, *c*, 1.42, showing the power of the diverging nozle for accelerating the velocity of discharge through its smallest area.

The application of the coefficient of delivery for the best form of orifice, as in Fig. 63, will reduce the theoretical to the actual value, and $1.5748 \times 0.956 = 1.5055$ cubic feet, the actual flow per second.

In order to turn the value of this flow into horse-power, multiply the volume by 60 (seconds in a minute) and by 62.5 (pounds in a cubic foot) and by 16 (feet in height); divide the cumulative product by 33,000 (pounds lifted 1 foot high per minute), the quotient equals the gross horse-power.

The expression will then be:

$$\frac{1.5055 \times 60 \times 62.5 \times 16}{33,000} = 2.737 \text{ horse-power}$$

The realization of work from this gross power requires a further reduction for the inefficiency of water-wheels, motors, or turbines, the value of which varies from 65 to 85 per cent. of the gross power of the spouting stream, taking the highest coefficient for a turbine or Pelton wheel and multiplying by the above gross power

$$0.85 \times 2.737 = 2.326$$

the actual horse-power available for work for the best form of water-wheels.

THE CONTRACTED VEIN

When a cylindrical ajutage is used, there will be a partial vacuum formed between the sides of the tube and the contracted vein, as shown in Fig. 68. If a pipe ascending from a reservoir of water is let into this part of the ajutage, the water will rise in the pipe; and if the height is not too great, the vessel may be emptied.

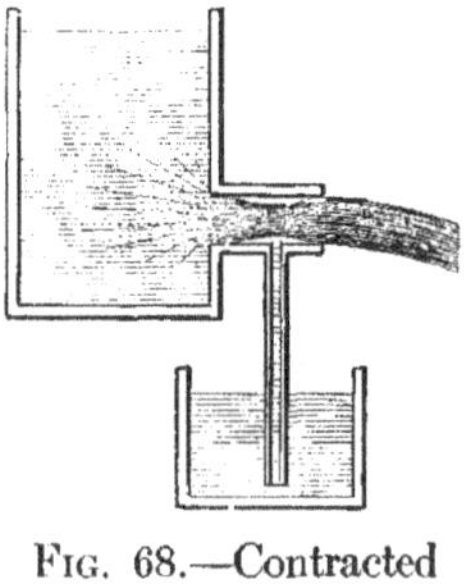

FIG. 68.—Contracted vein.

The contracted vein is caused by the converging motion of the water entering the ajutage.

The horse-power that may be obtained from the flow from an orifice or nozle, as shown in Table IV, requires a deduction for the coefficient of the form of the orifice or nozle (C V); tabular flow × coefficient of the orifice, for the actual flow in gallons per minute which, multiplied by the weight, 8.34 pounds per gallon, and by the height or head due to the pressure and the product divided by 33,000, equals the theoretical horse-power. From this a deduction must be made for the actual power of the water-wheel. See subsequent chapter on water motors, wheels, and turbines.

TABLE IV.—THEORETICAL DISCHARGE OF WATER FROM ORIFICES AND NOZLES, FROM WHICH THE COEFFICIENT OF FORM AND FRICTION MUST BE DEDUCTED FOR THE ACTUAL FLOW IN GALLONS. (ELLIS.) SEE FIGS. 65, 66, AND 67.

Head.		Velocity of discharge in feet per second.	Number of United States Gallons of 231 Cubic Inches Discharged per Minute.																		
Lbs.	Feet.		1/16	1/8	3/16	1/4	3/8	1/2	5/8	3/4	7/8	1	1 1/8	1 1/4	1 3/8	1 1/2	1 3/4	2	2 1/4	2 1/2	Lbs.
10	23.1	38.58	0.37	1.48	3.30	5.90	13.2	23.6	36.8	53.2	72.2	94.4	119	148	178	212	289	378	478	590	10
15	34.7	47.25	0.45	1.81	4.02	7.23	16.2	28.7	45.0	65.1	88.4	116	146	181	218	260	354	463	586	723	15
20	46.2	54.55	0.52	2.09	4.66	8.35	18.7	33.4	52.0	75.3	102	134	169	209	252	300	409	534	676	835	20
25	57.8	60.99	0.58	2.33	5.23	9.33	20.9	37.2	58.2	84.1	114	149	189	233	282	336	457	597	756	933	25
30	69.3	66.82	0.64	2.56	5.71	10.2	22.8	40.9	63.7	92.2	125	164	207	256	309	368	501	654	828	1022	30
35	80.9	72.16	0.69	2.76	6.16	11.0	24.7	44.2	68.8	99.6	135	177	223	276	334	397	541	707	895	1104	35
40	92.4	77.14	0.74	2.95	6.60	11.8	26.4	47.2	73.6	106	144	189	239	295	357	425	578	755	956	1180	40
45	104.0	81.83	0.78	3.13	6.99	12.5	28.0	50.2	78.1	113	153	200	253	313	378	450	613	801	1014	1252	45
50	115.5	86.26	0.82	3.30	7.37	13.2	29.5	52.8	82.3	119	161	211	267	330	399	475	646	845	1069	1320	50
55	127.1	90.46	0.86	3.46	7.73	13.8	30.9	55.4	86.3	125	169	221	280	346	418	498	678	886	1122	1385	55
60	138.6	94.49	0.90	3.62	8.08	14.5	32.3	57.8	90.1	130	177	231	293	362	437	520	708	925	1171	1446	60
65	150.2	98.35	0.94	3.77	8.40	15.1	33.6	60.2	93.8	136	184	241	305	377	455	542	737	963	1219	1506	65
70	161.7	102.06	0.97	3.91	8.73	15.6	34.9	62.5	97.4	141	191	250	316	391	472	562	765	999	1265	1561	70
75	173.3	105.65	1.01	4.04	9.03	16.2	36.1	64.6	101	146	198	259	327	404	488	582	792	1034	1309	1616	75
80	184.8	109.11	1.04	4.18	9.33	16.7	37.8	66.6	104	150	204	267	338	418	504	601	818	1068	1352	1669	80
85	196.4	112.46	1.07	4.31	9.62	17.2	38.5	68.8	107	155	210	275	348	431	520	620	843	1101	1394	1720	85
90	207.9	115.72	1.10	4.43	9.89	17.7	39.6	70.8	110	160	217	283	358	443	535	637	867	1133	1434	1770	90
95	219.5	118.89	1.13	4.55	10.2	18.2	40.7	72.8	113	164	223	291	368	455	550	655	891	1164	1474	1820	95
100	231.1	121.98	1.16	4.67	10.4	18.7	41.7	74.6	116	168	228	299	378	467	564	672	914	1194	1512	1866	100
105	242.6	125.00	1.19	4.78	10.7	19.1	42.8	76.5	119	172	234	306	387	478	578	688	937	1224	1549	1912	105
110	254.2	127.94	1.22	4.90	10.9	19.6	43.8	78.3	122	177	239	313	396	490	591	705	959	1253	1586	1957	110
115	265.7	130.82	1.25	5.01	11.2	20.0	44.8	80.1	125	181	245	320	405	501	605	720	980	1281	1621	2002	115
120	277.3	133.63	1.27	5.12	11.4	20.4	45.7	81.8	127	184	250	327	414	512	618	736	1001	1308	1656	2044	120
125	288.8	136.38	1.30	5.22	11.7	20.9	46.7	83.5	130	188	255	334	422	522	630	751	1022	1335	1691	2086	125
130	300.4	139.08	1.33	5.32	11.9	21.3	47.6	85.1	133	192	260	341	431	532	643	766	1042	1362	1724	2128	130

THE PROJECTED JET

A jet issuing from the side of a vessel describes, theoretically, a parabola, precisely as in the case of a solid projectile; for the impelling force and the force of gravity act upon the jet in the same manner, and the resultant force gives it the same direction. The range, or distance to which the jet is projected, is greatest when the angle of elevation is 45°, and is the same for elevations which are equally above or below 45°, as 60° and 30°. The resistance of the air, however, alters the results, and the statement is true only when the jet is projected into a vacuum.

If a vessel filled with water have orifices made in its side at equal distances in a vertical line from the top to the bottom, a stream issuing from an orifice midway between the surface and the bottom will be projected farther than any of the streams issuing from the orifices above or below. This may be demonstrated by the diagram, Fig. 69. Let a semicircle A F E be described on the side of a vessel of water, its diameter being equal to the height of a liquid. The range of a jet issuing from either of the orifices B, C, or D will be equal to twice the length of the ordinates B F, C I, or D K respectively; and therefore jets issuing from B and D will meet at a point H on a level with the bottom and twice the length of the ordinates B F and D K. Now, as the ordinate C I is the greatest, the range of the jet issuing from C will be greater than that of any other jet.

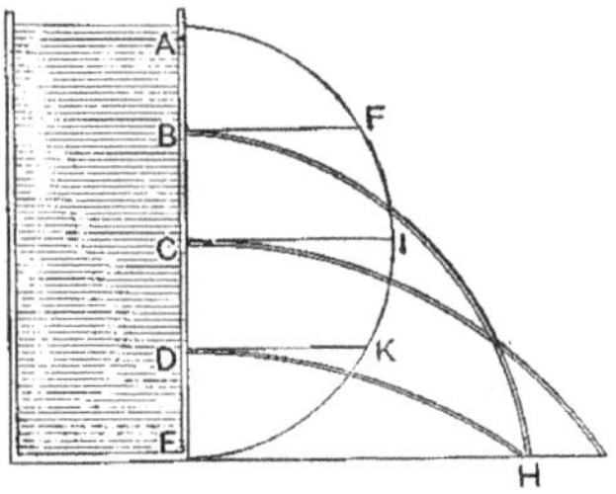

FIG. 69.—Horizontal jets.

The volume of water issuing from each jet in the same time is proportionally as the square root of its depth from the surface.

JETS AND FIRE STREAMS

The velocity of a jet at the orifice is the same at all angles from horizontal to vertical, and where the flow behind the orifice is unobstructed by friction, is equal to a coefficient for the form of the aperture multiplied by the square root of the product of twice gravity multiplied by the height in feet, for which the formula is

$$\text{C.} \quad \sqrt{2g \times h} = \text{velocity in feet per second.}$$

The greatest range of a horizontal jet on the floor line of a tank is when the orifice is placed in the middle of the water-head, as shown in Fig. 70, while jets from equal distances from the water-head and floor line will meet at the floor line, as shown in the cut.

The curved form of these jets is that of a section of a parabola from its axis, which corresponds with the front line of the tank.

Using the terms as expressed in Fig. 69, for the distance that a jet will touch the horizon of the bottom of the tank, the coefficient of the

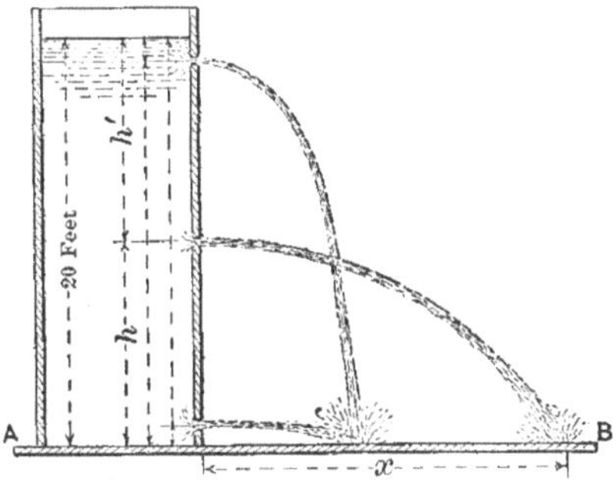

FIG. 70.—Horizontal jet.

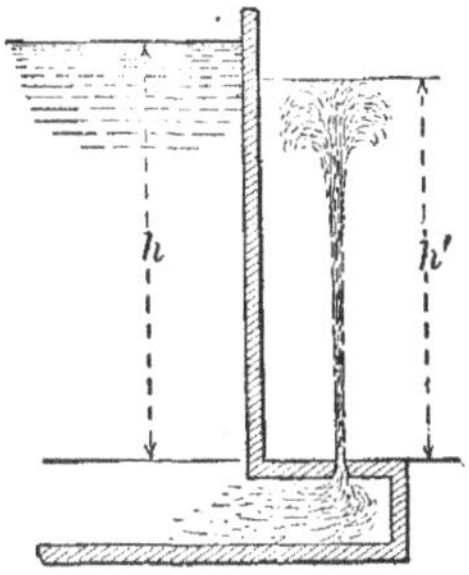

FIG. 71.—Vertical jet.

velocity multiplied by twice the square root of the pressure head h', multiplied by the height of the centre of the orifice above the floor, A B, equals the distance from the line of the tank that the jet will strike the floor—or, using the best form of orifice,

$$0.93 \sqrt{h' \times h} \times 2 = x,$$

the distance as in Fig. 70.

For example, with $h = 10$ feet, $h' = 10$, then

$$\sqrt{10 \times 10} = 10 \times 2 \times 0.93 = 18.6 \text{ feet},$$

at which point the jet touches the floor line at B.

In the same way, using the above formula, the jet from both the top and bottom orifices will touch the floor line at 8 feet from the front of the tank.

The quantity of water discharged by these jets will be

$$\text{coefficient} \times \frac{\text{area in square inches}}{144} \times \sqrt{2g \times h'} = \text{volume in cubic feet}$$

per second: as

$$\sqrt{2g} = 8.02$$

the square root of the head in feet from the centre of the orifice to the surface of the water may be used × by 8.02 for facilitating computation.

For example, taking the middle orifice in Fig. 70 at 1 inch in diameter, then the equation will be with the best form of orifice, and h′ = 10 feet,

$$0.93 \times \frac{1'' \times 0.7854}{144} \times 8.02 \times \sqrt{10} = 0.138 \text{ of a cubic foot per second.}$$

Let it be here understood that static height and pressure by gauge in pipes, tanks, and stand-pipes are convertible terms, so that, where gauge pressure is expressed in pounds per square inch, it should be multiplied by 2.3093 for its equivalent height of head in feet, and when head in feet is expressed, it should be multiplied by 0.433 for pressure in pounds per square inch. For vertical jets or fountains the coefficient of velocity may vary for various forms of orifice, as shown before, from 0.62 to 0.93 per cent. of the theoretical effect.

Taking the best form of nozle, the formula for velocity becomes

$$0.93\sqrt{2g \times h} = \text{velocity in feet per second,}$$

h being the height or head in feet of the source of supply, and not allowing for friction in intermediate pipes.

The height of the jet may be computed from the formula

$$\frac{V^2}{2\,g} = h'$$

or the square of the velocity divided by twice gravity equals the height of the jet.

For example, as illustrated in Fig. 71, with a head, *h*, of 64 feet and a nozle of best form, the computation becomes

$$0.93 \times \sqrt{64.33 \times 64}$$

the last terms of which may be multiplied together and the square root taken, or the square root of each member taken and multiplied, which reduced $=0.93\times 8.02\times 8=59.66$, the velocity of the jet at the nozle in feet per second, and

$$\frac{59.66^2}{64.33}=55.32 \text{ feet,}$$

the height of the jet from a head of 64 feet.

This does not include the friction in the reservoir or pipes leading to the nozle or orifice. By using the coefficients for various forms of nozle, the height of the vertical jet will be found to vary from 97 per cent. to 62 per cent. of the whole height, or, if the jet is taken from water mains, the gauge pressure multiplied by 2.3093 will represent the hydrostatic height to be used in the formulas.

The retardation by friction of the atmosphere affects height as given by the formulas to a considerable degree, small jets and any jets under high pressure being most retarded by air friction, as shown farther on in the table of actual heights from Freeman's experiments.

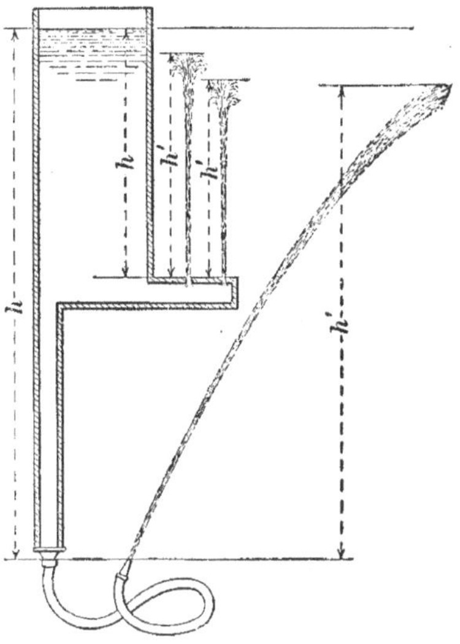

FIG. 72.—Vertical jets and fire stream.

TABLE V.—POUNDS PRESSURE LOST BY FRICTION, IN EACH 100 FEET of $2\frac{1}{2}$-INCH FIRE HOSE, FOR GIVEN DISCHARGES OF WATER PER MINUTE.

Diameter of Nozle. Inches.		PRESSURE AT HOSE NOZLE.								
	Head in lbs. per sq. in..	20	30	40	50	60	70	80	90	100
	Head in feet..........	46.2	69.3	92.4	115.5	138.6	161.7	184.8	207.9	231.0
1	Galls. discharged.......	110	134	155	173	189	205	219	232	245
	Rubber hose, lbs. lost...	4.35	6.40	8.40	10.20	12.80	14.80	17.00	19.20	20.50
	Leather hose, lbs. lost...	6.33	8.53	10.83	13.10	15.34	17.79	20.11	22.40	24.83
$1\frac{1}{8}$	Galls. discharged.......	139	170	196	219	240	259	277	294	310
	Rubber hose, lbs. lost...	6.79	10.16	13.60	17.05	20.59	24.00	27.00	30.00	33.00
	Leather hose, lbs. lost....	9.05	12.71	16.38	20.11	23.88	27.61	31.41	35.24	39.07
$1\frac{1}{4}$	Galls. discharged.......	171	210	242	271	297	320	342	363	383
	Rubber hose, lbs. lost...	10.28	15.64	20.85	25.46	29.50	39.00	43.81	49.42	55.00
	Leather hose, lbs. lost...	12.84	19.00	24.07	30.11	35.94	41.57	47.36	53.25	59.20
$1\frac{3}{8}$	Galls. discharged.......	207	253	293	327	358	387	413	439	462
	Rubber hose, lbs. lost...	15.00	22.96	29.40	40.50	48.20	55.70	64.70	72.00	79.26
	Leather hose, lbs. lost...	18.81	26.39	35.01	43.38	52.00	60.40	68.59	76.73	84.87

The form of nozle for fire streams has been the subject of experiment for many years, and has finally, in an exhaustive trial by John R. Freeman, C.E., resulted in a form probably the most perfect yet attained, having a coefficient of 0.976 for a $1\frac{1}{8}$-inch nozle, and a trifle greater coefficient with a $\frac{7}{8}$-inch and $\frac{3}{4}$-inch nozle of the same form on the same plug pipe, the advantage of the smaller stream being due to the less velocity friction between the pressure gauge at the foot of the butt and the nozle.

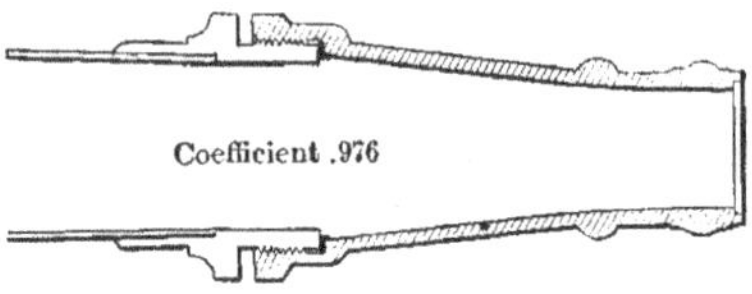

FIG. 73.—Nozle of best form.

This form is shown in its exact proportions in Fig. 73.

The loss of pressure in standard fire hose of $2\frac{1}{2}$-inch diameter with a fairly smooth inner surface is, in experiments by Freeman, from 13 to 14 pounds per square inch for each 100 feet in length, in ordinary fire service.

TABLE VI.—HORIZONTAL AND VERTICAL DISTANCE OF JETS WITH VARYING HEADS AND SIZES OF NOZLES.

Diameter of Nozle. Inches.		PRESSURE AT NOZLE.								
	Head in lbs. per sq. in..	20	30	40	50	60	70	80	90	100
	Head in feet..........	46.2	69.3	92.4	115.5	138.6	161.7	184.8	207.9	231.0
1	Galls. discharged.......	110	134	155	173	189	205	219	232	245
	Horizontal distance of jet................	70	90	109	126	142	156	168	178	186
	Vertical distance of jet..	43	62	79	94	108	121	131	140	148
$1\frac{1}{8}$	Galls. discharged.......	131	170	196	219	240	259	277	294	310
	Horizontal distance of jet................	71	93	113	132	148	163	175	186	193
	Vertical distance of jet..	43	63	81	97	112	125	137	148	157
$1\frac{1}{4}$	Galls. discharged.......	171	210	242	271	297	320	342	363	383
	Horizontal distance of jet................	73	96	118	138	156	172	186	198	207
	Vertical distance of jet..	43	63	82	99	115	129	142	154	164
$1\frac{3}{8}$	Galls. discharged.......	207	253	293	327	358	387	413	439	462
	Horizontal distance of jet................	75	100	124	146	166	184	200	213	224
	Vertical distance of jet..	44	65	85	102	118	133	146	158	169

The following table gives the extreme height of fire streams for various pressures and water-heads, as gauged at the butt, with best form of fire nozles of two sizes, the intermediate sizes being proportional to their areas. (From Freeman's experiments.)

TABLE VII.—EXTREME HEIGHT.

Pressure lbs. per square inch.	Head in feet.	Height of stream. ¾ nozle. Feet.	Height of stream. 1¼ nozle. Feet.
25	57.7	49	51
43.3	100	82.5	88
50	115.5	92.3	101.2
70	161.6	113.5	131.8
86.6	200	127	150.7
100	230.9	133.2	158.9

The loss of pressure or water-head by friction in long hose is very serious, so that to maintain a $\frac{7}{8}$-inch nozle jet of 100 feet in height, which would require about 60 pounds pressure if thrown from the engine, would require 65 pounds with 50 feet of rubber-lined hose; 69 pounds with 100 feet; 75 pounds with 200 feet; 89 pounds with 400 feet; and 116 pounds with 800 feet.

With a limited pump pressure the height of the stream would fall in proportion with the increase in length of hose. For standard $2\frac{1}{2}$-inch hose the difference in required pressure for a given height is considerably greater for nozles above $\frac{7}{8}$-inch diameter and less for those under $\frac{7}{8}$-inch diameter than the differences above stated, the experiments showing that for 70 pounds pressure with $1\frac{1}{8}$-inch nozle and 50 feet best rubber-lined hose an effective fire stream of 81 feet in height could be attained, while with 250 feet of hose it was found to be only 61 feet, and with 500 feet of hose a height of only 46 feet could be attained.

The table on pages 74 and 75 will be found very convenient for reference as to the theoretical velocity, discharge, and horse-power of nozles of various sizes from 1 inch to 4 inches, and for various heads from 1 foot to 1,000 feet.

The first column represents the height or head of water, the second column the theoretical velocity through nozles of good form, without deduction for friction in pipes or flumes leading to the nozles. The first column under the size of nozle is the flow in cubic feet per second, the second that of the theoretical horse-power, no allowance for the coefficient for water-wheels being made, which amounts to from 60 to 86 per cent. of the power here given.

TABLE VIII.—VELOCITY, DISCHARGE, AND HORSE-POWER OF NOZLES.

Head in feet.	Velocity per second in feet.	1 inch.		1½ inches.		2 inches.	
		Cubic ft.	H. P.	Cubic ft.	H. P.	Cubic ft.	H. P.
1	8.02	0.041	0.004	0.093	0.010	0.164	0.018
1½	9.83	0.050	0.008	0.111	0.019	0.200	0.034
2	11.35	0.058	0.013	0.130	0.028	0.232	0.052
2½	12.68	0.064	0.018	0.145	0.041	0.256	0.072
3	13.90	0.069	0.024	0.159	0.054	0.284	0.096
3½	15.01	0.076	0.030	0.171	0.068	0.304	0.120
4	16.05	0.081	0.037	0.183	0.083	0.324	0.148
4½	17.02	0.086	0.044	0.194	0.099	0.344	0.176
5	17.95	0.091	0.051	0.205	0.113	0.364	0.204
6	19.66	0.100	0.068	0.224	0.153	0.400	0.272
7	21.23	0.108	0.086	0.242	0.193	0.432	0.344
8	22.70	0.116	0.104	0.260	0.252	0.464	0.416
10	25.38	0.129	0.146	0.290	0.329	0.516	0.584
12½	28.37	0.144	0.204	0.324	0.460	0.576	0.816
15	31.08	0.158	0.269	0.355	0.505	0.632	1.08
17½	33.57	0.170	0.339	0.383	0.782	0.680	1.36
20	35.89	0.182	0.414	0.410	0.931	0.728	1.66
22½	38.07	0.193	0.494	0.435	1.11	0.772	1.98
25	40.13	0.204	0.578	0.458	1.30	0.816	2.31
27½	42.08	0.213	0.667	0.480	1.50	0.852	2.67
30	43.95	0.228	0.760	0.513	1.71	0.912	3.04
32½	45.75	0.232	0.857	0.522	1.93	0.928	3.43
35	47.47	0.241	0.958	0.542	2.15	0.964	3.83
40	50.75	0.257	1.17	0.579	2.63	1.03	4.68
45	53.83	0.273	1.40	0.614	3.14	1.09	5.60
50	56.75	0.288	1.64	0.648	3.68	1.15	6.56
60	62.16	0.315	2.15	0.709	4.84	1.26	8.60
70	67.14	0.341	2.71	0.766	6.10	1.36	10.8
80	71.78	0.364	3.31	0.819	7.45	1.46	13.2
90	76.13	0.386	3.95	0.864	8.88	1.54	15.8
100	80.25	0.407	4.63	0.916	10.4	1.63	18.5
125	89.72	0.455	6.47	1.02	14.1	1.82	25.8
150	98.28	0.499	8.60	1.12	19.1	2.00	34.0
175	106.1	0.539	10.7	1.21	24.0	2.16	42.8
200	113.5	0.576	13.1	1.29	29.4	2.30	52.4
250	127.1	0.644	18.3	1.45	41.1	2.58	73.2
300	139.0	0.705	24.0	1.59	54.0	2.82	96.9
350	150.1	0.762	30.3	1.71	68.1	3.05	121.0
400	160.5	0.814	37.0	1.83	83.2	3.26	148.0
450	170.2	0.864	44.2	1.94	99.3	3.46	176.0
500	179.4	0.910	51.7	2.05	116.0	3.64	206.0
550	188.2	0.955	39.7	2.10	134.0	3.82	238.0
600	196.6	0.999	68.0	2.23	152.0	3.99	272.0
700	212.3	1.06	85.7	2.46	192.0	4.36	342.0
800	226.9	1.15	104.7	2.58	235.0	4.60	418.0
900	240.7	1.22	124.9	2.75	281.0	4.88	499.0
1000	253.8	1.29	146.2	2.89	329.0	5.16	584.0

TABLE VIII.—*Continued.*

Head in feet.	Velocity per second in feet.	2½ inches.		3 inches.		3½ inches.		4 inches.	
		Cu. ft.	H. P.	Cu. ft.	H. P.	Cu. ft.	H. P.	Cu. ft.	H. P.
1	8.02	0.255	0.029	0.372	0.040	0.50	0.056	0.656	0.072
1½	9.83	0.312	0.053	0.444	0.076	0.61	0.105	0.800	0.136
2	11.35	0.360	0.082	0.520	0.116	0.70	0.160	0.928	0.208
2½	12.68	0.402	0.114	0.589	0.164	0.79	0.224	1.02	0.288
3	13.90	0.440	0.150	0.636	0.216	0.86	0.295	1.14	0.384
3½	15.01	0.475	0.189	0.684	0.272	0.94	0.370	1.22	0.480
4	16.05	0.507	0.231	0.742	0.332	1.02	0.452	1.30	0.592
4½	17.02	0.540	0.275	0.776	0.396	1.06	0.540	1.38	0.704
5	17.95	0.567	0.315	0.820	0.452	1.11	0.600	1.46	0.816
6	19.66	0.622	0.425	0.896	0.612	1.22	0.833	1.60	1.09
7	21.23	0.672	0.535	0.968	0.772	1.32	1.05	1.73	1.38
8	22.70	0.720	0.656	1.04	0.928	1.40	1.28	1.85	1.66
10	25.38	0.805	0.915	1.16	1.32	1.57	1.79	2.16	2.34
12½	28.37	0.897	1.28	1.30	1.84	1.76	2.50	2.30	3.46
15	31.08	0.985	1.68	1.42	2.42	1.93	3.29	2.53	4.32
17½	33.57	1.06	2.11	1.53	3.13	2.08	4.20	2.72	5.44
20	35.89	1.14	2.58	1.63	3.72	2.23	5.07	2.91	6.64
22½	38.07	1.21	3.08	1.74	4.44	2.36	6.05	3.09	7.92
25	40.13	1.27	3.61	1.83	5.20	2.54	7.08	3.26	9.24
27½	42.08	1.33	4.17	1.92	6.00	2.61	8.17	3.41	10.68
30	43.95	1.42	4.75	2.05	6.84	2.70	9.31	3.65	12.16
32½	45.75	1.45	5.35	2.09	7.72	2.84	10.50	3.71	13.72
35	47.47	1.51	5.98	2.17	8.60	2.95	11.71	3.86	15.32
40	50.75	1.61	7.31	2.32	10.52	3.15	14.33	4.12	18.72
45	53.83	1.71	8.23	2.46	12.56	3.34	17.10	4.36	22.40
50	56.75	1.79	10.2	2.59	14.72	3.52	20.03	4.60	26.24
60	62.16	1.97	13.4	2.84	19.36	3.86	26.32	5.04	34.40
70	67.14	2.13	16.9	3.06	24.40	4.17	33.17	5.42	43.36
80	71.78	2.27	20.6	3.28	29.80	4.46	40.55	5.84	52.96
90	76.13	2.44	24.6	3.46	35.52	4.73	48.37	6.16	63.20
100	80.25	2.54	28.9	3.66	41.64	4.98	56.67	6.52	74.08
125	89.72	2.84	40.4	4.08	58.20	5.57	79.20	7.28	103.5
150	98.28	3.11	53.1	4.48	76.48	6.10	104.1	8.00	136.0
175	106.1	3.36	66.8	4.84	96.28	6.60	131.5	8.04	171.2
200	113.5	3.59	81.7	5.10	117.7	7.06	160.2	9.20	219.6
250	127.1	4.02	114.0	5.87	164.5	7.88	223.9	10.3	292.8
300	139.0	4.40	150.0	6.36	216.3	8.63	294.3	11.2	284.0
350	150.1	4.76	189.0	6.84	272.6	9.33	371.2	12.2	484.8
400	160.5	5.09	231.0	7.30	323.0	9.97	453.2	13.0	592.0
450	170.2	5.40	276.0	7.76	397.4	10.5	541.0	13.8	707.0
500	179.4	5.69	323.0	8.20	466.0	11.1	627.0	14.5	827.2
550	188.2	5.96	372.0	8.40	536.8	11.6	731.0	15.2	955.2
600	196.6	6.23	475.0	8.92	611.0	12.2	832.7	16.9	1088.0
700	212.3	6.79	535.0	9.84	771.2	13.3	1051.0	17.4	1371.0
800	226.9	7.19	654.0	10.3	942.0	14.1	1282.0	18.4	1675.0
900	240.7	7.63	780.0	11.0	1124.0	14.9	1530.0	19.5	1998.0
1000	253.8	8.04	914.0	11.5	1316.0	15.7	1791.0	20.6	2339.0

MEASUREMENT OF FLOW OVER WEIRS

The method of weir measurement is more suitable for small and rapid streams, mountain torrents and springs.

For this purpose a rough estimate of the size of the weir must be made to enable you to make it ample to pass the whole stream. A small weir may be made by a simple rectangular notch in a board, which for convenience of measurement may be 8 to 10 times wider than the depth. Bevel the edge of the notch toward the downstream side of the weir, as shown in Fig. 74.

Set it across the stream so that the bottom of the notch shall be level; holding the board in position by stakes or stones, and packing

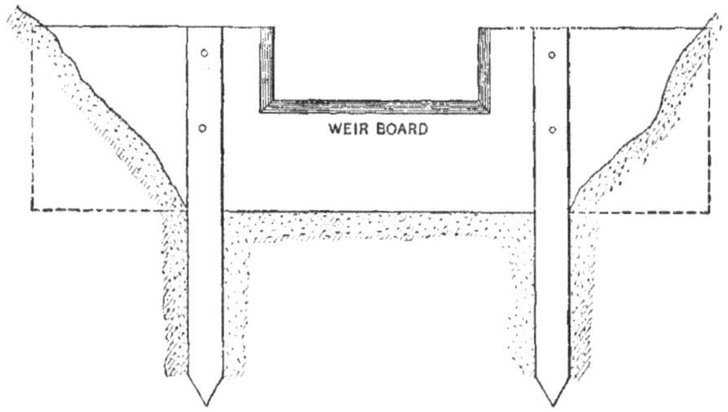

FIG. 74.—Weir board.

the under side and ends with any convenient material—as gravel or sand—and tightening by a layer of loam.

Drive a slender stake at A, Fig. 75, 2 or 3 feet upstream from the weir, and on one side of the notch, so that the top, B, shall be exactly level with the top of the weir board at C. The filling at the back should be kept a few inches below the spill at D, Fig. 75, to prevent obstruction or friction of the water as it approaches the spill.

When the stream pours over the weir, and the height at the stake E, Fig. 75, becomes stationary, measure the distance from the top of the stake at B to the surface of the water at E, and subtract it from the full depth of the notch C D. This will give the true depth of the spill.

As the slope of the water surface over the weir increases with the depth, caused by the velocity increasing as the water approaches the spill, it makes a direct measurement at the spill an uncertain exponent in the computation of volume; hence the necessity of levelling back to a stake in comparatively quiet water.

The levelling may be done with a carpenter's level on a straight edge, or by a water-level, as before described.

The difference between the actual depth at the spill and the properly measured depth as above may vary from 0.9 to 0.75 of the stake measure.

A coefficient of correction for the slope and friction must be used which by experiment has been found to vary from 0.63 for 1 inch in depth to 0.58 for 24 inches in depth, so that, for all ordinary pur-

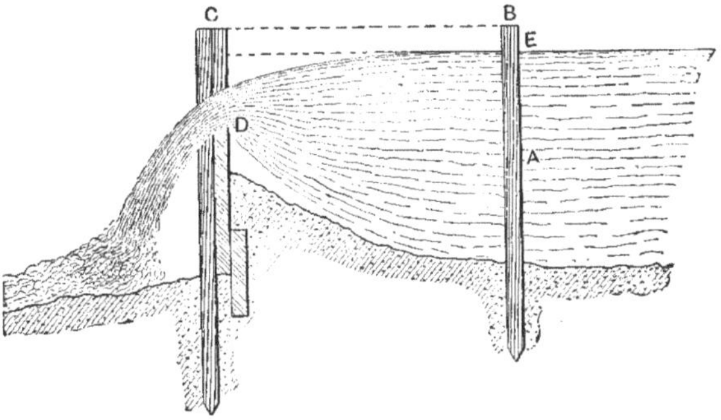

FIG. 75.—Stream profile.

poses, the coefficient C may be taken at 0.6. Then the equation for volume in cubic feet per second is

$$C \frac{2}{3} b, h, \sqrt{2 g, h}$$

For example, with a weir 6 inches deep, *h*, and 4 feet wide, *b*, the equation becomes when applied:

$$0.6 \times \frac{2}{3} \times 4' \times 0.5' \times \sqrt{64.32 \times 0.5}$$

The square foot of the last term

$$\sqrt{64.32 \times 0.5} = \sqrt{32.16} = 5.67$$

Then

$$0.6 \times \frac{2}{3} \times 4' \times 0.5 \times 5.67 = 4.536$$

cubic feet discharge per second. For the convenience of a more direct or easier computation, we give the following approximate table of depths in inches and decimals, with the corresponding depth in decimals of a foot, for depths up to one foot; which gives a close approximation for the flow in cubic feet per second, by simply multiplying the width of the weir in feet and decimals of a foot by the coefficient opposite to the measured depth of the weir as before described. Any measures between the depths given may be interpolated proportionately in all the columns. For example, with a weir as before, 4 feet wide by 6 inches deep, in the table opposite to 6 inches, or 0.5 of a foot, will be found 1.1295, which, multiplied by 4 feet, equals 4.518 cubic feet per second, which is near enough to the figures from the equation for all practical purposes:

TABLE IX.—WEIR TABLE FOR ONE FOOT IN WIDTH AND FOR DEPTHS UP TO ONE FOOT.

Depth inches and tenths.	Depth decimals of a foot.	Flow in cubic feet per second for one foot in width.	Depth inches and tenths.	Depth decimals of a foot.	Flow in cubic feet per second for one foot in width.
0.4	0.0333	0.01365	3.2	0.2666	0.44480
0.6	0.0495	0.03408	3.6	0.2995	0.53795
0.8	0.0666	0.05452	4.0	0.3333	0.63111
1.0	0.0833	0.08022	5.0	0.4165	0.88030
1.2	0.100	0.10592	6.0	0.5	1.12950
1.4	0.1165	0.13012	7.0	0.583	1.43795
1.6	0.1333	0.16616	8.0	0.666	1.74640
2.0	0.1666	0.22893	10.0	0.833	2.40040
2.4	0.200	0.29171	11.0	0.9166	2.76770
2.8	0.2333	0.36825	12.0	1.000	3.13500
3.0	0.2498	0.40653			

For further convenience we give the following weir table of larger scope than the one computed from the formula, which also gives a slightly larger volume, arising from the use of formulas for a different line of experiments, but near enough for all ordinary purposes.

It is arranged for depths of $\frac{1}{8}$ inch to 24 inches, and for a width of 1 inch; so that simply multiplying the quantity in the table at the junction of the unit and fractional part of an inch by the width of the spill in inches will give a near approximation to the volume or flow per minute.

In using this table of volume for computing the horse-power of a stream, 33,000 should be used as a divisor in the formula, which becomes:

TABLE X.—FLOW OF WATER OVER WEIRS FROM ⅛ INCH TO 24 INCHES IN DEPTH, FOR EACH INCH IN WIDTH, IN CUBIC FEET PER MINUTE.

Inches. Depth.	Even inches. Cubic feet.	⅛ inch.	¼ inch.	⅜ inch.	½ inch.	⅝ inch.	¾ inch.	⅞ inch.
0	....	0.01	0.05	0.09	0.14	0.20	0.26	0.33
1	0.40	0.47	0.55	0.65	0.74	0.83	0.93	1.03
2	1.14	1.24	1.36	1.47	1.59	1.71	1.83	1.96
3	2.09	2.23	2.36	2.50	2.63	2.78	2.92	3.07
4	3.22	3.37	3.52	3.68	3.83	3.99	4.16	4.32
5	4.50	4.67	4.84	5.01	5.18	5.36	5.54	5.72
6	5.90	6.09	6.28	6.47	6.65	6.85	7.05	7.25
7	7.44	7.64	7.84	8.05	8.25	8.45	8.66	8.86
8	9.10	9.31	9.52	9.74	9.96	10.18	10.40	10.62
9	10.86	11.08	11.31	11.54	11.77	12.00	12.23	12.47
10	12.71	12.95	13.19	13.43	13.67	13.93	14.16	14.42
11	14.67	14.92	15.18	15.43	15.67	15.96	16.20	16.46
12	16.73	16.99	17.26	17.52	17.78	18.05	18.32	18.58
13	18.87	19.14	19.42	19.69	19.97	20.24	20.52	20.80
14	21.09	21.37	21.65	21.94	22.22	22.51	22.79	23.08
15	23.38	23.67	23.97	24.26	24.56	24.86	25.16	25.46
16	25.76	26.06	26.36	26.66	26.97	27.27	27.58	27.89
17	28.20	28.51	28.82	29.14	29.45	29.76	30.08	30.39
18	30.70	31.02	31.34	31.66	31.98	32.31	32.63	32.96
19	33.29	33.61	33.94	34.27	34.60	34.94	35.27	35.60
20	35.94	36.27	36.60	36.94	37.28	37.62	37.96	38.31
21	38.65	39.00	39.34	39.69	40.04	40.39	40.73	41.09
22	41.43	41.78	42.13	42.49	42.84	43.20	43.56	43.92
23	44.28	44.64	45.00	45.38	45.71	46.08	46.43	46.81
24	47.18	47.55	47.91	48.28	48.65	49.02	49.39	49.76

Cubic feet per minute × 62.5 lbs. × height in feet divided by 33,000 = the gross horse-power of the stream.

From this quotient, 20 per cent. should be deducted for the available power that can be obtained by the best turbines and water-wheels, although 85 per cent. is claimed by some makers.

THE MINER'S INCH

A miner's inch is the measure of water established in the early days of hydraulic mining, and may be defined to be the quantity of water that will flow from a square orifice in the side of a flume, 1 inch in diameter, through a 2-inch plank, the centre of the orifice being $6\frac{1}{2}$ inches below the surface of the watre or a $6\frac{1}{2}$-inch head.

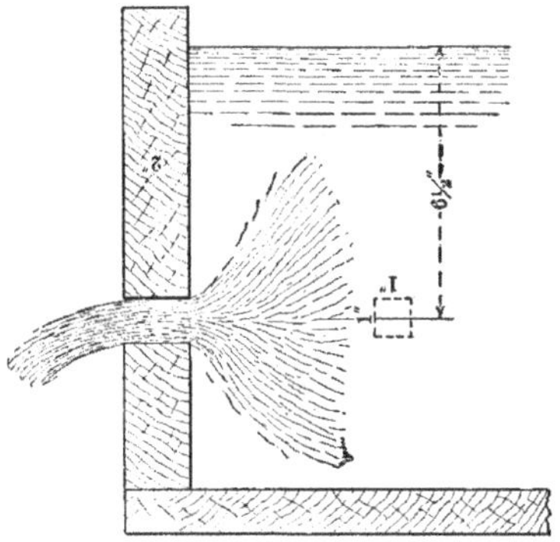

FIG. 76.—The miner's inch.

The ratio or coefficient of actual to theoretical discharge varies somewhat in the trials of experimenters, from 0.623 to 0.616, so that the actual flow may be from the rule for one miner's inch, Fig. 76.

RULE

Area in decimals of a square foot × coefficient × square root of twice gravity × square root of height in feet, or

$$A \times C \times \sqrt{2g \times h}$$

or

$$0.006944 \times 0.623 \times 8.02 \times \sqrt{\frac{6.5}{12}} = 0.0255 \text{ cubic foot per second, and}$$

$0.0255 \times 60'' = 1.53$ cubic feet per minute = a miner's inch of water.

The miner's inch varies in quantity in various districts of California and other mining States; the method of testing its measured flow also varies greatly from the plan illustrated in Fig. 76 to apertures larger than 1 square inch, long sluices and weirs, all of which give variable results.

See chapter on Irrigation.

The actual value or flow seems to vary considerably at various elevations, and values are quoted at from 1.25 to 1.75 cubic feet per minute for actual discharge, the less in quantity for the greater height.

From experiments with exact measurements of orifice, height, time and flow at moderate elevations in California, the mean flow of a miner's inch was found to be 1.5744 cubic feet per minute, with an ascertained coefficient of 0.616.

The variation of water-level in flumes makes a uniform flow somewhat difficult of control, and a resort to gates regulated by floats is made where uniform flow is required.

The available horse-power, from one or any number of miner's inches, may be readily computed by multiplying the number of miner's inches by their volume per minute (1.5744 × number of inches), and by the weight of a cubic foot of water (62.5), and by the fall in feet that can be obtained, dividing the last product by 33,000, which gives the value of the horse-power of the water. From this a deduction of 20 per cent. should be made for the loss by transmission of the power through the motor, which is also a variable and is the mean of various kinds ranging in efficiency from 75 to 85 per cent.

The most approved forms of turbines and impact wheels are found to yield from 80 to 86 per cent. of the primary power of the water.

Thus as an example: For 100 miner's inches and a fall of 100 feet, the figures of the equation become

$$100 \times 1.5744 \times 62.5 \times 100 = \frac{984,008}{33,000} = 29.8 \text{ h.-p.}$$

and 29.8 less 20 per cent. = 23.84 available horse-power.

CHAPTER V

FLOW OF WATER IN PIPES

WHILE the theoretical flow of water through long pipes depends mainly upon the fundamental formula of $\sqrt{2g\,h}$, the friction of the water upon the walls of the pipe, its length and diameter, as well also as some other conditions, such as foulness of the walls by fixed rust nodules or silt, require a coefficient varying with its velocity and the diameter of the pipe which is assignable for clean, smooth surfaces, but for obstructions can be estimated only from some known general condition of the interior surface.

The laws in relation to the flow may be briefly summed up as follows:

1. The loss in friction is proportional to the length of the pipe, with equal velocities.
2. The friction increases nearly as the square of the velocity.
3. The friction decreases with the increase of diameter of the pipe, for a given length.
4. It increases with the roughness of the interior surface.
5. It is independent of the pressure.
6. In wooden pipes the friction is 1.75 greater than in metal pipes.

In order to utilize the full flow of water through pipes by gravity or pressure, the entering end should be bell-mouthed or curved out in a long cone, 8 or 10 times the diameter of the pipe in length, the end of the cone or bell being no larger than twice the area of the pipe.

From the variety of formulas on water-flow friction by different experimenters, we have selected those giving average results.

With clean, smooth pipe, the coefficient c, named also the friction factor, f, varies inversely with the velocity and size of the pipe as shown by inspection of Table XI.

To determine the loss of head in feet or in pounds pressure, we have the formula $f\frac{l}{d}\times\frac{V^2}{2g}=h'$, loss of head in feet, which multiplied by .433 equals the loss in pounds pressure.

For example, length 1,000 feet, diameter 6 inches, velocity 4 feet per second; we find in Table XI, for a velocity of 4 feet and .5 foot diameter, that the friction factor is .022. Then,

$$.022\times\frac{1000}{.5}\times\frac{16}{64.32}=10.94 \text{ feet loss of head by friction.}$$

A further deduction must also be made for the form of the entrance to the pipe, which if bell shape of best form may be no more than 2 per cent.; while with a straight end flush with the walls of the reservoir 18 per cent., and with the end of the pipe projecting into the reservoir as much as 25 per cent. must be added to the above loss of head.

TABLE XI.—FRICTION FACTORS FOR VELOCITY OF FLOW AND SIZE OF PIPE.

Diameter in feet.	Velocity in Feet per Second.						
	1	2	3	4	6	10	15
0.1	.038	.032	.030	.028	.026	.024	.023
0.25	.032	.028	.026	.025	.024	.022	.021
0.50	.028	.026	.025	.023	.022	.020	.019
0.75	.026	.025	.024	.022	.021	.019	.018
1.00	.025	.024	.023	.022	.020	.018	.017
1.25	.024	.023	.022	.021	.019	.017	.016
1.50	.023	.022	.021	.020	.018	.016	.015
1.75	.022	.021	.020	.018	.017	.015	.014
2.00	.021	.020	.019	.017	.016	.014	.013
2.50	.020	.019	.018	.016	.015	.013	.012
3.00	.019	.018	.016	.015	.014	.013	.012
3.50	.018	.017	.016	.014	.013	.012	.011
4.00	.017	.016	.015	.013	.012	.012	.011

A general formula for the velocity in feet per second is,

$\sqrt{\frac{2g\,h}{1+.5\,f\frac{l}{d}}}$ = velocity; in which f may be taken as a mean factor of .02 and finally corrected from the value of f, due to the approximate value in the formula, as shown in Table XI. Then taking the last example of 1,000 feet of 6-inch pipe with an assumed head of $16\frac{1}{2}$ feet, we have,

$\sqrt{\frac{64.32\times16\frac{1}{2}}{1+.5\times.02\times\frac{1000}{.5}}}=\frac{1061.28}{60}=\sqrt{17.7}=4.21$, the approximate velocity. The factor f for near this velocity in the table is .023, the true one for the above formula. The coefficient for the loss of head at the entrance of a long pipe due to the contracted vein when the end is flush with the inside of the reservoir is $\left(\frac{1}{.82^2} - 1_1\right)\frac{V^2}{2g}$, which for a computed velocity of 4 feet per second = .121, which should be added to the friction loss, and as in the first case 10.94 + .12 = 11.06 feet, total loss of head.

The volume of flow through long pipes varies slightly by the formulas of different experimenters; from the velocity flow as above, the area of the pipe of 6-inch diameter is .19635 × 4.21 = .826 cubic foot per second.

By the experiments of D'Arcy and others, for gravity flow in pipes, the velocity may be obtained from the formula $C\sqrt{r}\ \sqrt{\text{slope}}$ in which the constant C is variable with the size of the pipe; $\sqrt{r}$ = square root of the hydraulic radius or $\sqrt{\frac{\text{diameter}}{4}}$ and the slope = $\frac{\text{head}}{\text{length}}$, all in feet.

The coefficients in the D'Arcy formula may be taken at 92 for sizes less than 6-inch diameter; 93 for 6 inches; 94 for 2 feet, and 95 for 4 feet.

Examples with a head of 16½ feet and 1,000 feet in length the slope is $\frac{16.5}{1000} = .0165$ and $\sqrt{.0165} = .128$. Then for a 6-inch pipe we have $\frac{\sqrt{.5}}{4} = \sqrt{.125} = .3535$.

The formula may then be stated as, 93 × .3535 × .128 = 4.20 feet velocity per second; and the velocity, 4.20 × area (.19635) = .82467 cubic feet per second.

From a formula by Eytelwein $\sqrt{\frac{d^5\ h}{l}} \times 4.71$ = volume in cubic feet per minute; by which the square root of the fifth power of the diameter of the pipe in inches, multiplied by the factor 4.71, is made a tabular number for pipe sizes from ½ inch to 48 inches in Table XIII, which requires only $\sqrt{\frac{l}{h}}$ as the divisor of the tabular number for the discharge in cubic feet per minute.

LOSS OF HEAD BY BENDS AND ELBOWS

The loss of head by long bends, say of a radius of 10 diameters or more, is very small and practically needs no consideration. The loss of head, however, like pipe friction, increases as the square of the velocity. The formula of Weisbach gives approximate values for the head required to overcome the friction in bends and elbows.

$\frac{a}{180} \times \left[.131 + 1.847 \left(\frac{d}{2r} \right)^{\frac{7}{2}} \right] \times \frac{V^2}{2g}$ = friction head of bend or elbow, in which a = angle of curve, d = diameter of pipe, r = radius of curve, h = height or head, all in feet.

The following table comprises the values of the bracketed terms of the formula from which the other terms are easily computed:

TABLE XII.—COEFFICIENTS OF CURVES.

$\frac{d}{2r}$	.1	.131	.25	.145	.4	.206	.6	.44	.75	.806	.9	1.408
	.15	.133	.3	.158	.45	.244	.65	.54	.8	.977	.95	1.674
	.2	.138	.35	.178	.5	.294	.7	.66	.85	1.177	1.00	1.978

Then for a 90° bend of 2-inch pipe with a radius of 10 inches, $\frac{2.}{20}=.1$ and in column 3 opposite .1 is the coefficient sought, .131.

Then $\frac{90^\circ}{180^\circ} \times .131 \times \frac{V^2}{2g}$ = the friction head of the bend in feet.

For a velocity of 4 feet per second, for example, $\frac{90}{180} \times .131 \times \frac{16}{64.32} =$ .01624.

With an ordinary cast-iron elbow with $\frac{d}{2r} = \frac{2}{2} = 1.$ and in the table opposite to 1. is 1.978; then as before $.5 \times 1.978 \times .248 = .2452$ of a foot. With a return bend this may be doubled and in a stack as in a return bend coil the friction may be manifold the above amount.

In Figs. 77 and 78 are illustrated the conditions for the following formula and table in which the height is the vertical elevation of the surface of the water in the reservoir or pond above the point of discharge, and the length of pipe along its line; not the horizontal distance.

This apparently simple formula for the flow of water through long pipes may be expressed thus

$\sqrt{\frac{d^5 h}{\text{length}}} \times 4.71 =$ volume in cubic feet per minute; where d is the internal diameter of the pipe in inches, raised to its fifth power, or

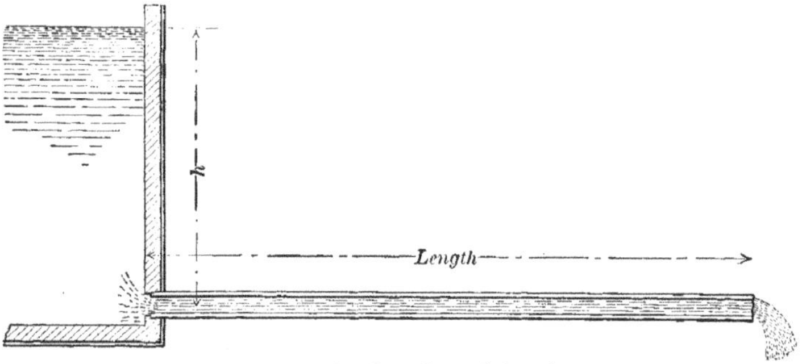

FIG. 77.—Head and straight pipe.

multiplied 4 times upon itself and its sums. Thus the fifth power of 2 is $2 \times 2 = 4 \times 2 = 8 \times 2 = 16 \times 2 = 32$. The fifth power of 3 multiplied in the same manner is 343. h is the head or height of water-supply above the orifice; l is the total length of the pipe; and 4.71

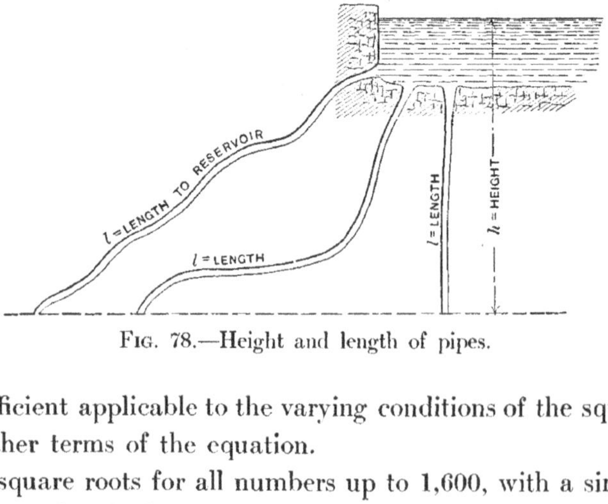

FIG. 78.—Height and length of pipes.

is a coefficient applicable to the varying conditions of the square root of the other terms of the equation.

The square roots for all numbers up to 1,600, with a simple rule for any number higher, and for decimals, may be obtained from Haswell's or Kent's engineers' pocket-books.

For example: What will be the flow from a 2-inch pipe 1,000 feet long, with a fall of 25 feet? The formula will then be:

$$\sqrt{\frac{2^5 \times 25}{1000}} \times 4.71 = \text{volume in cubic feet per minute.}$$

The fifth power of 2 is $32 \times 25 = \frac{800}{1000} = 0.8$ and square root of 0.8 is $0.894 \times 4.71 = 4.21$ cubic feet per minute, and $4.21 \times 7\frac{1}{2} = 31.57$ gallons per minute.

TABLE XIII.—CONTAINING THE EQUIVALENT OF THE SQUARE ROOT OF THE FIFTH POWER BY INVERSION OF THE OTHER TERMS FOR ORDINARY SIZES OF IRON PIPE FROM ½ INCH TO 48 INCHES IN DIAMETER. THIS REQUIRES ONLY THE FACTORS OF HEAD AND LENGTH OF PIPE, WITH THE TABULAR NUMBER FOR SIZE, TO COMPLETE THE EQUATION FOR A DISCHARGE OF CUBIC FEET PER MINUTE FOR ROUND TUBES.

Diameter inches.	Tabular number.	Diameter inches.	Tabular number.	Diameter inches.	Tabular number.
½	0.83	4½	194.84	20	8,449.0
¾	2.29	5	263.87	22	10,722.0
1	4.71	6	416.54	24	13,328.0
1¼	8.48	8	854.99	26	16,278.0
1½	13.02	10	1,493.5	28	19,592.0
2	26.69	12	2,356.0	30	23,282.0
2½	46.67	14	3,463.3	36	36,725.0
3	73.55	15	4,115.9	42	53,995.0
3½	108.14	16	4,836.9	48	75,392.0
4	151.02	18	6,493.1		

The quantities in the table, opposite the sizes of pipe, represent the square root of the fifth power of the diameter of the pipe in inches, multiplied by 4.71 from Eytelwein's formula.

By the use of this table the more simple formula of

$$\frac{\text{Tabular number}}{\sqrt{\frac{\text{Length of pipe}}{\text{Head or height}}}} = \text{volume in cubic feet per minute.}$$

Take the terms of the last example for a 2-inch pipe, the tabular number for a 2-inch pipe is 26.69. The equation then becomes

$$\frac{26.69}{\sqrt{\frac{1000}{25}}}, \text{ then } \frac{1000}{25} = 40 \text{ and } \sqrt{40} = 6.324, \text{ and } \frac{26.69}{6.324} = 4.22 \text{ cubic}$$

feet per minute.

For the velocity of flow in pipes, where the area and quantity are both known, the simple equation is

$$\frac{\text{Volume in cubic feet}}{\text{Area in square feet}} \text{ or}$$

$$\frac{\text{Volume in cubic feet} \times 144 \text{ square inches per foot}}{\text{Area in square inches}} \text{ or}$$

$$\frac{V \times 144}{D^2 \times 0.7854} = \text{velocity in feet per minute.}$$

When D is the diameter in inches and V is the volume in cubic feet per minute, or if in gallons, divided by 7.48 for cubic feet.

Again, taking the preceding example of a 2-inch pipe discharging 4.22 cubic feet per minute,

$$\frac{4.22 \times 144}{3.1416} = 193.34 \text{ feet per minute velocity.}$$

If the diameter of the pipe is given in feet and decimals of a foot, the equation becomes simply

$$\frac{\text{Volume}}{D^2 \times 0.7854} = \text{velocity in feet per minute.}$$

Gate valves should be used wherever greatest efficiency of flow is required. The entrance heads of all pipes, especially where the largest efficiency is required, as for power purposes, should have a long bell mouth, and, if possible, with an increasing curve, for the purpose of avoiding the loss by the contracted vein at entrance head. This may be no longer than from 6 to 10 times the diameter of the pipe. For entrance heads at springs and reservoirs, the bell mouth may be made of galvanized sheet iron, and neatly inserted in the socket of the wrought-iron pipe. For cast iron, they are made under the name of reducing pieces, with the small end to fit the socket end of the cast-iron pipe.

The following table, showing the computed loss of head by friction in pipes from 1 to 14 inches in diameter, and for a length of 100 feet, will be found very convenient, the loss of head being proportionable for any multiple of 100 feet, friction being in proportion to the water surface of the pipe, also the discharge in cubic feet for different velocities, and the velocity of a given discharge for various size pipes from 1 inch to 14 inches in diameter:

TABLE XIV.—LOSS OF HEAD BY FRICTION OF WATER IN PIPES. COMPUTED FOR PIPES 100 FEET LONG.

Velocity through pipe in feet per second.	Inside Diameter of Pipe in Inches. 1		2		3		4		5		6		8		10		12		14	
	Discharge per minute in cubic feet.	Loss of head in feet by friction.	Discharge per minute in cubic feet.	Loss of head in feet by friction.	Discharge per minute in cubic feet.	Loss of head in feet by friction.	Discharge per minute in cubic feet.	Loss of head in feet by friction.	Discharge per minute in cubic feet.	Loss of head in feet by friction.	Discharge per minute in cubic feet.	Loss of head in feet by friction.	Discharge per minute in cubic feet.	Loss of head in feet by friction.	Discharge per minute in cubic feet.	Loss of head in feet by friction.	Discharge per minute in cubic feet.	Loss of head in feet by friction.	Discharge per minute in cubic feet.	Loss of head in feet by friction.
1	0.327	0.588	1.30	0.294	2.95	0.196	5.22	0.147	8.17	0.118	11.77	0.098	20.88	0.074	32.70	0.059	47.10	0.049	64.08	0.042
2	0.654	1.97	2.60	0.989	5.89	0.659	10.44	0.494	16.34	0.395	23.54	0.329	41.76	0.247	65.40	0.198	94.20	0.164	128.16	0.141
3	0.981	4.05	3.90	2.02	8.83	1.35	15.67	1.02	24.51	0.815	35.32	0.679	62.64	0.509	98.15	0.407	141.30	0.330	192.24	0.291
4	1.30	6.84	5.20	3.42	11.80	2.28	20.89	1.71	32.69	1.37	47.09	1.14	83.52	0.856	130.85	0.685	188.40	0.570	256.32	0.489
5	1.63	10.2	6.50	5.14	14.70	3.43	26.12	2.57	40.87	2.05	58.87	1.71	104.40	1.28	163.50	1.03	235.40	0.855	320.40	0.735
6	1.96	14.3	7.80	7.17	17.70	4.78	31.34	3.59	49.05	2.87	70.64	2.39	125.28	1.79	196.20	1.43	282.50	1.20	384.48	1.03
7	2.29	19.1	9.10	9.52	20.60	6.35	36.57	4.77	57.22	3.81	82.41	3.18	146.16	2.39	228.90	1.90	329.60	1.59	448.57	1.36
8	2.61	24.4	10.40	12.2	23.56	8.14	41.79	6.11	65.40	4.89	94.19	4.07	167.04	3.06	261.60	2.45	376.70	2.04	512.66	1.75
9	2.94	30.3	11.70	15.2	26.51	10.12	47.02	7.59	73.57	6.07	105.97	5.06	187.92	3.79	294.29	3.03	423.80	2.53	576.75	2.17
10	3.27	37.0	13.0	18.5	29.45	12.32	52.24	9.24	81.75	7.39	117.74	6.16	208.80	4.62	327.00	3.70	470.90	3.02	640.84	2.64
11	3.59	44.1	14.3	22.0	32.40	14.71	57.47	11.03	89.92	8.82	129.52	7.36	229.68	5.52	359.70	4.41	518.00	3.68	704.93	3.15
12	3.92	51.9	15.6	26.0	35.34	17.31	62.70	12.98	98.10	10.38	141.30	8.67	250.56	6.49	392.39	5.19	565.10	4.32	769.02	3.71
13	4.25	60.0	16.9	30.0	38.33	20.10	67.92	15.08	106.27	12.06	153.07	10.05	271.44	7.54	425.09	6.03	612.20	5.03	833.10	4.30
14	4.58	69.0	18.2	34.0	41.23	23.12	73.15	17.34	114.45	13.87	164.85	11.56	292.32	8.67	457.79	6.93	659.30	5.78	897.18	4.95
15	4.90	79.0	19.5	39.0	44.20	26.32	78.38	19.74	122.62	15.79	176.63	13.16	313.20	9.87	490.49	7.90	706.35	6.38	961.27	5.64
16	5.23	88.0	20.8	44.0	47.12	29.72	83.60	22.29	130.80	17.83	188.40	14.86	334.08	11.15	523.18	8.92	753.45	7.43	1,025.36	6.37
17	5.56	100.0	22.1	49.0	50.05	33.33	88.83	25.00	138.97	20.00	200.18	16.67	354.96	12.50	555.88	10.00	800.50	8.33	1,089.45	7.15
18	5.89	111.0	23.4	55.0	53.0	37.14	94.05	27.86	147.15	22.29	211.96	18.57	375.84	13.93	588.58	11.14	847.60	9.29	1,153.54	7.96
19	6.21	123.0	24.7	61.0	55.95	41.12	99.28	30.84	155.32	24.67	223.73	20.56	396.72	15.42	621.28	12.34	894.70	10.28	1,217.63	8.81
20	6.54	136.0	26.0	68.0	58.89	45.32	104.50	33.99	163.50	27.19	235.51	22.66	417.60	17.00	653.98	13.60	941.75	11.33	1,281.72	9.71

The above table will also be found convenient for engineers using pumping machinery, for ascertaining the work of the pump due to friction; thus, in pumping through 250 feet of 1-inch pipe at the rate of 3.27 cubic feet per minute, as per table, the velocity is found to be 10 feet per second, and the friction 37 feet for 100 feet of pipe; $37 \times 2\frac{1}{2}$ hundred feet equals $92\frac{1}{2}$ feet head, and $92\frac{1}{2} \times 0.433 = 40$ pounds pressure per square inch, due to friction alone. Any quantity in discharge between the figures as given in the table will have equal proportions in each of the columns for velocity and friction.

TABLE XV.—FRICTION LOSS IN POUNDS PRESSURE PER SQUARE INCH. FOR EACH 100 FEET OF LENGTH IN DIFFERENT SIZE CLEAN IRON PIPES, DISCHARGING GIVEN QUANTITIES OF WATER IN GALLONS PER MINUTE; VELOCITY OF FLOW IN PIPE, IN FEET PER SECOND.

G. A. ELLIS.

Gallons discharged per minute.	½ Inch.		¾ Inch.		1 Inch.		1¼ Inch.		1½ Inch.		2 Inch.		2½ Inch.		3 Inch.		4 Inch.		6 Inch.		Gallons discharged per minute.
	Veloc. in pipe per second.	Friction loss in pounds.	Veloc. in pipe per second.	Friction loss in pounds.	Veloc. in pipe per second.	Friction loss in pounds.	Veloc. in pipe per second.	Friction loss in pounds.	Veloc. in pipe per second.	Friction loss in pounds.	Veloc. in pipe per second.	Friction loss in pounds.	Veloc. in pipe per second.	Friction loss in pounds.	Veloc. in pipe per second.	Friction loss in pounds.	Veloc. in pipe per second.	Friction loss in pounds.	Veloc. in pipe per second.	Friction loss in pounds.	
5	8.17	24.6	3.63	3.3	2.04	0.84	1.31	0.31	0.91	0.12		0.04									5
10	16.33	96.0	7.25	13.0	4.08	3.16	2.61	1.05	1.82	0.47	1.02	0.12									10
15			10.9	28.7	6.13	6.98	3.92	2.38	2.73	0.97		0.27									15
20			14.5	50.4	8.17	12.3	5.22	4.07	3.63	1.66	2.04	0.42									20
25			18.1	78.0	10.2	19.0	6.53	6.40	4.54	2.62			1.63	0.21	1.13	0.10					25
30					12.3	27.5	7.84	9.15	5.45	3.75	3.06	0.91		0.30		0.12					30
35					14.3	37.0	9.14	12.4	6.36	5.05				0.42		0.14					35
40					16.3	48.0	10.4	16.1	7.26	6.52	4.09	1.60		0.51		0.17					40
45							11.7	20.2	8.17	8.15				0.62		0.27					45
50							13.1	24.9	9.08	10.0	5.11	2.44	3.26	0.81	2.27	0.35	1.28	0.09			50
75							19.6	56.1	13.6	22.4	7.66	5.32	4.90	1.80	3.40	0.74					75
100									18.2	39.0	10.2	9.46	6.53	3.20	4.54	1.31	2.55	0.33	1.13	0.05	100
125											12.8	14.9	8.16	4.89	5.67	1.99					125
150											15.3	21.2	9.80	7.00	6.81	2.85	3.83	0.69	1.70	0.10	150
175											17.1	28.1	11.4	9.46	7.94	3.85					175
200											20.4	37.5	13.1	12.48	9.08	5.02	5.11	1.22	2.27	0.17	200
250													16.3	19.66	11.3	7.76	6.39	1.89	2.84	0.26	250
300													19.6	28.06	13.6	11.2	7.66	2.66	3.40	0.37	300
350															15.9	15.2	8.94	3.65	3.97	0.50	350
400															18.2	19.5	10.2	4.73	4.54	0.65	400
450															20.4	25.0	11.5	6.01	5.11	0.81	450
500															22.7	30.8	12.8	7.43	5.67	0.96	500

TABLE XVI.—FRICTION LOSS IN POUNDS PRESSURE FOR LARGER PIPE AND FLOW THAN IN THE PRECEDING TABLES, FOR EACH 100 FEET IN LENGTH.

Galls. per min.	Size of Pipes.											
	3	4	6	8	10	12	14	16	18	20	24	30
500	30.8	7.43	0.96	0.25	0.09	0.04	.017	.009	.005	.....	.....	.....
600	.....	9.54	1.41	0.38	0.14	0.07	.026	.014	.007	.....	.....	.....
700	.....	14.32	1.89	0.47	0.18	0.08	.034	.017	.010	.002	.....	.....
800	.....	.....	2.38	0.61	0.22	0.09	.045	.023	.012	.004	.....	.....
900	.....	.....	2.60	0.78	0.27	0.11	.055	.028	.015	.008	.001	.....
1,000	.....	.....	3.88	0.94	0.32	0.13	.062	.036	.020	.012	.005	.002
1,250	.....	.....	.....	1.46	0.49	0.20	.091	.049	.028	.021	.009	.003
1,500	.....	.....	.....	2.09	0.70	0.29	.135	.071	.040	.030	.014	.004
2,000	.....	.....	.....	.....	1.23	0.49	.234	.123	.071	.042	.020	.006
2,500	.....	.....	.....	.....	.....	0.77	.362	.188	.107	.064	.032	.009
3,000	.....	.....	.....	.....	.....	1.11	.515	.267	.150	.091	.047	.012
3,500	.....	.....	.....	.....	.....	.....	.697	.365	.204	.124	.055	.016
4,000	.....	.....	.....	.....	.....	.....	.910	.472	.263	.158	.067	.022
4,500	.....	.....	.....	.....	.....	.....	.....	.593	.333	.189	.082	.027
5,000	.....	.....	.....	.....	.....	.....	.....	.730	.408	.244	.102	.035

TABLE XVII.—LOSS OF HEAD BY FRICTION OF WATER IN LARGE PIPES AT VELOCITIES FROM 1 TO 30 FEET PER SECOND. CALCULATED FOR PIPES 100 FEET LONG.

Velocity of water through pipe in feet per second.	Inside Diameter of Pipe in Inches.									
	15		16		17		18		20	
	Discharge per minute in cubic feet.	Number of feet loss of head due to friction.	Discharge per minute in cubic feet.	Number of feet loss of head due to friction.	Discharge per minute in cubic feet.	Number of feet loss of head due to friction.	Discharge per minute in cubic feet.	Number of feet loss of head due to friction.	Discharge per minute in cubic feet.	Number of feet loss of head due to friction.
1	73.58	.039	83.68	.037	94.56	.035	106.00	.033	130.87	.029
2	147.16	.132	167.36	.123	189.12	.116	212.00	.110	261.74	.099
3	220.74	.272	251.04	.255	283.68	.239	318.00	.225	392.61	.204
4	294.32	.457	334.72	.428	378.24	.403	424.00	.380	523.48	.343
5	367.90	.683	418.40	.640	472.80	.601	530.00	.570	654.35	.515
6	441.48	.957	502.08	.895	567.36	.841	636.00	.795	785.22	.715
7	515.07	1.27	585.76	1.19	661.92	1.12	742.00	1.06	916.09	.950
8	588.66	1.63	669.45	1.53	756.48	1.44	848.00	1.36	1,046.96	1.23
9	662.25	2.02	753.14	1.89	851.04	1.78	954.00	1.68	1,177.83	1.51
10	735.84	2.46	836.83	2.31	945.60	2.18	1,060.00	2.06	1,308.70	1.85
11	809.43	2.94	920.52	2.76	1,040.16	2.59	1,166.00	2.45	1,439.57	2.21
12	883.02	3.46	1,004.21	3.24	1,134.72	3.05	1,272.00	2.89	1,570.44	2.59
13	956.60	4.02	1,087.90	3.77	1,229.28	3.55	1,378.00	3.35	1,701.31	3.02
14	1,030.18	4.62	1,171.59	4.33	1,323.84	4.08	1,484.00	3.86	1,832.18	3.47
15	1,103.77	5.26	1,255.28	4.93	1,418.40	4.65	1,590.00	4.38	1,963.05	3.95
16	1,177.36	5.94	1,338.96	5.58	1,512.96	5.25	1,696.00	4.96	2,093.92	4.46
17	1,250.95	6.67	1,422.64	6.25	1,607.52	5.88	1,802.00	5.55	2,224.79	5.00
18	1,324.54	7.43	1,506.32	6.97	1,702.08	6.55	1,908.00	6.19	2,355.66	5.57
19	1,398.13	8.22	1,590.00	7.71	1,796.64	7.26	2,014.00	6.86	2,486.53	6.17
20	1,471.72	9.06	1,673.68	8.50	1,891.20	8.00	2,120.00	7.56	2,617.40	6.80

TABLE XVIII.—FREE FLOW OF WATER IN CUBIC FEET PER SECOND THROUGH CLEAN PIPES, WITH A UNIFORM SLOPE IN FEET PER MILE AND IN INCHES PER ROD.

Fall per mile in feet.	Fall per rod in inches.	Bore of Pipes and Cubic Feet Discharged.													
		8 in. cu. ft.	10 in. cu. ft.	11 in. cu. ft.	12 in. cu. ft.	14 in. cu. ft.	16 in. cu. ft.	18 in. cu. ft.	20 in. cu. ft.	22 in. cu. ft.	24 in. cu. ft.	27 in. cu. ft.	30 in. cu. ft.	33 in. cu. ft.	36 in. cu. ft.	40 in. cu. ft.
2.64	.10	......	......	......	......	......	......	......	......	......	......	8.27	10.24	12.92	16.35	22.08
3.70	.14	......	......	......	......	......	2.25	3.10	4.07	5.25	6.64	9.09	11.90	15.14	19.76	26.27
4.75	.18	......	......	......	......	1.83	2.59	3.49	4.68	6.01	7.56	10.26	13.48	17.58	22.30	29.80
5.28	.20	......	......	......	1.26	1.91	2.72	3.66	4.92	6.32	7.95	10.74	14.21	18.74	23.47	31.46
6.33	.24	......	.87	1.12	1.40	2.11	3.02	4.06	5.40	6.94	8.75	11.93	15.81	20.28	26.12	34.68
7.39	.28	......	.96	1.22	1.48	2.27	3.28	4.40	5.82	7.51	9.47	12.96	17.18	22.20	28.24	37.57
8.44	.31	.57	1.04	1.32	1.63	2.44	3.49	4.75	6.27	8.05	10.13	13.98	18.58	23.76	30.29	40.54
9.50	.35	.61	1.11	1.39	1.72	2.59	3.69	5.03	6.65	8.55	10.77	14.81	19.66	25.22	32.48	43.07
10.56	.39	.63	1.19	1.49	1.82	2.72	3.92	5.30	7.05	9.07	11.43	15.63	20.79	26.94	34.49	45.20
11.62	.43	.66	1.26	1.58	1.94	2.88	4.12	5.63	7.42	9.55	12.05	16.44	21.80	28.27	36.15	48.12
12.67	.47	.70	1.32	1.65	2.02	3.02	4.32	5.87	7.79	10.01	12.61	17.23	22.83	29.02	37.74	50.48
13.72	.51	.73	1.37	1.72	2.11	3.15	4.51	6.18	8.14	10.48	13.23	18.01	23.93	31.06	39.40	52.67
14.78	.55	.76	1.42	1.78	2.20	3.29	4.68	6.38	8.48	10.91	13.79	18.75	24.86	32.28	40.86	55.04
15.84	.59	.82	1.47	1.85	2.29	3.42	4.87	6.64	8.77	11.29	14.25	19.50	25.87	33.62	42.28	56.33
18.48	.68	.87	1.58	1.99	2.46	3.62	5.31	7.17	9.49	12.25	15.50	21.13	27.96	36.17	45.95	61.09
21.12	.79	.93	1.68	2.13	2.66	3.99	5.67	7.65	10.16	13.12	16.62	22.62	29.84	38.57	48.83	65.41
26.40	.99	1.04	1.86	2.39	3.02	4.46	6.39	8.66	11.43	14.78	18.71	25.34	33.55	43.12	54.89	73.09
31.68	1.18	1.15	2.05	2.63	3.31	4.91	7.02	9.54	12.59	16.20	20.42	27.74	36.79	47.40	59.95	80.32
36.96	1.38	1.26	2.22	2.85	3.60	5.37	7.66	10.33	13.66	17.53	22.05	29.96	39.66	51.35	65.17	86.70
42.24	1.58	1.34	2.38	3.06	3.85	5.77	8.16	11.09	14.66	18.78	23.61	31.99	42.39	54.91	69.80	92.58
47.52	1.78	1.42	2.51	3.23	4.07	6.11	8.64	11.71	15.54	19.93	25.07	33.97	45.23	58.20	74.33	98.00
52.80	1.98	1.49	2.66	3.41	4.30	6.44	9.10	12.37	16.47	21.06	26.42	35.89	47.71	61.62	78.46	103.99
63.36	2.37	1.64	2.93	3.76	4.72	7.00	9.95	13.65	17.99	23.07	29.03	39.76	52.92	68.00	82.84	......
73.92	2.77	1.78	3.21	4.01	5.09	7.60	10.87	14.75	19.49	24.68	31.49	43.22	57.65	73.95	......	......
84.48	3.16	1.91	3.45	4.39	5.48	8.17	11.63	15.84	21.03	26.97	33.90	46.57	......	......	......	......
95.04	3.56	2.03	3.67	4.67	5.83	8.93	12.43	16.90	22.45	29.70	36.18	48.06	......	......	......	......
105.60	3.96	2.15	3.85	5.25	6.16	9.26	13.14	17.85	23.56	31.15	38.45	......	......	......	......	......

THE MEASUREMENT OF WATER FLOWING IN PIPES

THE VENTURI METER

The principle of action of the Venturi meter is founded on the well-known property of a Venturi tube to exercise a sucking action through holes bored into its narrowest section. The construction of the meter, as shown by the accompanying cut, is merely a contraction of the main pipe, to which two ordinary pressure gauges are connected—one at any convenient point before contraction of pipe begins; the other at the smallest section. When any flow in the pipe occurs the pressure on throat gauge will fall, if the flow becomes sufficiently rapid, all pressure at the throat may disappear and a vacuum obtain. The other gauge, however, will continue to indicate the pressure due to the supply. By mathematical calculation and experimental confirmation, a formula, based on the different pressures on the gauges, has been obtained, which accurately indicates the velocity of flow through the throat of the meter. An ordinary self-recording differential gauge may be used to obtain a diagram of these variations in pressure, from which both the velocity at any given time, and the total quantity passed in any interval, may be readily determined.

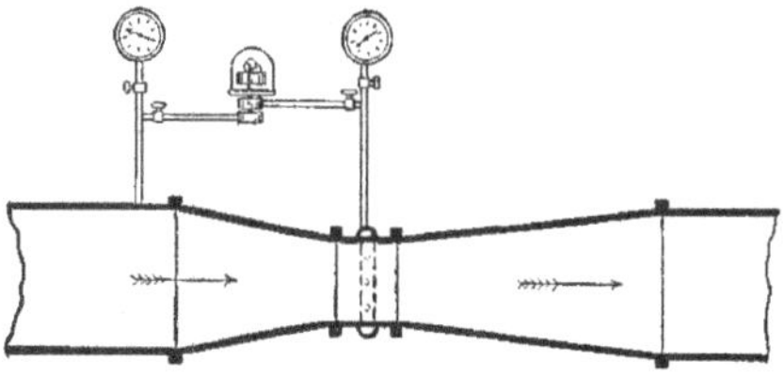

FIG. 79.—Venturi meter.

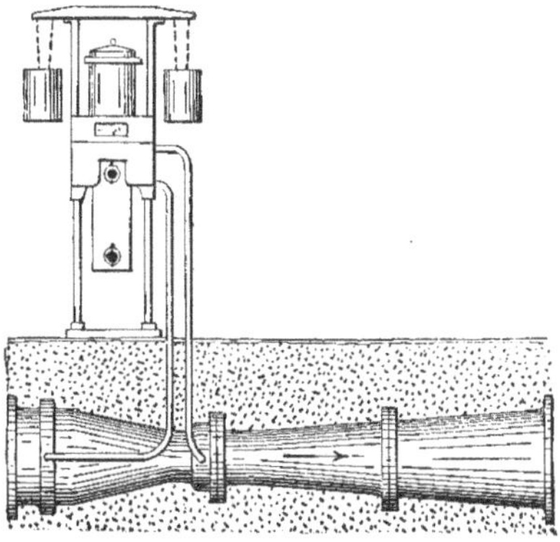

FIG. 80.—Registering Venturi meter.

Another form of Venturi meter is in use by means of pilot tubes with open mouths within the Venturi tube and connected through a registering water-meter; Fig. 80.

The differential velocity of the water in the main pipe and in the throat of the double conical tube produces a differential pressure in the small tubes with their mouths turned in opposite directions, which is used for registering the amount of water flow in the main pipe by the flow through the small pipes.

A small water-meter records a continuous run by dial or counter, and by an attachment of a chart recorder the mean volume of flow may be measured for any length of time.

The Thomson water-meter is shown in Fig. 81. The displacing or measuring member consists of a flat disk, having a ball-and-socket bearing, and is adapted to oscillate in a chamber, comprised of two sections joined together, in which each of the inside faces approximates the frustum of a cone, the exterior confining wall assuming the form of a circular zone. The disk has a single slot projecting radially from the ball, which embraces a fixed metallic diaphragm, set within and crosswise of one side of the chamber, the disk being thus prevented from rotating; but when it is caused to oscillate in contact with the cone frustums, the chamber, by these means, is divided into subcompartments or measuring spaces. Now, if the ports of ingress and egress are properly disposed on opposite sides of the diaphragm, the disk will act as its own valve. The course of the flow through the meter is as follows: entering the compartment, formed by the upper and lower caps, the current passes on all sides of the chamber, to and through the inlet port; thence through the measuring chamber (causing the oscillation of the disk), then through the outlet port, to the outlet spud and the pipe.

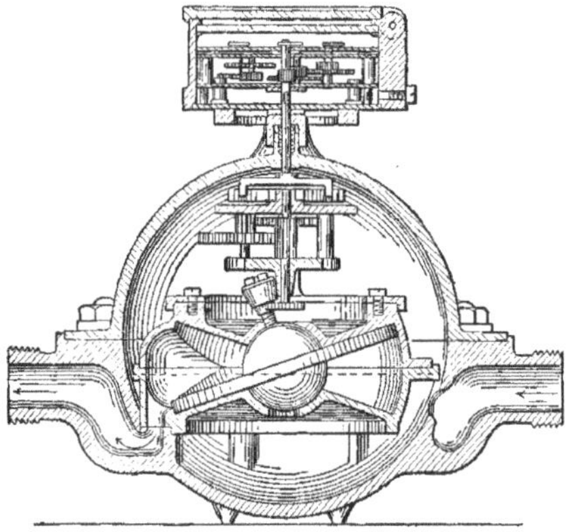

FIG. 81.—Thomson water-meter.

The oscillation of the disk produces in its central axis, at a right angle to the plane of the disk, circular motion. Advantage is taken of this to control its proper relative action in respect to the cone frustums, by mounting a conical roller upon a spindle fixed in the ball. This roller impinges upon and rolls around the fixed conical stud or hub, formed on the inner side of the gear frame. The roller turns upon a conical sleeve which is screwed upon the disk spindle.

This circular motion of the spindle is also utilized to drive the registering mechanism by means of an arm secured to the primary pinion of the train, the arm impinging upon and being driven by the lower extension of the roller. The trend of the motion of the disk is to thrust the edge of the slot constantly against the outlet side of the diaphragm.

The water-pressure regulator is a most useful device for reducing the pressure in the service pipes from high pressure water-works. Fig. 82 represents one of the several models in use.

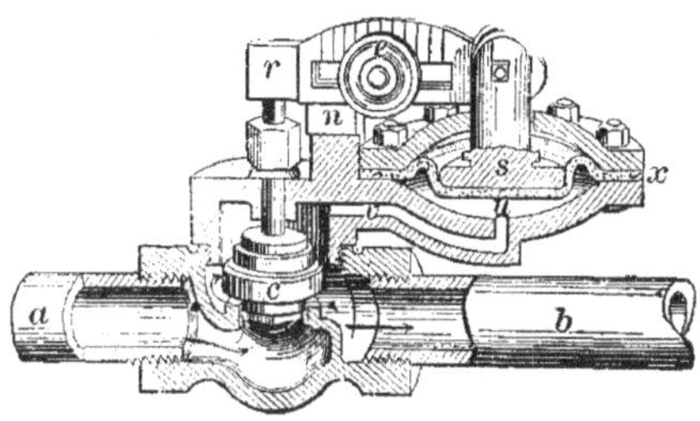

FIG. 82.—Water-pressure regulator.

The diaphragm *x* and plunger *s* operate the valve *c* through the lever *r*, and the relative pressures are regulated by the movable fulcrum *e*. *a*, high-pressure pipe; *b*, low-pressure pipe; *v*, passage to diaphragm from low pressure.

CHAPTER VI

THE SIPHON AND THE WATER RAM

THE origin of the siphon like that of the pump is lost in antiquity. Its earliest use has come to us through the channels of early civilization as recorded on the walls of Egyptian temples and tombs.

One of the earlier uses of the siphon is supposed to be for drawing the clear water from the settling jars of the turbid water of the Nile for domestic use.

In Fig. 83 is illustrated the use of the siphon as depicted on a tomb at Thebes of the age of 1500 years B.C.

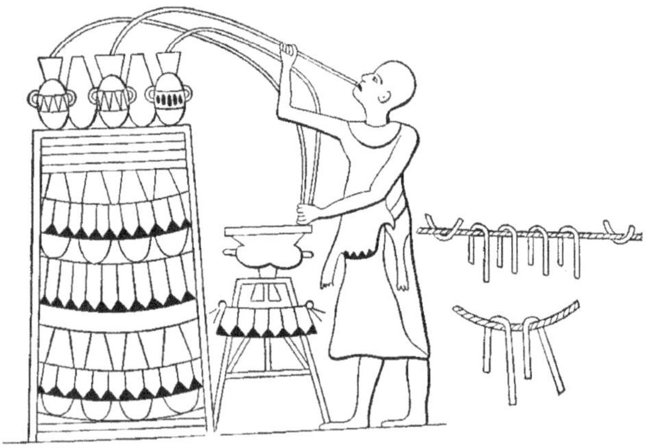

FIG. 83.—Egyptian siphons.

Hanging on lines at the right is shown the small siphons used in the Egyptian kitchens of that early period.

A variety of forms of siphons for drawing off liquids and especially for acids and other liquid chemicals have been devised which are illustrated in the following figures:

Fig. 84.—A bottle siphon which is charged by blowing through the short tube in the cork.

Fig. 85.—A rarefied bulb siphon; the bulb is heated by a spirit lamp, the long leg closed by the finger with the short leg in the liquid; remove the lamp, when the cooling air will charge the siphon.

Fig. 86.—Liquid charging siphon, or trompe.

Fig. 87.—Sucking siphon with safety bulb.

Fig. 88.—Constant charge siphon.

Fig. 89.—Constant charge tipping siphon.

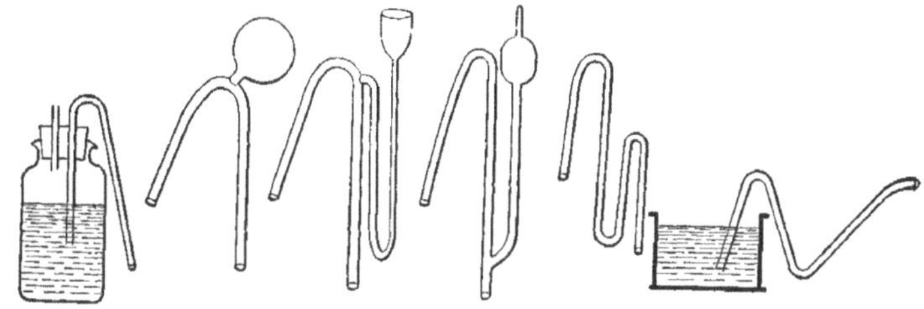

FIGS. 84 85 86 87 88 89

Chemical siphons.

PUMP, DIFFERENTIAL LEVEL, JUMP AND BLOWING SIPHONS

Fig. 90.—Pump charging siphon. The chamber is half full of water, when by lifting the invert tube with the finger closing the long leg of the siphon, it is charged.

Figs. 91 and 92 are suction charging of plain siphons by withdrawing a vessel of water which seals the long leg. The latter figure also acts as a regulator to the flow of the siphon.

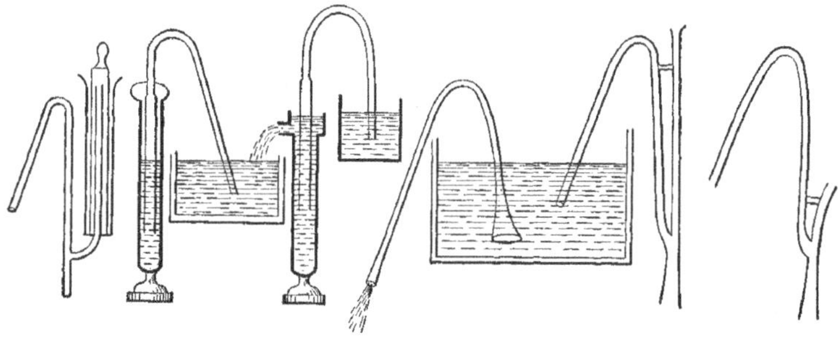

FIGS. 90 91 92 93 94 95

Fig. 93 is a jump charger, by which the momentum of the liquid caused by the plunging of the bell-shaped short leg in the fluid, sends it over the apex of the siphon sufficient in quantity to charge the other leg.

Figs. 94 and 95 are blowing siphons, which are charged by blowing through the tubes connected to the discharging legs. The draught in the conical mouth-piece raises the liquid to the apex of the siphon and so charges it.

The inverted form, which is not a true siphon, although so named in engineering phrase, has reached giant proportions as a conveyer of water across great valleys in Europe and our Western States.

The fact that all water in its natural condition contains air in mechanical combination, which is liberated by removing the atmospheric pressure, is the only and great drawback to the continuous flow of water through it, up to a height near the atmospheric or vacuum limit of about thirty-three feet.

The conditions of partial vacuum, or negative pressure, or, rather, loss of pressure, commence to liberate the air at a few feet elevation, and increase up to the limit at which the quantity of air breaks the continuity of the stream, when the operation stops; the rarefied air remaining at the apex, unless by some mechanical means it can be withdrawn from the siphon.

When siphons have but small lift, say from 8 to 10 feet, and can be so arranged as to discharge enough lower than the water at the inlet end as to create a strong current, a siphon may run continuously by carrying the liberated air along with the water, but with heights of from 15 to 25 feet the air accumulates in quantity sufficient to cut the stream at the apex of the siphon and stop the flow; often in a few hours, when the apex is short and holds but little air.

The essential features of a good siphon are illustrated in Fig. 96, which is simple, cheap, and easily managed, but requiring frequent removal of the air by closing the cocks at B and C and refilling the chamber at A and the apex of the siphon with water through the plug, D, in the funnel. The chamber at A may be made quite large, to hold the air of a whole day's run.

The top of the pipe A should have a reducing socket with a funnel of galvanized sheet iron soldered to it, as shown in the cut, to form a water seal over the plug, for convenience of filling, as well as to insure an air-tight joint at the plug, which, by frequent use, may not be perfectly tight, and here we should say that absolute tightness against leakage in of air is a most essential feature of the successful working of siphons.

We also illustrate, in Fig. 97, an arrangement for replacing the accumulating air in the siphon with water without stopping the flow.

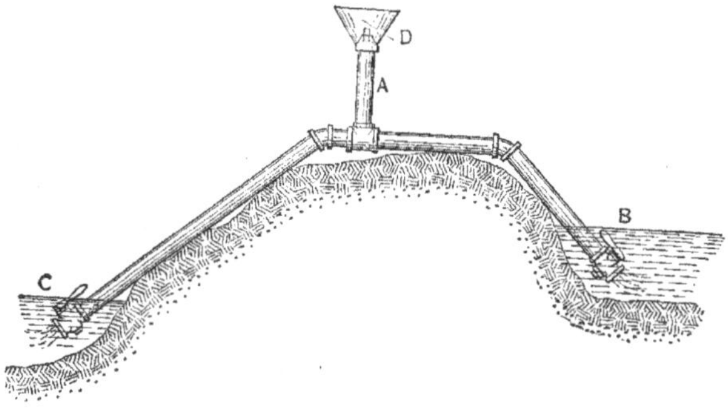

FIG. 96.—Siphon filled at apex.

In this plan make the apex section at A of a size larger pipe than the other parts of the siphon. This allows the liberated air from the rising leg to separate and float along the upper side of the pipe and exchange with the water from the reservoir B.

The slower motion of the water along the apex, and the placing of the tee piece very near the point of descent, allows of the separation

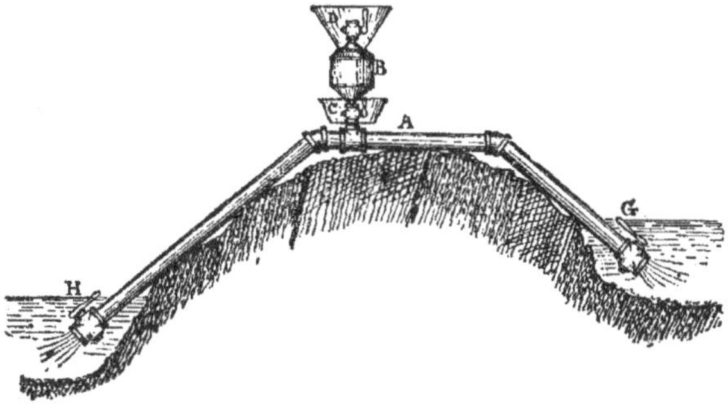

FIG. 97.—Air-chambered siphon.

of the air and prevents its accumulation beyond the outlet at C, and thereby choking the circulation.

The two cocks C and D should have water seals as shown to insure against leakage of air into the siphon.

The reservoir B, and the seals may be made of galvanized sheet iron with soldered joints, and attached by soldering to galvanized pipe couplings. To start the siphon, close the cocks H and G, open the cocks C and D, and fill the siphon and reservoir with water to the top of the funnel D, and also fill funnel C as a seal. Close the cock D, and open the cocks G and H, when the water will flow with a velocity due to the difference of water-level at the ends, less the friction of pipe and fittings.

When the chamber at B has filled with air, which may be known by rapping on it with the knuckles, close cock C, open D, and fill with water, closing D first, and open C. After a few trials a constant siphon will become a time-keeper, so that the filling of the chamber may be made at stated intervals.

In regard to the sizes, for a 2-inch siphon the cocks C and D may

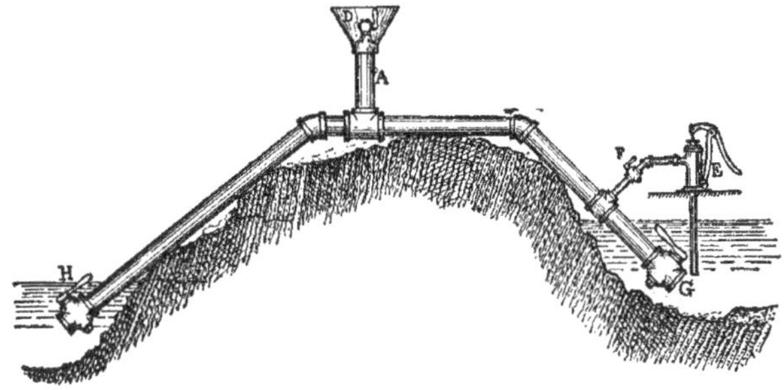

FIG. 98.—Pump-charged siphon.

be $\frac{3}{4}$ inch, and the chamber from 1 to 2 cubic feet, according to height of siphon.

A siphon arranged as illustrated in Fig. 98 is a most convenient one, where the apex is at considerable distance from the water-supply; and for quarries where the water drainage is limited and taken from sumps, which are exhausted by a short run of the siphon.

This is essentially like Fig. 96, with the addition of a common plumber's force pump at E, connected to the siphon behind the inlet cock, G, with a cock in its force pipe, F; or, if the stuffing-box of the pump is air-tight, no cock will be needed at F. A small air-cock in the funnel at D completes the arrangement ready for operating.

By closing cocks G and H, and opening the air-cock in the funnel at A, the whole siphon may be easily filled by the pump at E, also filling the funnel full. The air-cock at D should be closed first, to keep it sealed by the water in the funnel. The cocks G and H are then to be opened and the cock F closed. The siphon will then flow until the chamber, A, and part of the horizontal apex pipe becomes filled with air; when the operation of filling may be repeated. There is no special limit to the capacity of the air-chamber, which may be made of a larger pipe than the siphon, and extended at a small angle sidewise to give capacity for holding air without increasing the total elevation of the water-head in the siphon. The air-cock at D should never exceed an elevation of 32 feet above the cock at G.

There is another way of starting and keeping up the flow of a siphon by an air-pump attached directly to the apex with a cock in the position of the air-cock at D, Fig. 98. The operation of this plan can be made with a plain siphon, without cocks, at H and G, and the pump used only to pump out the air when required. The starting of the siphon requiring no manipulation of terminal cocks, nor traversing over the distance of a long siphon. The only specialty is a good air-pump.

We do not recommend the use of cast-iron pipe for siphons, the difficulty of keeping it absolutely air-tight becoming an expensive item in its maintenance. Wrought-iron black or galvanized pipe, with cast-iron screw fittings, tarred or galvanized, make the only reliable outfit for siphonage. There is probably no limit to the size of a siphon pipe, within ordinary engineering requirements, with adequate facilities for removing the air.

SIPHONAGE IN LONG SUCTION PIPES

Obstruction to the flow of water in long suction pipes that are not laid in a straight slope are sometimes quite serious obstacles to the full action of the pump, or they may entirely stop the flow with high lifts, or very long pipes. The inverts along the line of pipe hold back the air that is liberated by the negative almospheric pressure from the pump draught and thus increase the height of the lift by their cumulative action. This is illustrated in Fig. 99, where it is shown that the uptakes of the inverts are filled with water which is balanced only with air on their other sides. The condition shown is extreme, but gradually takes place after a suction pipe has been fully charged with water. It has been known to completely interrupt the flow. The necessity for straight slopes is also applicable to long drainage siphons, which have been known to cease their flow from the above conditions.

Notwithstanding the difficulties that have been met in the instalment of town and city water-works in which long siphon suction pipes have been used, the remedy for air obstruction has been found by tapping the apex of each siphon and connecting them to an air-pump in the power-house with small pipe; in this way a long suction pipe may be kept fully and constantly charged.

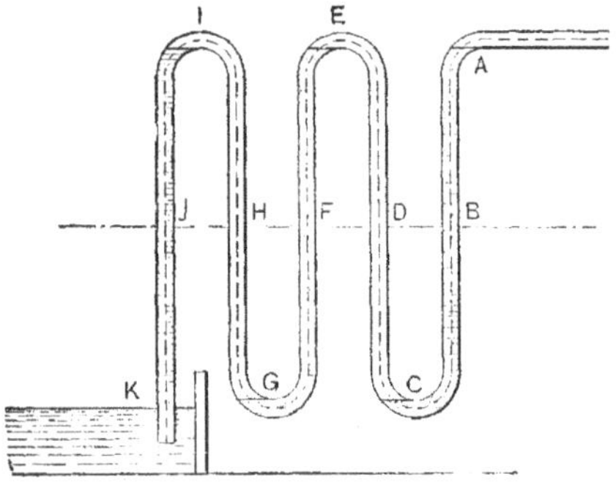

FIG. 99.—Invert siphon suction.

The velocity and volume of flow in a siphon with the discharge end submerged in the tail water and with the apex within the practical height of continual flow, is the same and depends upon the same conditions as that of other pipes between two water-heads, as shown by examples in Chapter V.

THE HYDRAULIC OR WATER RAM

The principle of the water ram may be briefly stated to be found in the power of impact of a column of water, when its motion is instantly arrested by the closing of a waste valve, and may be compared with the power of a blow from a hammer upon an anvil, which will crush a piece of metal that would require the static weight of thousands of such hammers if applied without the power of motion.

The relative conditions of the head of water that can be utilized, its quantity, and the height of required elevation, are the principal factors for determining the quantity of water, apart from the friction of the water in long pipes, that can be utilized from the

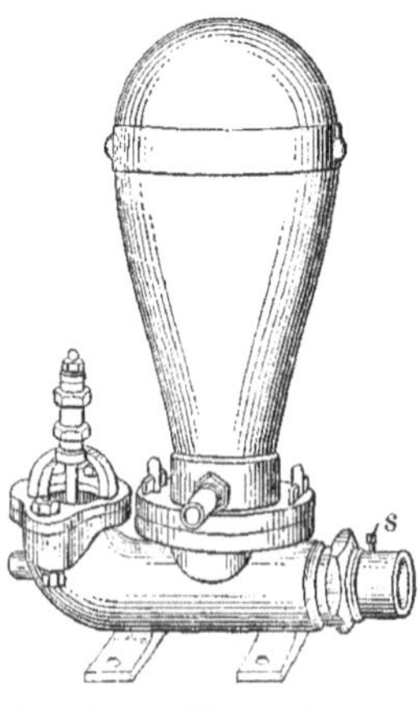

FIG. 100.—Hydraulic ram.

flow of a spring or stream, with the best proportioned apparatus. The number of strokes of the valve per minute and the weight of the valve is a matter of regulation for the utmost power of any given size ram.

Where the water-supply is limited, as from springs, the utmost efficiency is required, and a possibility of 85 per cent. in useful effect has been and may be obtained by the best adjustment of flow and discharge in improved rams.

The ordinary efficiency varies from 50 to 67 per cent.

The useful effect may be readily ascertained by measuring the waste valve flow in gallons per minute, also the discharge in gallons per minute; the height of flow or feed pipe, and the height of the point of discharge, above the waste valve of the ram. Then the equation becomes:

$$\frac{\text{Height of discharge}}{\text{Height of feed head}} \times \frac{\text{quantity discharged}}{\text{quantity wasted}} =$$

the useful effect. This, when abbreviated to formula, is $\frac{h}{h'} \times \frac{q}{q'} =$ coefficient of ram; in which

h = height of discharge.

h′ = height of feed head.

q = quantity of discharge in gallons per minute.

q′ = quantity of waste in gallons per minute.

Thus, for example, a fairly efficient ram having a fall of 10 feet, and forcing the water 50 feet high, if found forcing 5 gallons per minute and wasting 40 gallons per minute, will have an efficiency of $\frac{50}{10} \times \frac{5}{40} = 0.625$.

For a given fall the efficiency increases inversely as the height of discharge, or the discharge largely increases with a decrease of height.

A formula for regulating the weight of the waste valve to produce the maximum supply for the ordinary form of hydraulic rams has been proposed and formulated by an Austrian engineer, which may be of value for the adjustment of rams that are furnishing a deficient supply.

The formula is stated as the area of the waste valve in inches × height of supply above the valve in feet × length of drive pipe in feet, and the product divided by 500 will give the weight, including the valve, for best effect, when the diameters of the drive and discharge pipes are of proper proportions. The formula may be expressed as $\frac{a'' \times h' \times l'}{500}$ = weight in pounds.

For example, for a ram having a waste valve 2 inches diameter, in the clear opening, the area will be $2^2 \times 0.7854 = 3.14$ square inches, a fall of 5 feet, and drive pipe 40 feet long, then $\frac{3.14 \times 5 \times 40}{500} = 1.25$ or $1\frac{1}{4}$ pounds for the weight of the valve.

The drive or feed pipe should not be of less length than 6 times the height for a fall of from 6 to 10 feet; and for a less fall than 6 feet, from 8 to 10 times the fall in length, for best efficiency.

A fall of 2 feet in the drive pipe is as low as can be made to give satisfaction, unless by special adjustment and low discharge.

The discharge pipe should be from one-third to one-half the diameter of the driven pipe, according to length. There seems to be no reasonable limit to the distance that a medium or large size ram will discharge—even up to 2,000 or 3,000 feet, with a moderate elevation. For very long distances, it is better to have the discharge pipe larger than one-half the diameter of the drive pipe, to lessen friction.

The coupling attaching the drive pipe to the ram should have a snifting-hole from $\frac{1}{16}$-inch to $\frac{1}{8}$-inch diameter, as shown at S, Fig. 100, drilled in a slanting direction toward the ram (if not provided by the makers), for the purpose of keeping the air-chamber supplied by air. The momentary relief from the impact of the water in the drive pipe draws a little air in through the snifting-hole and discharges it at the next impact into the air-chamber. This is necessary to the continuous action of the ram, as the air under pressure in the air-chamber is gradually absorbed and the efficiency of the ram is thereby destroyed.

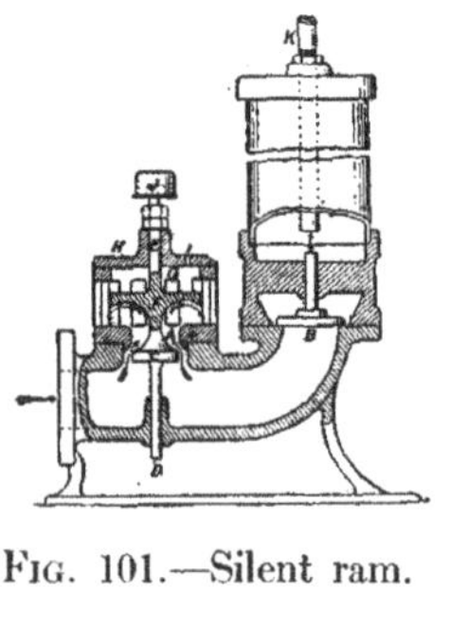

Fig. 101.—Silent ram.

The largest impact rams now made will under favorable circumstances supply from 60 to 70 gallons of water per minute, with elevation under 100 feet; say 86,000 gallons per day; enough for the largest stock farms, or for irrigating from 20 to 30 acres of land.

For irrigation purposes on a larger scale, see chapter on Irrigation.

The noise from the action of a hydraulic ram has been overcome in rams in use in Europe, the construction of which is shown in Fig. 101. The principle is in the use of a piston-valve disk, stopped by an air-cushion in a cage with a curved reaction disk and detailed as follows:

The curved reaction disk, F, serves to lift the piston valve, C, quickly without shock. The air-cushion at G stops the lift at the moment of closure of piston valve, C. J, a stop set-screw; H, valve-cage; B, force-valve; K, force-pipe; I, vent-hole to air-cushion.

The siphon water ram is a novelty of French origin, showing possibilities in this class of water-raising devices that can be located

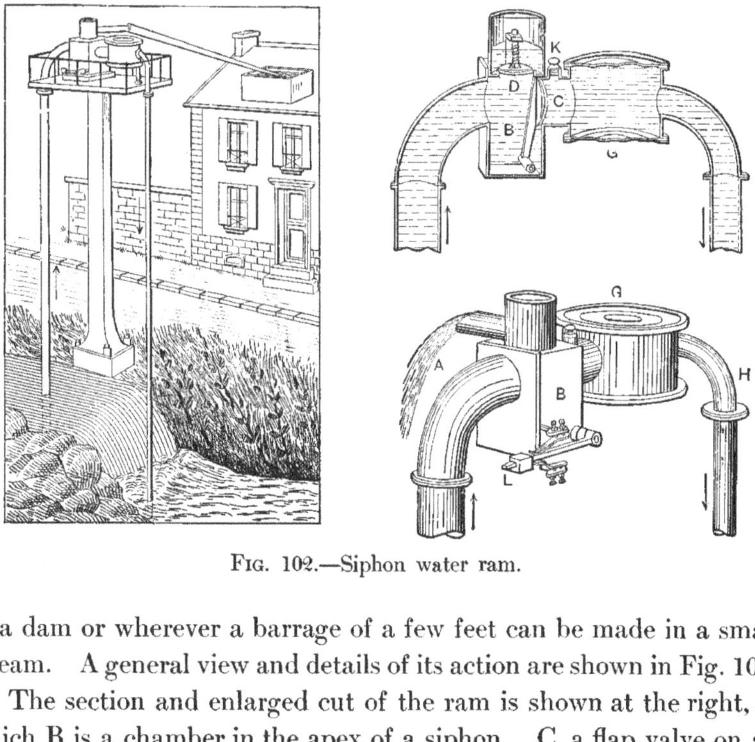

FIG. 102.—Siphon water ram.

at a dam or wherever a barrage of a few feet can be made in a small stream. A general view and details of its action are shown in Fig. 102.

The section and enlarged cut of the ram is shown at the right, in which B is a chamber in the apex of a siphon. C, a flap valve on an arm and spindle extending to outside of chamber and held open by the lever and weight, L, with its movement adjusted by the springs above and below the lever. D, discharge valve. G, a chamber with elastic heads or diaphragms of thin corrugated metal, for an air-chamber and to prevent hammer. K, plug for filling the siphon with water or by the suction of an air-pump.

The height of ram may be made convenient up to 14 feet above the barrage with a water fall of 6 feet or more and is claimed to deliver one-third of water passing the siphon at its own height and in proportion for higher or long delivery.

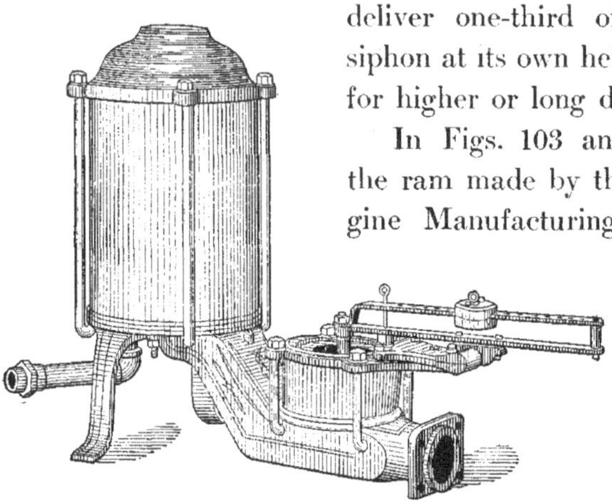

FIG. 103.—Rife hydraulic ram.

In Figs. 103 and 104 are illustrated the ram made by the Rife Hydraulic Engine Manufacturing Company, in sizes for ordinary domestic supply to that of town and irrigation requirements.

The principal novelty in the construction of this ram is the balanced waste valve, so arranged that the valve can be exactly weighted to the amount for its best work under varying conditions of fall and discharge. This is done by merely sliding the weight on the lever to the proper place and clamping it with a set-screw.

Its rubber-faced valve is also an important factor in suppressing the hammer or jar of metallic-faced valves.

The supply of water from a pure spring at only a slight elevation of two or more feet, by the power of the water of an undesirable stream, is a peculiar function of the hydraulic ram.

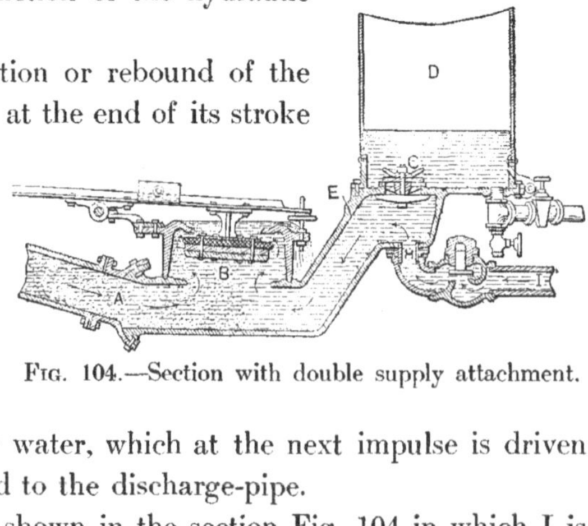

FIG. 104.—Section with double supply attachment.

The momentary reaction or rebound of the driving column of water at the end of its stroke suspends the pressure due to its fall or produces a slight vacuum which allows the small pressure from the spring to supply the end of the drive pipe beneath the air-chamber with spring water, which at the next impulse is driven into the air-chamber and to the discharge-pipe.

This arrangement is shown in the section Fig. 104 in which I is the inlet for spring water; H, a check valve and opening directly under the air-chamber valve; E, the snifting-hole close to the valve. The lever is on an outside fulcrum and allows the weight to be moved to decrease or increase the weight of the waste valve.

By adjusting the counterweight so that the valve is nearly balanced, the valve comes to its seat very quickly after the flow past it begins. The result is that the ram makes a great number of short, quick strokes, which are much easier on the mechanism than slower and heavier strokes. Of course the stroke must be sufficiently powerful to act efficiently in overcoming the head in the delivery pipe. The adjustable weight permits this to be effected with great nicety.

FIG. 105.—Irrigation ram, or hydraulic engine.

The question of efficiency of hydraulic rams has been much discussed. The Rife Hydraulic Engine Co., however, uses Rankine's formula in calculating efficiency, which is

$$E = \frac{q\ h}{(Q-q)\ H}$$

where Q is the quantity of water flowing per second in the drive pipe; *q*, the quantity flowing per second to the stand pipe through the discharge pipe; H, the height from the waste valve to the level of the water feeding the drive pipe; and *h*, the difference in the level of the water-supply reservoir and the water in the stand pipe. D'Aubisson's formula for efficiency $E = \frac{q\ (H-h)}{Q\ H}$ is the correct one, considering the energy only at both ends of the machine.

The Rife hydraulic engine illustrated in Fig. 105 is designed for town water-supply or for irrigation, for which it is well suited where there is sufficient running water and fall to supply its power demand. It weighs approximately 2,800 pounds; the capacity of the air-chamber is $20\frac{3}{4}$ cubic feet; diameter of drive pipe, 8 inches; diameter of the waste valve, 18 inches; weight of waste valve, 50 pounds; diameter of delivery pipe, 4 inches; height to top of air-chamber, $7\frac{1}{2}$ feet.

A general formula for the possible amount of water-supply that may be obtained from a spring or stream is, $\frac{S\,F}{H} \times 40 =$ gallons per hour; in which S = supply of water from spring or stream in gallons per minute; F = fall in feet of the drive pipe; H = the elevation of discharge above the ram. If the discharge pipe is of great length its friction factor should be added to the height.

TABLE XIX.—GALLONS DELIVERED PER 24 HOURS BY RIFE ENGINE FOR 1 GALLON PER MINUTE SUPPLIED

Power head in feet.	Pumping Head in Feet.												
	15	20	30	40	50	60	70	80	90	100	120	140	160
2	150	112	.....	.....	.....	.....	.....	.....	.....	.....	.....	.....	.....
3	225	168	112	84	.....	.....	.....	.....	.....	.....	.....	.....	.....
4	300	225	150	112	.....	.....	.....	.....	.....	.....	.....	.....	.....
5	375	281	187	140	112	.....	.....	.....	.....	.....	.....	.....	.....
6	450	337	225	168	135	112	.....	.....	.....	.....	.....	.....	.....
7	525	394	262	197	157	131	112	.....	.....	.....	.....	.....	.....
8	.....	450	300	225	180	150	128	112	.....	.....	.....	.....	.....
9	.....	506	337	253	202	168	145	126	112	.....	.....	.....	.....
10	.....	562	375	281	225	187	160	140	125	112	.....	.....	.....
12	.....	.....	450	337	270	225	192	168	150	135	112	.....	.....
14	.....	.....	525	394	315	262	225	196	175	157	131	112	.....
16	.....	.....	.....	450	360	300	257	225	200	180	150	128	112
18	.....	.....	.....	506	405	337	290	253	220	202	168	144	126
20	.....	.....	.....	562	450	375	321	281	250	225	187	160	140
22	.....	.....	.....	.....	506	412	353	308	275	247	206	176	154
24	.....	.....	.....	.....	532	450	385	337	300	270	225	192	168
26	.....	.....	.....	.....	.....	488	417	365	325	292	244	208	182
28	.....	.....	.....	.....	.....	525	450	394	350	315	263	224	196
30	.....	.....	.....	.....	.....	562	482	422	375	337	281	240	210

THE PEARSALL HYDRAULIC ENGINE

The general principle of the action of these engines is that water from a stream or river is allowed to flow through a pipe into a tail race at lower level, through an open valve in the engine, and the water in the pipe having thus acquired a certain velocity the valve is closed, and the momentum of the mass of water in the pipe causes some of it to pass into an air vessel, from which the pressure of the air in the air vessel causes it to flow out by a pipe to any height desired. Several novel features are, however, embodied in these engines to attain the perfect utilization of this principle which has hitherto had only a very partial application.

Where the water power is to be used to compress air, the action is similar, and the construction is only slightly modified.

The vertical section, Fig. 106, shows the construction of the engine; C is a pipe from the head race to the engine; D is the main valve actuated by the rod E and the cam F, and alternately opening and closing the annular opening; B is a chamber above the main valve; K K are valves opening from the chamber B into the chamber L, which communicates with an air vessel; a small air motor of ordinary construction turns the shaft T and cam F, and thus opens and closes the annular valve D.

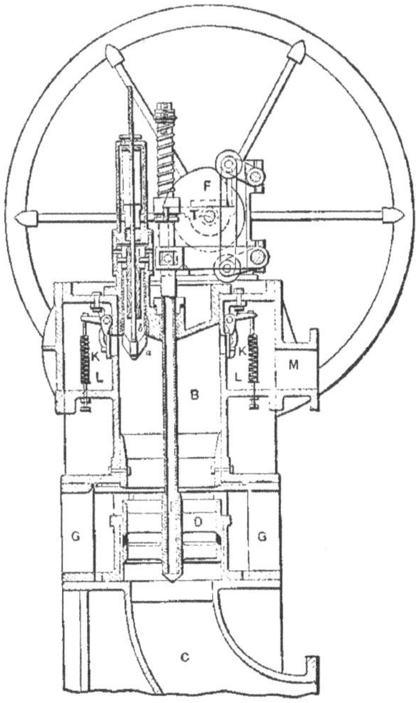

FIG. 106.—Pearsall's hydraulic engine.

The shaft T is turned by the motor at a speed of about 25 revolutions per minute, and, by means of the cam, F, actuates the main valve, D, and alternately opens and closes the orifice G at regular intervals. Immediately the orifice G is opened, the water flows in the pipe C, escaping freely into the tail race, until by the continued revolution of the shaft T, the cam again closes G. At the closing of the valve water rises in the chamber B, driving out the air, which escapes from the chamber through the air-valve. Almost immediately after the closing of the main valve, the water, still rising in the chamber, reaches the lower edge of the tube *b*, up which it flows, touches the float attached to the air-valve and closes it. The momentum of the column of water then compresses the small quantity of air contained in the air-chamber above the lower edge of the tube, *b*, opens the valves K K, and enters the chamber L (and thence the air-vessel), first the air and then some water following it. Having thus expended its energy the water comes to rest. The valves K K then close and the continued revolution of the shaft T and its cam, F, reopens the main valve, D. The flow in the pipe C recommences, and at the same time the water in chamber B also escapes into the tail race through the same opening, atmospheric air entering through the air-valve, which falls open as soon as the high pressure ceases. This completes the cycle of operations, the delivery of water at high pressure from the air-vessel being continuous.

The annular sliding valve D is peculiar in that its closing is completed by a slight movement of the seat, which consists of a rubber ring in the form of a curtain. The valve is moved up to, but not quite touching it. Immediately there is any internal pressure the edge of the rubber curtain is forced against the valve, and so completes its closing.

The motor is actuated by the small quantity of air compressed at each stroke and enters the air-vessel with the water. The depth of the tube *b* dipping into the antechamber is regulated by a screw, so that the quantity of air compressed may be made equal to the requirements of the motor.

The efficiency of the engine from its actual work is 71 per cent.

The engine was designed by Howard D. Pearsall and made in England for the Isabella furnaces, Chester County, Pa.

The means for raising large volumes of water from as low initial heads as 1½ or more feet has been brought out in the Niagara hydraulic engine, built at Chester, Pa. Its capacity is from 20,000 to 5,000,000 gallons per day and in sizes with drive pipes from 12 to 48 inches diameter with delivery pipes one-half the diameter of the drive pipes and a limit of lift from 500 to 800 feet in height.

Its great capacity under low heads as found in sluggish rivers makes it a most economical means of water-supply for irrigation, town water-works, and for mining. Fig. 107 shows a section of this hydraulic engine.

The driving water column enters engine at D and finds a free flow through the opening B and cylindrical impact valve A, thus insuring the greatest possible speed of driving column in the shortest period of time. The friction of the flowing water on cylindrical impact valve A causes it to rise and brings the flanges of valve against their rubber seats, thus causing the suddenly increased pressure or momentum of driving column to be exerted under the air-chamber valves E, which allows the water to enter the air-chamber where the ramming action is absorbed by a large air-cushion relieving the engine of unnecessary shocks or vibrations and insuring a uniform and comparatively low speed of delivery through the outlet G.

When the cylindrical impact valve A is closed, the blow or ramming action is delivered against the valve hood, H; the same not

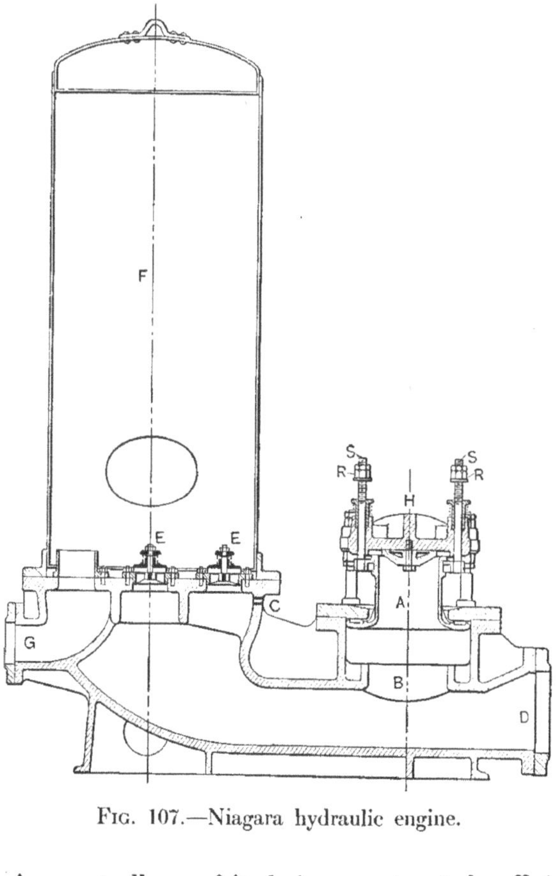

FIG. 107.—Niagara hydraulic engine.

being a moving part allows of its being constructed sufficiently heavy to withstand any desired pressure.

By means of the regulating nuts, R, the drop of impact valve can be readily adjusted so as to make the engine use the proper quantity of water within the limits specified for each size engine.

The inflowing driving column is guided in a straight upward direction parallel through the walls of impact valve by means of the port, B, so that the circular impact valve is relieved from unbalanced side pressure, allowing its motion to be naturally vertical.

The impact valve has no points of contact or friction to overcome. This allows it to use gritty or sandy water without in any way affecting the engine.

The two stems, S, are for regulating the amount of valve drop rather than acting as valve guides.

The air-chamber cushion is kept constantly supplied by means of the automatic air-feed C.

The action of water as a jet for the purpose of propelling vessels has been a failure in many trials from its low efficiency; but for many special purposes, such as the drainage of sumps and cellars, where water can be obtained under sufficient pressure, the hydraulic ejector may be made a most useful and reliable device.

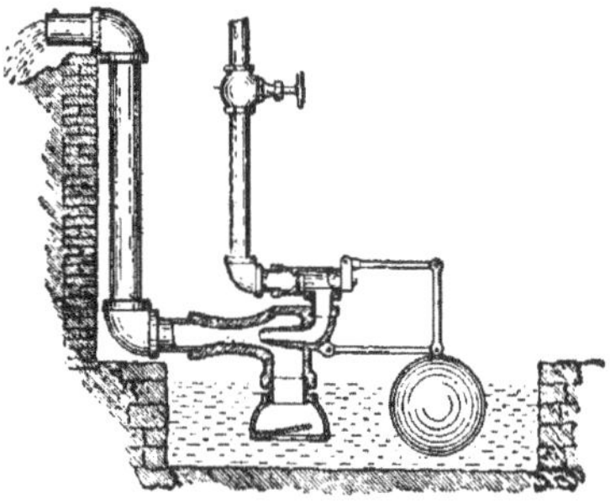

FIG. 108.—Automatic ejector.

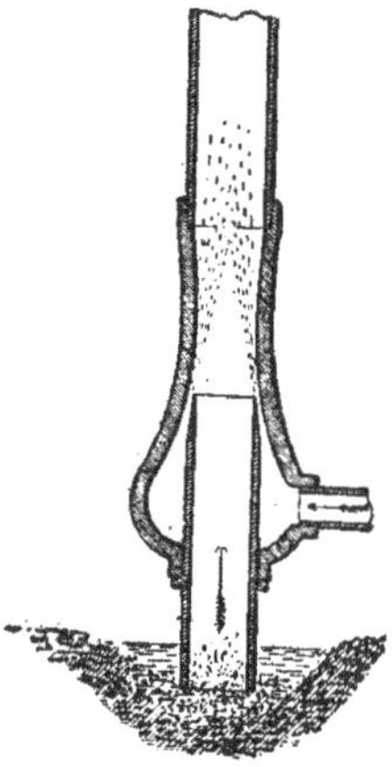

FIG. 109.—Sand ejector.

Fig. 108 illustrates the design of an automatic water ejector in which the jet is started and stopped by a float in the sump, so arranged with a loose or sliding joint that the jet starts when the sump is full and stops when it is empty, thus avoiding waste of water.

A useful device, invented by Captain Eads, is the hydraulic sand ejector used for discharging sand from the caissons of the St. Louis bridge.

Fig. 109 shows a section of the sand ejector, the action of which consists of a thin annular jet of water under high pressure, 40 to 50 pounds per square inch, ejecting sand and water from a sump and discharging at a high elevation.

A device involving the principle of the impact of a swift body or scoop against water is illustrated in Fig. 110, consisting of a long water trough between the tracks. A movable spout in the

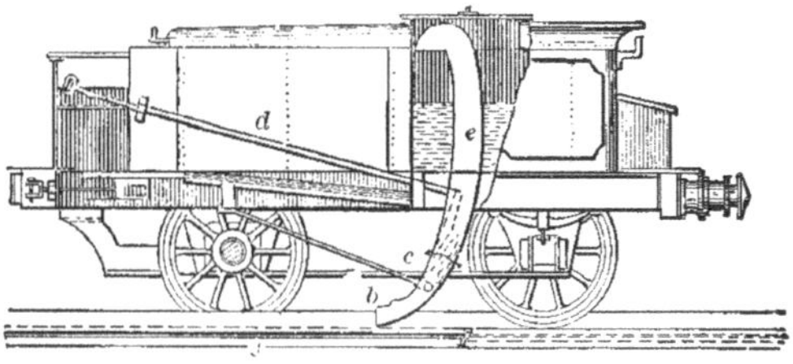

FIG. 110.—Railway water-scoop.

tender is dropped into the water trough at an angle to scoop up the water and propel it into the tank by the speed of the train.

An example of the power of water under pressure and velocity is shown in the hydraulic system of pile-driving and in the driving of pipe wells.

By this means large iron piles are driven to great depths by merely pumping water down through the pipe.

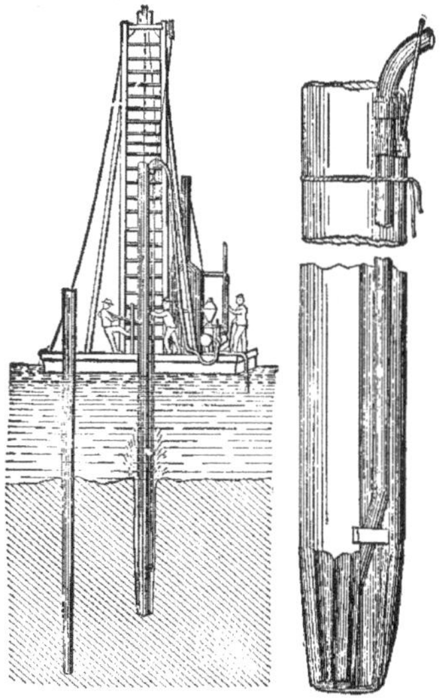

FIG. 111.—Hydraulic pile-driving.

Fig. 111 illustrates the method of driving wooden piles by the hydraulic system. A pile with a groove on its side in which a pipe is laid to the bottom of the pile, loosely clipped in place to enable its withdrawal after the pile is set. A strong stream of water from a pump excavates a passage for the pile to the required depth. No hammer is needed—only a steady pressure.

CHAPTER VII

DAMS AND RESERVOIRS

THE storage of water at higher levels than that of the local flow of streams by means of dams and reservoirs, and for the purposes of irrigation, city and domestic use was a recourse of the earliest civilization, of which the most notable example was Lake Mœris in Egypt, of 150 square miles in area, with embankments 30 feet high, which can still be traced for many miles. The canal for supplying this vast reservoir was about 120 miles in length, running along the west side of the valley of the Nile, and received its water from the floods of the Upper Nile. Its sluices irrigated about 1,200 square miles.

The present great barrage of the Nile at Philæ is the greatest work of modern times for irrigation.

One of the greatest reservoirs of modern times is that of Alicante in Spain, finished in 1594. The dam is curved, convex upstream, is 140 feet high, 36 feet thick at the top and 112 feet at the bottom, and built of solid limestone masonry.

It has stood against the greatest floods of the past 300 years, although the spileways have been flooded and the crest overflowed to a depth of 8 feet.

Its reservoir capacity is 131,000,000 cubic feet.

The tanks of Ceylon are among the wonders of hydraulic engineering, some of which are 20 miles in circumference, with embankments of massive masonry, which seem to have defied the hand of time. Their age is unknown; they are a part of a vast system of irrigation. Similar structures are found in southern India and Arabia and point to the occupation of these countries by a civilized people older than the Arabs and Hindoos.

The construction of dams in the United States has had a progressive history, with many failures, from the rude log dams of the early settlers to the enormous masonry dams required for the present storage of water for the requirement of our great cities and manufacturing industries.

Wooden dams are usually of cribwork, of either rough round logs with the bark on, or of hewn timber—in either case about a foot through. These timbers are merely laid on top of each other, forming in plan a series of rectangles, with sides of about 7 to 12 feet. They are not notched together, but simply bolted by 1-inch square bolts (often ragged or jagged) about 2 to $2\frac{1}{2}$ feet long, through every timber at every intersection.

The cribs are usually, but not always, filled with stone. In triangular dams, disposed as in Fig. 112, this stone filling is not so essential as in other forms, because the weight of the water, and of the gravel backing, tends to hold the dam down to its base. Still,

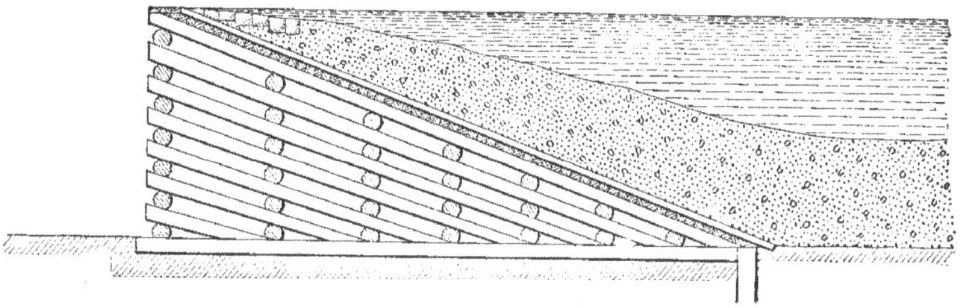

FIG. 112.—Log and timber dam.

even in these, when the lower timbers are not bolted to a rock bottom, or otherwise secured in place, some stone may be necessary to prevent the timbers from floating away while the work is unfinished and the gravel not yet deposited behind it. On rock, the lowest timbers are often bolted to it, to prevent them from floating away during construction; and when the water is some feet deep, this requires coffer-dams.

In this method of construction the overfall should impinge upon solid rock for the safety of the dam. When built upon a soft, sandy, or gravel bottom, ample provision should be made by trenching and sinking below the bottom of the stream with piling for anchorage and a long, deep, back fill to prevent seepage and undermining. The planking should run lengthwise of the dam and battened.

Fig. 113 shows a type of dam with the frame or crib made of logs and covered with a timber or plank backing laid crosswise of the dam and battened. If built on a sand or gravel bottom, great care should be taken to sink the cribwork as deeply as possible and anchor it with stone filling or piles. Such a dam should have an overfall

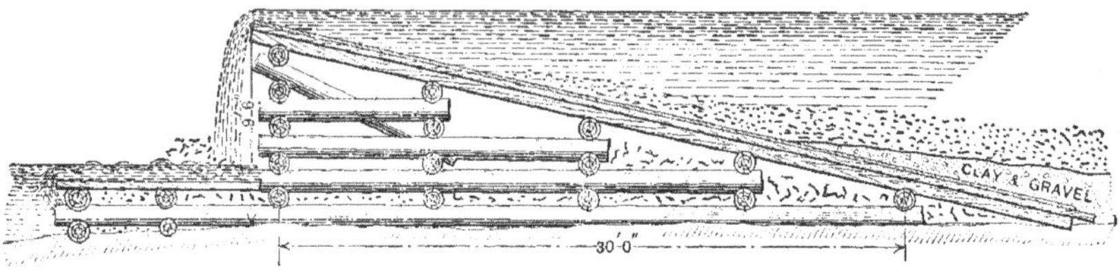

FIG. 113.—The log dam.

floor made on an extension of the lower tier of logs as far from the face of the dam as its height.

Although this type of construction is rude, it answers its purpose where logs are plenty and other material scarce, as in lumber districts, and in view in the old dams in the Eastern States. In Fig. 114 is shown a section of the crib dam across the Schuylkill River at the Fairmount water-works, Philadelphia. It measures 1,600 feet from bank to bank, forming an angle of about 45° with the direction of the stream. By this extension of length the rise above the top of the dam is lessened during high water.

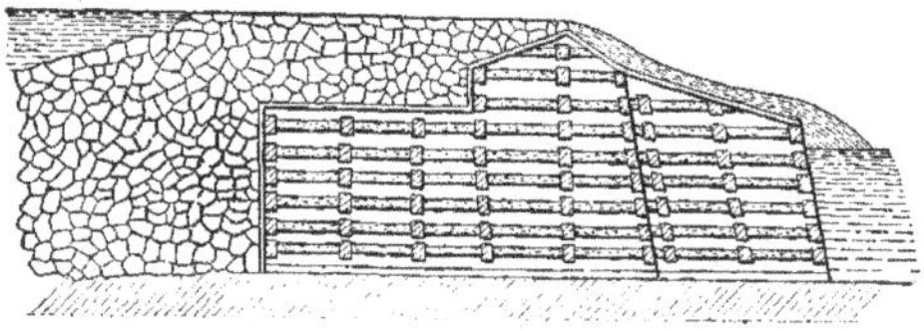

FIG. 114.—Dam of the Fairmount water-works.

The slack water above the dam extends about six miles, and a canal and locks are provided for overcoming the rise. A part of the

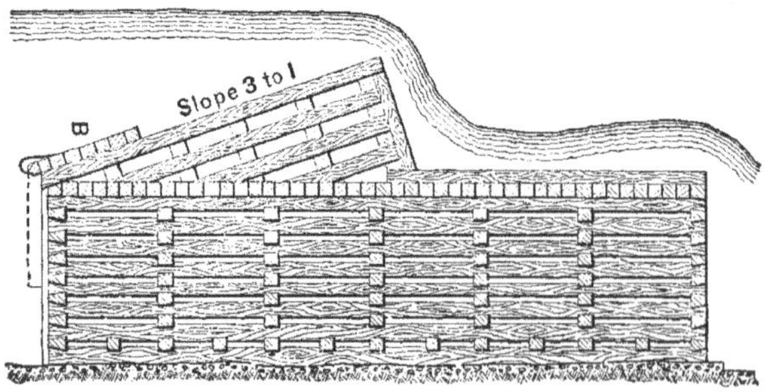

FIG. 115.—Timber crib dam.

bottom consisted of mud, and upon this portion, 270 feet in length, a foundation of rubble was laid, and covered with earth. This portion is 150 feet broad at the base and 12 feet on top, being incased with large stones. The overfall dam is 1,204 feet in length, founded on the bare rock, the deepest portion having a depth of 24 feet below low tides.

In Fig. 115 is shown a section of a timber crib dam of the Ottawa River type. A crib framing of timber filled in with stone, topped by a slope frame of 3 to 1, and apron with its apex at half the width of

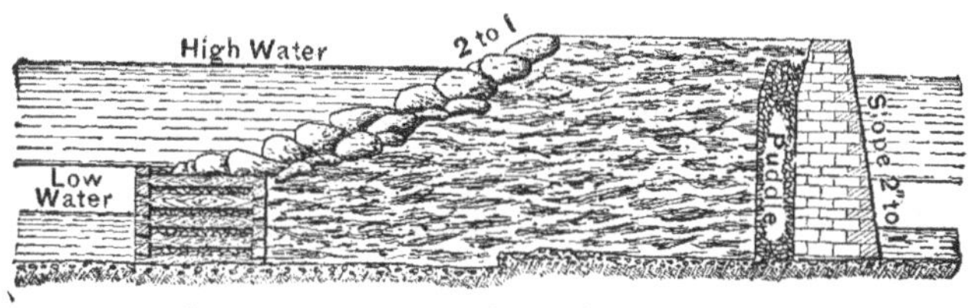

FIG. 116.—River and canal embankment.

the crib to divide the total fall of the water. B, cross planking on top and back. Back filling of stone and earth.

In Fig. 116 is shown a section of the embankment of the Ottawa River navigation system, in which a solid stone wall is built on the canal side backed by a clay puddle wall to near its top with a back filling of earth against a timber cribwork on the river side, filled with stone and the whole bank riprapped with large stone. Where rock bottom is not accessible the wall and puddle is carried quite a distance below the bottom of the canal.

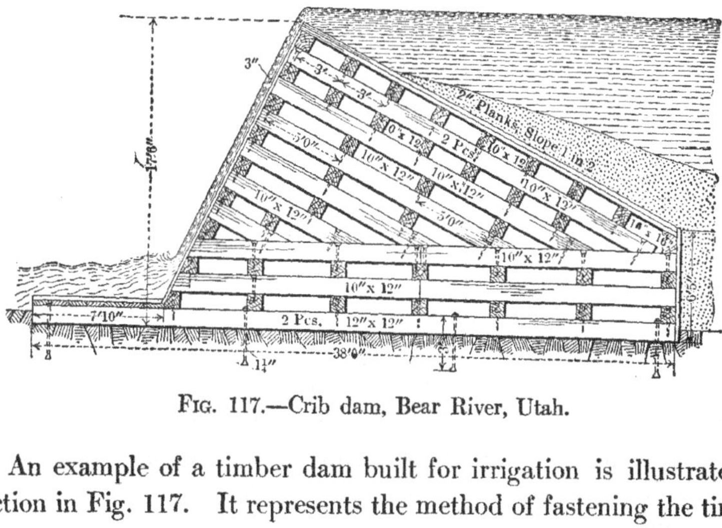

FIG. 117.—Crib dam, Bear River, Utah.

An example of a timber dam built for irrigation is illustrated in section in Fig. 117. It represents the method of fastening the timber work with long spikes and the fastening of the crib to the rock bottom with anchor bolts, 1½ inch, 3 feet in length with fox-wedges. The entire cribwork is filled with stone; the downstream face and apron are planked, while the rear slope is covered with two layers of 2-inch plank.

This dam has an overflow of 370 feet in length and for 100 feet under one end rested on a bed of gravel and bowlders which was washed out soon after the dam was finished; at the next low water a solid concrete wall was underlaid and the back fill increased so that it is now in a safe condition to withstand the largest floods.

Its height, 17′ 6″, is about the limit for wooden dams for safety and permanence, which the history of the Holyoke dam, of 30 feet in height, has shown.

VIBRATION FROM VERTICAL FACE DAMS

One of the greatest annoyances that can occur in the vicinity of a large dam with a vertical face or one so nearly vertical that air is enclosed behind the sheet of falling water, is the vibration set up by the elasticity of the confined air which produces a synchronous vibration in mills and dwellings for miles around. The vibration from the old Holyoke dam has rattled the windows at Springfield eight miles distant.

The remedy has been found in breaking the continuity of the falling water at a number of places.

FISHWAYS

As falls or dams of over 6 feet in height are a bar to the migration of fish upstream, especially where a rocky bottom, without an overfall basin, prevents their leaping propensities, one of the needed adjuncts of every dam or waterfall is a fishway. This needed appliance is practically enforced to a limited extent by the fish commissioners under the laws of many States.

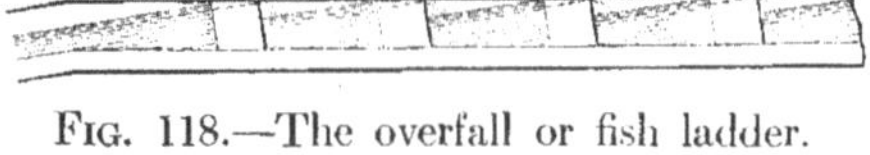

FIG. 118.—The overfall or fish ladder.

The principle of a practical fishway consists in retarding the velocity of the water in an inclined sluiceway by obstructions, so that the mean velocity will be no more than 8 feet per second with resting places made by the nature of the obstructions. Such a sluice should be open to the light and may be a series of basins with overfalls of one foot or less, which by returning upon itself in a zigzag way may be made to terminate near the foot of the dam where the fish naturally gather.

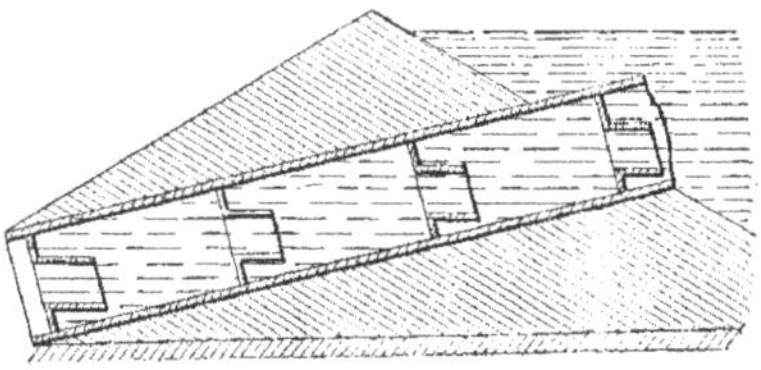

FIG. 119.—Fishway.

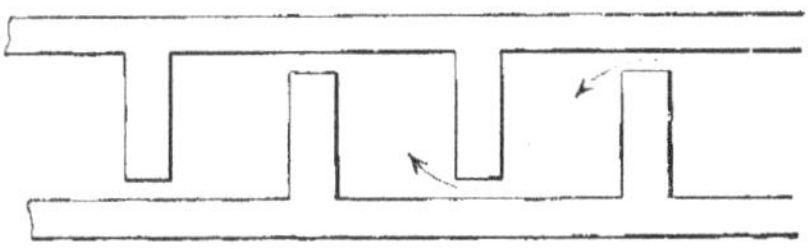

FIG. 120.—Square dam fishway.

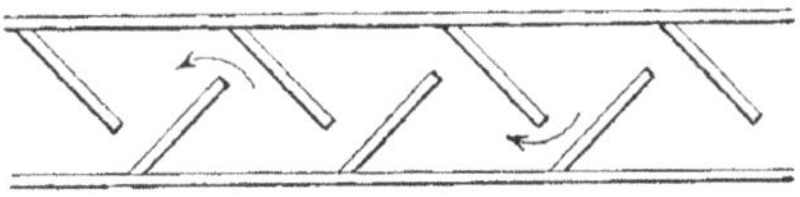

FIG. 121.—Oblique dam fishway.

A series of stepped basins over which the water descends, turning a fall into a cascade, and sometimes known as a fish ladder; or it may consist of a chute with a sinuous track for diminishing the velocity and assisting the passage of the fish to the level above the dam. In it is an inclined chute having a series of chambers containing comparatively still water, the current being confined to a relatively smaller space.

In Figs. 120 and 121 are shown other ways of constructing fish chutes of approved design. Square and oblique dams across a part of a sluiceway.

DAMS OF MASONRY AND CONCRETE

The progress and experience during the past half-century in the building of dams for stability has advanced in the line of all industries, and the art of building great structures of this class has settled into definite formulas for permanence and lasting qualities.

The vast heights made necessary for the ample storage of water for the increasing population of cities, for power and for irrigation, require absolute safety in these structures, which nothing but solid impervious rock and hydraulic cement can insure.

No reliance should be placed on the tensile strength of stone or cement in restraint of the pressure of the water to overturn the dam or to move it forward. The outline of its sections at all points should be so formed as to bring the resultants, or centres of pressure and weight as a factor of safety, at points in the base not less than one-third of its length from the downstream toe; or, in technical phrase, within the middle third of the base, and also at all horizontal levels above the base.

The highest conception in regard to the rising lines of both faces of a high dam indicates curves to exactly meet the safe factor of the pressure and gravity forces at all levels in its structure.

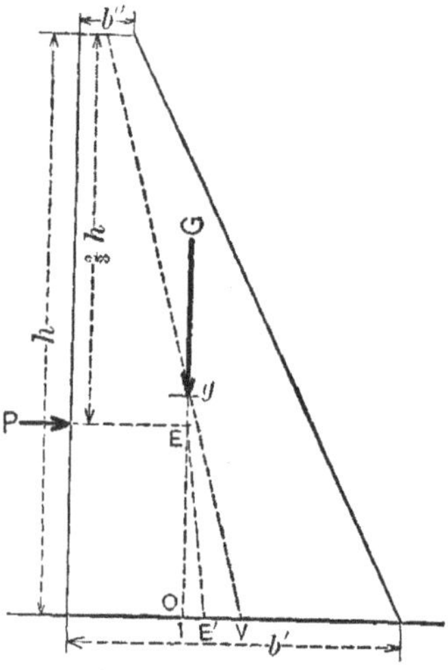

FIG. 122.—Trapezoidal dam.

In Fig. 122 is shown a dam or retaining wall of trapezoidal profile in which the resultants of pressure and gravity are detailed, and a computation is made for one foot in length of the dam.

The dotted line b'', g V, is the equal division of the profile through the centre of gravity and the top and base lines.

The lines of force, gravity G, and pressure P, meet at E and their resultant cuts the base at E'. The sum of their moments about E' should be zero, i. e., $P \times \frac{1}{3}h = G \times O\,E'$ for an exact balance. For the computations the following letters will designate the conditions in the formula:

h = height of dam; b'' = length across top.

b' = length of base; g = centre of gravity of profile.

G = weight of the section 1 foot thick.

P = point of mean pressure of the water = $\frac{1}{3}$ height.

E = horizontal pressure at vertical of centre of gravity.

E' = resulting point of pressures at the base line.

n = ratio of the distance E' from the middle of the base, to the whole length of the base.

y = weight of water per cubic foot.

y' = weight of masonry per cubic foot.

Assigning for the height of the dam to be 100 feet with the water even with the top with a width of section at top of 10 feet and at the bottom, assumed from the formula $b' = \frac{h}{\sqrt{3}} = \frac{100}{1.73} = 57.7$ feet.

The weight of the masonry may be assumed as 150 pounds per cubic foot (y′) and the weight of the entire section, one foot thick;

$$G = \frac{b' + b'' \times h}{2} \times 1 \times y' \text{ and with the figures,}$$

$$\frac{57.7 + 10 \times 100}{2} \times 1 \times 150 = 507{,}750 \text{ pounds.}$$

The pressure of the water at P, the mean centre, per square foot is $4{,}335 \times 144 = y \times \frac{2}{3} h = 41{,}574$ pounds.

The centre of gravity above the base is, $\frac{h}{3} \times \frac{b' + 2\,b''}{b' + b''}$ and as figured $\frac{100}{3} \times \frac{57.7 + 20}{57.7 + 10} = 38.19 \text{ feet} = g\ O = \bar{y}$.

The horizontal distance of the centre of gravity O, from the foot of the base is, $b' + b'' - \frac{b' \times b''}{b' + b''}\,\frac{1}{3} = 19.72$ feet.

V b″ is the division line cutting the centres of base and top through the centre of gravity.

$\bar{y} = g\ O = \frac{h}{3} \times \frac{b' + 2\,b''}{b' + b''} = \frac{100}{3} \times \frac{57.7 + 20}{57.7 + 10} = 38.19$ feet from base to centre of gravity.

$$O\ V = \frac{\bar{y}}{h} \times \frac{b' - b''}{2} = \frac{38.19}{100} \times \frac{47.7}{2} = 9.11 \text{ feet.}$$

$$O\ E' = \frac{\bar{y} \times P}{G} = \frac{38.19 \times 41{,}574}{507{,}750} = 3.127$$

$E'\ V = O\ V - O\ E = 9.11 - 3.127 = 5.98$ feet, the resultant of the forces P and G from the centre V toward the upstream face of the dam and far within the middle third section of the base, which is assumed to be the limit of safety. In all cases for dams with a shorter base, the resultant moment of the pressures P and G should be confined to the middle third, and we find that the limit of the base cannot be trusted at less than one-third the height of the dam with no overflow.

THE QUAKER BRIDGE DAM

The best conditions for safety under the effects of flood, shock, and time, have been worked out in the profile and terminals of the Quaker Bridge Dam of the Croton water system of New York; the highest barrage masonry in the world, Fig. 123.

Its profile has been made a study of the best engineering talent, which has added safety to the plain trapezoidal type with economy

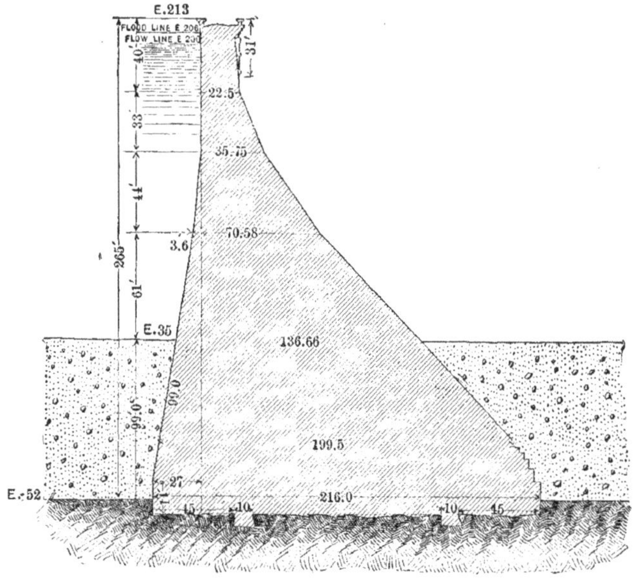

FIG. 123.—Quaker Bridge Dam.

of cost, by curves on both faces, extending its base to exceed the requirement of $\frac{h}{\sqrt{3}}$ by more than a third; the upper part conforming more nearly to the above formula.

The 18-foot driveway along the crest of the dam forms a connecting link between two macadamized roads, which follow the shore of the new Croton Lake, and form a continuous ride over forty miles in extent; the total height of masonry from the foundation to the crest being just under 300 feet, or to be exact 297 feet. At the foundations of the dam in the centre of the valley the masonry is 200 feet in thickness, and it narrows symmetrically to a thickness of 18 feet

Fig. 124.—Crest, spillway, and bridge.

VIEW OVER THE SPILLWAY, QUAKER BRIDGE DAM.

at the crest. The total length of the dam from the southerly abutment to the bridge is 1,168 feet, and the length of the spillway from the bridge to its terminus up the valley is 1,000 feet, making a total length of masonry of 2,168 feet. The 1,000 feet of spillway provides complete security against damage by sudden floods. As the waters flow over the spillway they enter a wide channel blasted out of the rocky side of the hill, and they are led beneath the steel arch bridge down to a new artificial channel, which ultimately directs them into the old bed of the Croton River.

The total storage of the Croton reservoir system is 102,000,000,000 gallons.

An abnormal thickness of a masonry dam, over that due to the constructive pressures P and G, is required from considerations of the known fact of the porosity of the cement mortar, which in a short time becomes saturated with water at its full pressure on the upstream face and through the horizontal seams, relieved only by friction and its exit at the downstream face of the dam.

This has been proved by the experiments of Mr. J. B. Francis, to indicate an upward pressure at the base, diminishing with the height and toward the downstream face, thus contributing to lift the weight of the masonry and to the overturning of a thin section dam.

Opinions seem to vary among engineers in regard to the value of a curve in the alignment of dams. This curve may be necessary in a

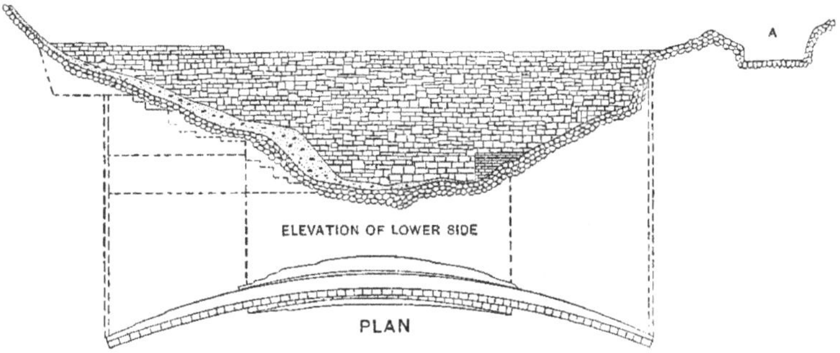

FIG. 125.—Elevation and curve of Bear Valley Dam.

dam of very thin section, but the stiffness or rigidness of long walls is subject to shear with slight irregular pressures, and although such dams are cheap to build, in material, they require more labor in perfecting the vertical joints to prevent compression. A curved dam is liable to differences of compression by varying stages of water and with this to a movement on its rocky base that is disintegrating to its structure.

Such dams are, however, in use and declared trustworthy when built of moderate length and abutting against rock walls.

We illustrate a dam of this form, the Bear Valley Dam, San Bernardino County, Cal.

This dam has a curve of 335 feet radius, about 260 feet in length, and abutting against rock walls with a spill at A, cut in the solid rock.

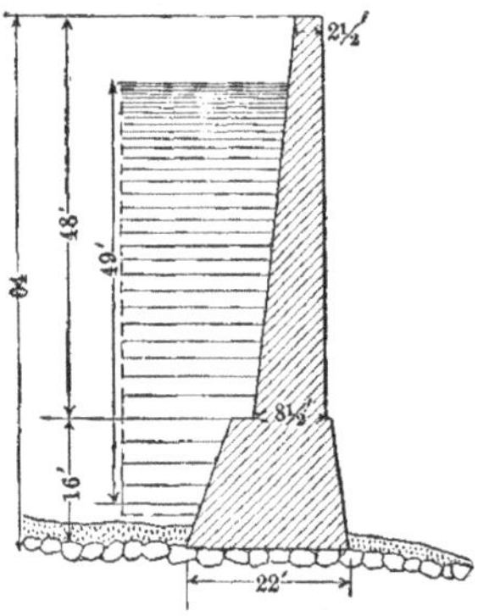

FIG. 126.—Section.

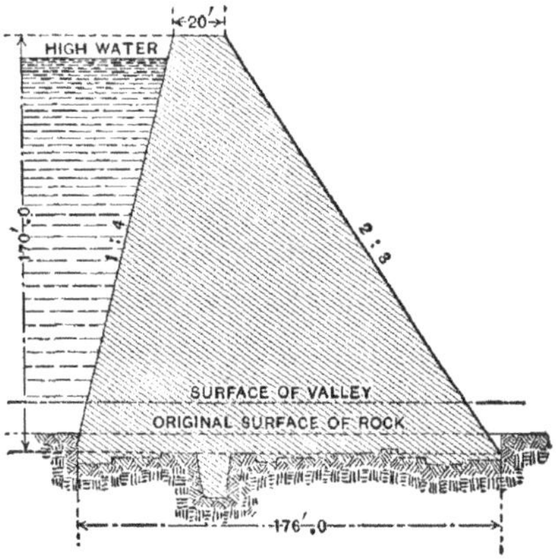

FIG. 127.—San Mateo Dam.

Fig. 126 is a profile section at the centre of the dam. The masonry has a granite ashlar facing with rubble heart all closely laid in cement. The drainage area is small, 56 square miles, and the dam is not subject to overflow.

The dam of the San Francisco water-works near San Mateo is curved on a radius of 637 feet with an exceptionally strong profile and base greater than its height. Its outlet is tunnelled through the solid rock with gates in a shaft.

FLOODS AND STABILITY OF MASONRY DAMS

The disintegrating effect of floods by deep overflow on the top of dams has been the cause of disasters in many instances by undermining, lifting, and pushing the dam downstream; the experience with which has led to the practice of greatly extending the base of the profile for absolute safety, especially where rock bottom is friable or shaly and to great depth where rock bottom is not available.

In order to give a dam a wide margin of safety from flood flow over the top, we give a graphic example of the forces and figured resultant for an overflow of a possible 10 feet in height over the top of a trapezoidal dam of masonry, 50 feet high with a base of 25 feet and a crest 10 feet wide, which may be considered a fairly safe dam on a rock bed.

Let $X = \frac{h}{3}\left(1 + \frac{h'}{h+2h'}\right)$ figured, $= 16.66 \times \left(1. + \frac{10}{50+20}\right) = 18.44$ ft.

$G = \frac{25+10\times 50}{2} \times 150 = 131{,}250$ pounds, weight of the dam 1 foot long.

$P = 4.335 \times 144 \times (h + h' - X) = 624.24 \times 41.56 = 25{,}945.4$ pounds pressure.

Then 131,250 : 25,945 :: 18.44 : 3.64 feet the distance of the resultant of the pressures P and G from the centre of gravity, O, of the dam and well within the middle third.

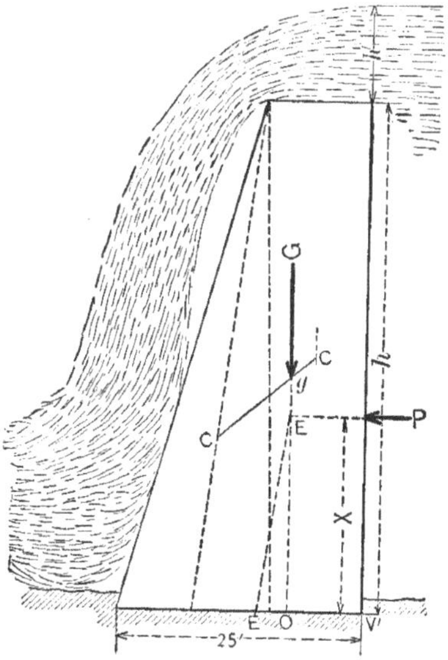

Fig. 128.—Overflow dam.

The protection of the toe of an overflow dam by a curved apron is an essential feature of safety from the disintegrating effect of floods in carrying over ice and lumber that tends to wear away the rock base or the riprap that was intended to supply the place of a properly built apron.

In Fig. 129 is shown the well-designed contour of the profile of the Holyoke Dam, designed to meet all flood conditions. Its length is 1,017 feet and 30 feet high. The entire facing of this dam is of granite with the crest and reverse apron dowelled. The interior is of rubble masonry.

It will be noticed that in this design are the best elements of stability, that in its trapezoidal outline the base is greater than the height and that the extension of the curved apron gives the base practically a length of one and three-quarters of the height.

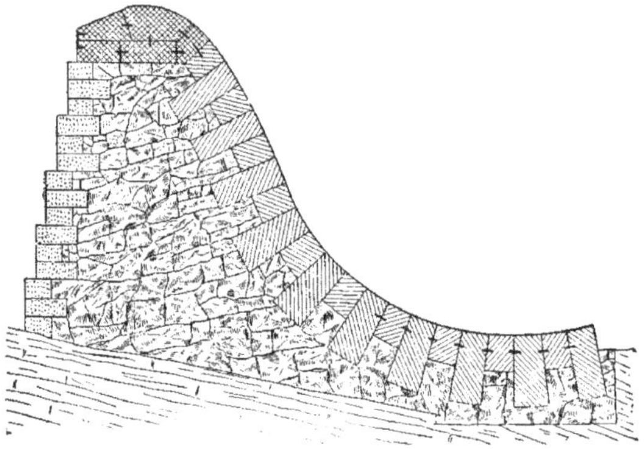

FIG. 129.—Section of Holyoke Dam.

The rapid advance in the use of cement concrete in building and experience with its weathering qualities and resisting properties in hydraulic structures has led to its adoption in the construction of dams.

In Fig. 130 is shown a section of the concrete dam at Mechanicsville, N. Y., with abutments also of concrete. The base is re-

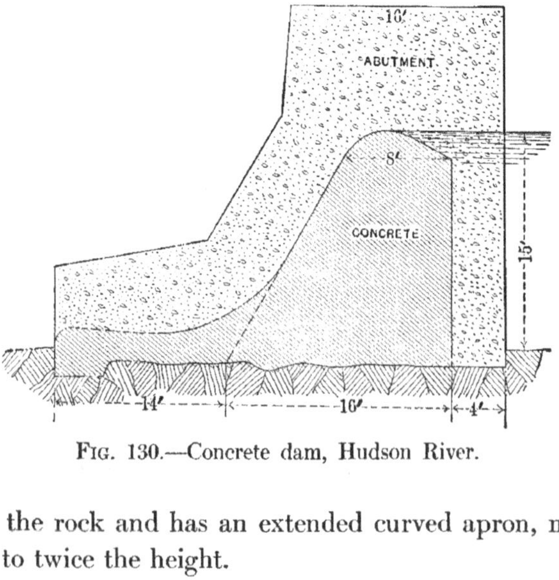

FIG. 130.—Concrete dam, Hudson River.

cessed into the rock and has an extended curved apron, making the base equal to twice the height.

The dam is 800 feet long.

RESERVOIR EMBANKMENTS

The embankments for impounding water only, such as reservoirs for city and town water-supply, where the water-level is limited by gates or overflows and which are only subjected to pressures at their high-water level or less, are generally of far more simple construction than of dams subject to flood action.

In Fig. 131 is shown a section of a reservoir embankment of moderate height, say 10 feet depth of water with the top 3 feet above the overflow, which may be a pipe laid sloping through the bank and to beyond its foot.

Such reservoirs on gravel, sand, or loam beds should be puddled on their entire bottom and extending to and beyond the clay puddle

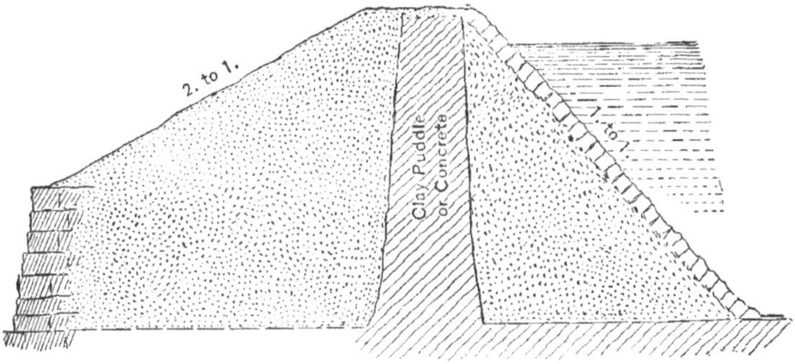

FIG. 131.—Clay puddle embankment.

wall and closely amalgamated with it. The puddle wall should be one-third its height in thickness at the bottom, tapering to one-quarter at the top, be well rammed and carried up with the slope filling. The inside or wet slope should be not less than 1 to 1 and faced with stone; coarse pebble, gravel, and bowlders are in use. The back filling may have a slope of not less than 2 to 1 and supported by a low foot wall for safety and appearance. A breadth at the top of 10 feet or more is recommended.

Concrete for bottom lining and walls is much in favor, although failures from cracks have occurred.

For elevations of less than 10 feet, earth embankments without puddle walls have been made a permanent feature where clay or cement was not available; where the footing of both slopes may abut against rough stone walls without mortar.

In Fig. 132 is shown a section of an earth embankment which should have a loam filling on the wet face and bottom of the reservoir with a pebble top dressing. The back filling may be of any soil available and compacted by ramming wet, or if of low height, may be rammed by the traversing of the filling teams.

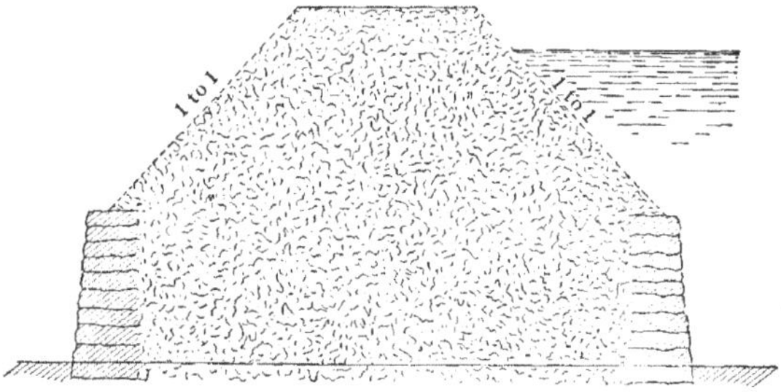

FIG. 132.—Earth embankment.

Face puddling is in use and has an advantage over the puddle wall in strengthening the embankment by placing the resistance of all the material behind the puddled face. A face puddle should have a layer of broken stone and gravel to prevent wash, and for best effect the face should have a slope of $1\frac{1}{2}$ to 1.

The laying of the outlet pipe is always a matter of some concern as to its stability and freedom from seepage; it should be an extra heavy pipe provided with anchorage bands and laid on a concrete foundation with a gate valve at the outer end, or in large works gate chambers are built on the inner slope of the embankment.

CHAPTER VIII

CITY AND TOWN WATER-SUPPLY

RESERVOIRS for city and town water-supply are of far greater importance than the mere service of domestic and manufacturing needs.

A fire service is an absolute requirement for safety from fire calamity; the water pressure for all purposes, save fire, need be no greater than from 30 to 40 pounds per square inch; although much higher pressures are in use, often to the detriment of house plumbing.

The size of a reservoir is a matter of judgment, comparable with the volume and constancy of supply; but in view of the variation in seasonal rainfall, it should be ample for a week's supply in case of floods bearing muddy water, a drought, or the breakdown of pumping machinery.

In great cities with scant water resources that are subject to droughts, the reservoir volume requires special attention as to storage capacity and resources, such as now confronts the Greater City of New York. Nearly all large cities are centres of manufacturing industry, and on that account require much more water than the requirement for domestic service. For domestic service alone, from 20 to 40 gallons per capita per day about covers the range of consumption; but in manufacturing cities and towns the consumption and waste range all the way up to near 300 gallons per day for each individual of the population.

Where a natural water-supply by gravity cannot be made available, the pumping system is in order, and for cities and large towns a well-organized steam pumping plant is preferable. There are numerous small towns throughout the United States and in other countries that should consider with favor the gas-, gasoline-, or oil-engine proposition with a view to economy in a combined distribution and fire system. The ready condition of the explosive pumping engine for starting makes it a valuable means for fire service; especially at night, when pumping is usually suspended and with expenses nil; while with steam power it takes time to start the pumps, or banked fires and watchfulness. Not only is a point in economy gained in the general running expenses by explosive power, but a constant readiness for putting a greater pressure on the pipe distribution by shutting off the reservoir or tower and pumping directly into the main pipe supply.

STAND-PIPES AND ELEVATED TANKS

Where elevated positions for reservoirs are not available, a stand-pipe or a tank on an elevated trestle is always a recourse with the advantage derived from freedom of location, for which the economy in piping may in a way compensate in cost.

Stand-pipes are generally built in plain vertical tubes

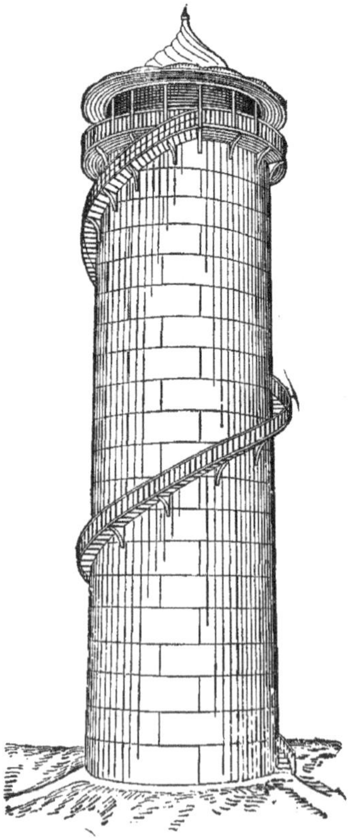

FIG. 133.—Des Moines stand-pipe.

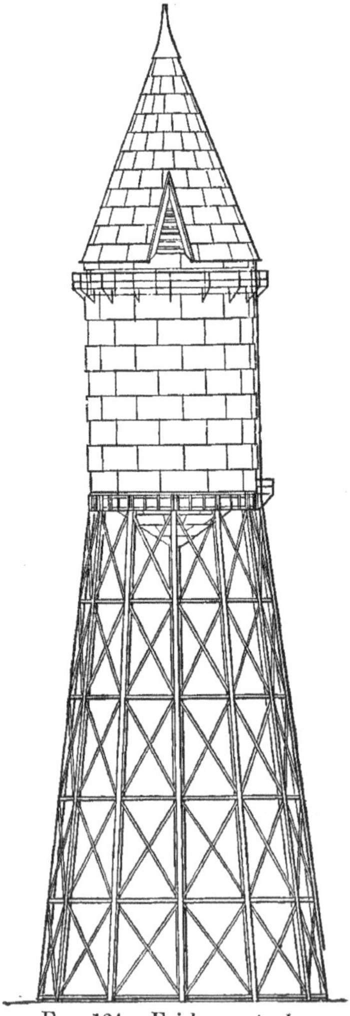

FIG. 134.—Fairhaven tank.

with open tops; but this plan is not recommended, because of the exposure of the water to sunlight and its pollution by organic life.

In Fig. 133 we illustrate an artistic design of a stand-pipe built for the Des Moines water-works, having a winding outside stairs, a gallery and observation balcony covered with a metallic roof.

It is 100 feet in height and 30 feet diameter.

In Fig. 134 is illustrated a most interesting model of an elevated tank from its extreme height and capacity.

It is a cylindrical tank 35 feet diameter and 50 feet high, resting on a steel skeleton base 100 feet high and covered with a conical roof 50 feet high, making a total height of 200 feet; a sightly object of the Fairhaven, Mass., water-works.

The bottom of the tank is of conical form and provided with a slip-joint connection with the flow pipe to avoid strain from differential expansion and compression. The expansion joint is an essential feature of every elevated tank. The one used on the Fairhaven tank is shown in Fig. 135, and consists of a cast-iron flanged stuffing-box, riveted to the bottom of the tank; a composition gland through which works a composition flanged piston bolted to the 10-inch rising main. The detailed make-up of stand-pipes, tanks, and their framework are matters of engineering experience too varied for the scope of this work.

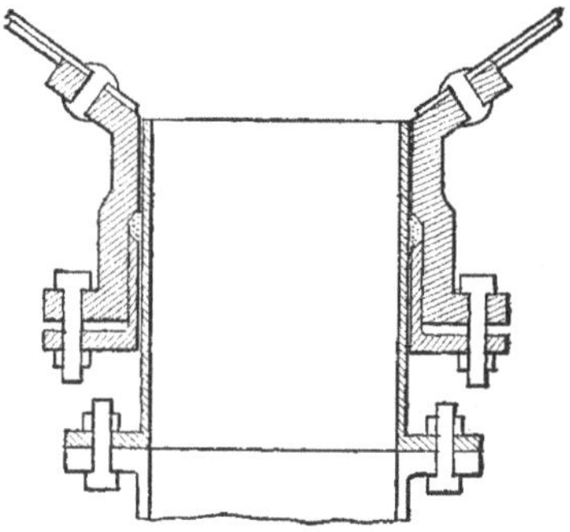

FIG. 135.—Expansion joint.

The system of charging for water by meter measurement has vastly improved the conditions as to the waste in units of the per capita consumption; sometimes as much as 70 per cent. has been saved.

A curious revelation in regard to the actual use of water by meter in different families, was given as authentic in the following statement:

"The water used by 2,553 families in Providence, R. I., was measured for one year. Five persons counted to a family, make a total of 12,765 people. By actual measurement

"167 families used but 6.15 gallons to a person per day.
"237 families used but 8.20 gallons to a person.

"361 families used but 10.25 gallons to a person.
"445 families used but 12.30 gallons to a person.
"446 families used but 14.35 gallons to a person.
"462 families used but 16.40 gallons to a person.
"435 families used but 18.27 gallons to a person."

And yet we find American cities publishing the information, year after year, that their works are actually supplying 150 to 300 gallons each day to each person in the city!

No less diversity of water consumption is shown by different cities than by the different families in a single city as noted in the following report of the per-capita consumption for different cities per day:

	Gallons.		Gallons.
New York	100	Louisville	80
Chicago	100	Milwaukee	80
Cincinnati	116	Glasgow, Scotland	64
St. Louis	102	St. Paul, Minn	44
Syracuse	101	New Orleans	37
Paris, France	98	Liverpool, England	31
Minneapolis	85	Berlin, Germany	21
Montreal, Canada	81		

The variation in quantity of water used per capita in private and apartment houses in New York is very great; from 40 to 90 gallons per day with an average of 50. The large draught due to the supply is caused by steam and factory use, stables, street watering, supply of shipping, and the leakage from street mains; which with other unknown wastage is above 119,000,000 gallons per day.

The comparative effect from metering the water-supply is shown in the following data:

	Per cent. services metered.	Gallons per capita.
Lexington, Ky	99	47
Utica, N. Y	96	53
Fall River, Mass	94	36
Buffalo, N. Y	1.6	233
Pittsburg, Pa	.6	231
Camden, N. J	1.4	280
Washington, D. C	....	200

For the dwelling part of a town, provision should be made for at least 60 gallons per capita per day, and the reservoir capacity, where the supply is limited, should be as much greater as the location and resources will permit, to the extent of from four to six times the daily requirement.

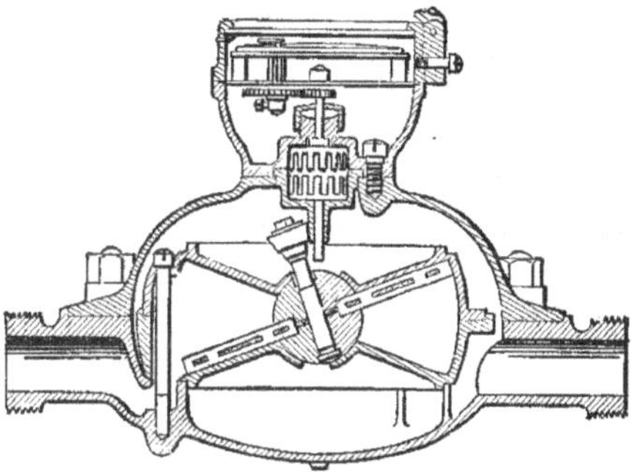

FIG. 136.—Thompson meter.

The distributing system of a town should be a subject of judgment and engineering scrutiny in the allotment of sizes of pipes according to the plan of the town, its street levels and slopes; but in no case should less than a 4-inch pipe be used.

The service distribution is a matter belonging to the expert plumbing trade, suited to dwelling and factory plans, and is not suitable for detail in this work.

The meter question, however, is in agitation and is being adopted in many places where the flow is not meeting the needs of population and waste.

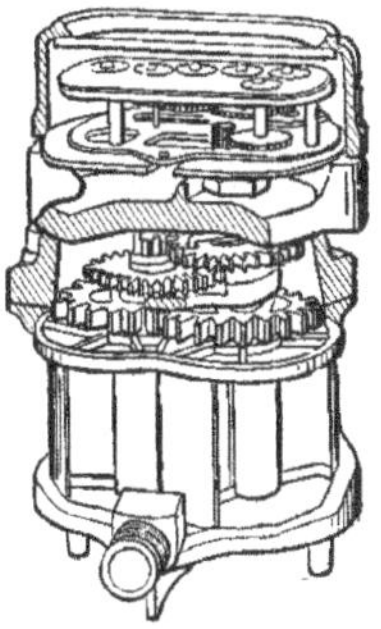

FIG. 137.—Union meter.

Fig. 136 shows a section of the Thompson meter, the measuring element consisting of a swinging disk movement on a ball socket, operated by the flow of water, which rotates a vertical crank, spindles, and gear train and index hand.

In the Union meter, Fig. 137, the water flows through a rotary motor with equalizing gear, from which the dial pointers are driven by a clock train and counter.

For measuring the flow in large pipes the Venturi meter is in favor; it is illustrated in Chapter V.

WATER HAMMER IN SERVICE PIPES

The increased pressure in house service pipes, caused by water impact when faucets are suddenly closed and a water hammer in the pipes may be heard all over the house, is a source of danger and trouble from the expansion and final bursting of lead pipes and the splitting of iron pipes and fittings, and shows the necessity of ample air-chambers at every faucet. The following tables, the result of experiments, show the impact pressure by the sudden closing of various sized orifices:

TABLE XX.—FORCE OF RAM GENERATED IN A LINE OF PIPE CONSISTING OF 111 FEET OF 6-INCH, 58 FEET OF 2-INCH, 99 FEET OF 1½-INCH, AND 4 FEET OF 1-INCH, WHICH WAS THE OUTLET PIPE.

Orifice of discharge in inches.	Velocity in the 1-inch pipe.	Impact. Ram in pounds per square inch.						
		1-inch.	1½-inch. Distance from 2-inch.			6-inch.	Dead end.	
			32 ft.	49 ft.	67 ft.		2½-inch.	The same with air-chamber.
3/16	2.57	72.8	72.4	70.8	70.3	14.5	18.8	....
1/4	5.36	129.3	125.9	127.0	128.8	23.5	42.2	....
1/2	10.05	....	....	....	....	51.7	88.3	16.8
5/8	19.23	....	....	....	....	....	....	49.3

The dead end referred to in the above table was a continuation of the pipe beyond the point in the 6-inch, where the 2-inch was connected, and consisted of 70 feet of 6-inch, 66 feet of 4-inch, and 4 feet of 2½-inch, which was at the extremity, and located, measuring on the line of pipe, about 322 feet from the orifice of discharge.

TABLE XXI.—FORCE OF RAM GENERATED IN A LINE OF PIPE CONSISTING OF 111 FEET OF 6-INCH, 58 FEET OF 2-INCH, 48 FEET OF 1½-INCH, 3 FEET OF 3-INCH, 48 FEET OF 1½-INCH, AND 4 FEET OF 1-INCH, WHICH WAS THE OUTLET PIPE.

Orifice of discharge in inches.	Velocity in the 1½-inch pipe.	Ram in Pounds per Square Inch.		
		1½-inch. 16 ft. from 3-inch.	Centre of 3-inch.	1½-inch. 16 ft. beyond 3-inch.
3/16	1.19	75.3	64.5	61.2
1/4	2.37	126.4	120.8	113.8
5/16	3.00	150.2	150.5	138.9
3/8	4.47	203.3	206.8	195.3

TABLE XXII.—FORCE OF RAM GENERATED IN A LINE OF PIPE CONSISTING OF 182 FEET OF 6-INCH, 66 FEET OF 4-INCH, 4 FEET OF 2½-INCH, 1 FOOT OF 2-INCH, 7 FEET OF 1½-INCH, AND 6 FEET OF 1-INCH, WHICH WAS THE OUTLET PIPE.

Orifice of discharge in inches.	Velocity in the 1-inch pipe.	Ram in Pounds per Square Inch.					
		1-inch.	1½-inch.	2½-inch.	6-inch.	With air-chamber. 2½-inch.	With air-chamber. 6-inch.
¼	5.39	66.7	49.4	22.2	4.8	....	....
$\frac{5}{16}$	6.71	76.1	61.5	35.6	6.6	....	....
⅜	10.02	106.3	81.8	52.0	15.8	14.0	12.3
½	20.94	177.5	121.5	99.0	36.8	38.7	25.6
1	43.9	....	....	183.0	80.1	105.8	65.6

Velocities in feet per second.

ON THE PREVENTION OF THE GROWTH OF ALGAE IN WATER SUPPLIES

The fouling of reservoirs and conduits in water-works by various growths is a problem that has frequently to be dealt with by the water engineer, and although it has not attracted general attention this contamination of large volumes of water is often a serious matter, and several commissions have had to deal with this subject. The contamination frequently consists of one, or several, of the numerous algæ, many of which, besides causing the clogging of valves, and an unsightly appearance, are capable of imparting disagreeable odors and tastes. Trials have been made to destroy or inhibit the growth of the plants by the addition of a minute quantity of a germicidal substance to the water. Various reagents have been employed to prevent the dense, unsightly growths which have been found to consist largely of desmids and other unicellular algæ allied to *Protococcus*.

On a large scale metallic salts such as those of iron or copper have been used. The employment of the latter in the form of crystallized copper sulphate has met with a fair amount of success, the poisonous copper being subsequently precipitated as a basic salt, the coagulating effect of which often renders the finished water very clear and bright.

In the experiments in this direction by Dr. G. T. Moore, of the

United States Agricultural Department, the quantity of crystallized sulphate required varied from 1 part per 1,000,000 to 1 per 50,000,000, while some waters did not appear to be amenable to the copper treatment. He found that the worst tastes and odors are most susceptible. In conjunction with Kellerman he recommends the following dilutions per 1 part of copper sulphate for the destruction of each organism named:

Clathrocystis	8,000,000
Cœlospherium	3,000,000
Microcystis	1,000,000
Oscillatoriæ	5,000,000

However, it has been found that the growth of a conferva (a variety of *Spirogyra*) was inhibited, but by no means entirely destroyed, by the addition of 1 part per 1,000,000, in an English lake water. Some forms of algæ appear to flourish especially in calcareous waters, and also in those containing quantities of free CO_2.

In the addition of the germicide to large reservoirs it must be remembered that the solution will diffuse downward and horizontally, any upward movement being too slow. At Newport (Mon.) reservoirs were successfully treated by trailing bags of copper sulphate across the surface of the water, the rate of solution being controlled by the thickness of the material enclosing the salt. Similar experiments at Gloucester water-works have been successful in removing the spring growth from the reservoir.

Aeration is a powerful agent in treating waters suffering from tastes and odors of growth and disintegration. Its effect has been most frequently noted with impounded reservoir waters which are later exposed to air and sunlight in another reservoir before the water is used by consumers. This beneficial effect appears to be accomplished partly by a marked reduction in free carbonic-acid which is one of the sources of food for organic growths, and partly by agitating the water so that the living organisms then contained in the water are either killed or prevented from subsequently multiplying.

Aeration of this type of water will also remove tastes and odors of decay present in the water when aerated. It will reduce but not necessarily remove satisfactorily the odors of disintegration and odors of decay resulting from the organisms in the water at the time of aeration.

It should be noted that as in the case of carbonic-acid far more vigorous and thorough aeration is required to remove tastes and odors than is necessary to saturate a water with oxygen, and there are, no doubt, some tastes and odors which could not be adequately removed by any practical amount of aeration. The violence of the contact between the air and water seems to be important; and a short, vigorous agitation with air often accomplishes that which indefinite exposure of the surface of still or flowing water to the air fails to accomplish.

Aeration affects the water in a number of ways. It practically saturates the water with oxygen, and this is, of course, one of the primary benefits which is availed of in connection with the removal of iron from ground waters. This same feature also comes into prominence in connection with the discolorization of stagnant reservoir water. In the stagnant layer of a deep reservoir fermentations take place, especially in summer, which result in the solution of considerable amounts of organic matter and also of iron in the ferrous state. This water can be aerated and filtered, thereby entirely freeing it from objectionable odors of decay, and utilizing the iron to a substantial extent as a coagulant to remove the color and other organic matter in the water, and so produce a water better and more acceptable in every way than could possibly have been obtained by filtration of the same water before stagnation and fermentation. It is only within the past few years that the practical significance of this proposition has been fully appreciated.

SOURCES OF WATER FOR TOWN SUPPLY

The most important point in the proposed instalment of a town water-works is the source and condition of the water. Lakes, ponds, rivers, springs, and wells have their special features of purity and quantity which require the first consideration in the designing of a water-works; lakes are generally uniform in quality, but ponds are liable to be infested with vegetable growth and a river supply is not reliable in purity; its high-water stages are muddy and carry the storm-wash of its watershed; in droughts it is sluggish and abounds in vegetable and primordial life. Springs and wells, properly cared for, are assumed to furnish acceptably pure water.

Aeration and filtration with in some cases a settlement reservoir built in two sections for cleaning are the means of obtaining acceptable purity from infected sources of supply.

It is well known now, among hydraulic engineers, that an ample aeration of water in reservoirs and tanks will prevent stagnation, check the growth of algæ, remove the disagreeable odor from decomposing vegetable matter, and deposit the salts of iron that sometimes pervade waters from iron soils or that have traversed long lines of iron pipe.

Fig. 138 represents the pipe plan for aerating a tank 62 feet in diameter, 59 feet high, holding 1,300,000 gallons, at Brockton, Mass. On the bottom of the tank are three 2-inch galvanized iron pipes branching from the main air-pipe with 39 branches of $\frac{1}{4}$-inch and $\frac{3}{8}$-inch brass tubes perforated with $\frac{1}{8}$-inch holes, 3 feet apart; a check valve in the main pipe prevents the water going back to the air-pump when not in operation. A duplex air-compressor $7\frac{1}{2}\times9\times9$ inches, furnishes 172,000 cubic feet of free air in 24 hours. By this means the water is thoroughly agitated and aerated, doing away with its former odor and taste. Another method of aeration of water is by pumping air directly into the main at the pumping-station or into the delivery main from a reservoir.

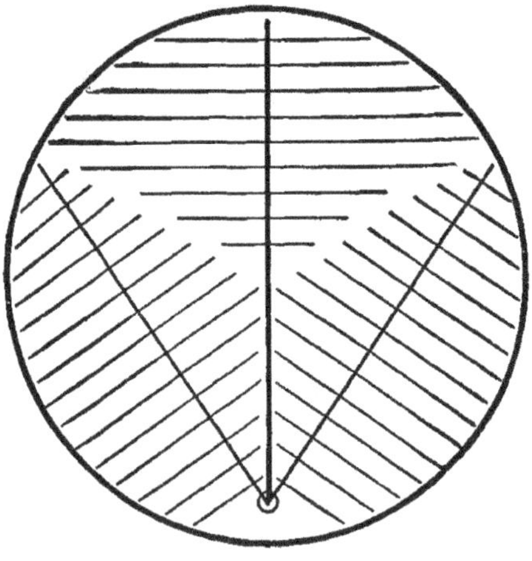

Fig. 138.—Water-tank aeration.

Cascades, fountains, the introduction of air to conduits, artificial falls, thin films of water passing over large surfaces—in fact any device that will permit the air to mingle with the water, give new life to the waters and death to organisms.

The plan adopted by the Utica, N. Y., Water Company is on the fountain principle, discharging the water under pressure through a series of pipes in a shallow reservoir, shown in Figs. 139 and 140.

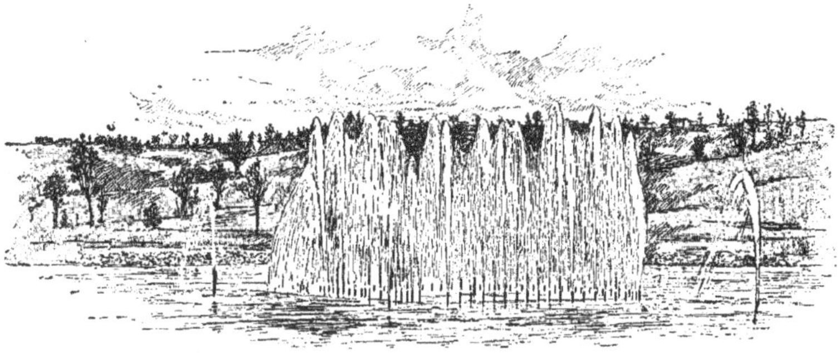

FIG. 139.—Aeration at the Utica water-works.

A similar system with four jet pipes equal to the capacity of the supply pipe is in operation at Fresh Pond reservoir, Cambridge, Mass.

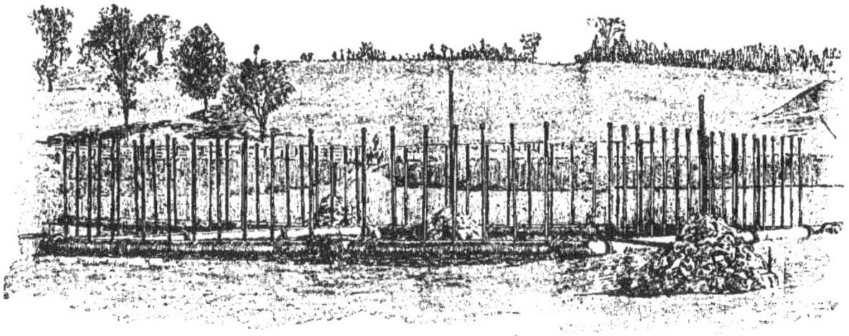

FIG. 140.—Aeration pipe system, Utica water-works.

The jets are thrown 40 feet into the air and by so atomizing the water it becomes throroughly aerated.

SETTLING BASINS

When the source of water-supply is from streams subject to storm-wash, its turbidity is mostly composed of fine sand, loam, and silt raised from the deposit in the stream at low-stage flow. In such cases a settling reservoir of an elongated plan, or a canal, is needed that the influx may have time to deposit its holding by a slow passage through it. A ditch large enough to convey the required supply of water slowly, say 1 foot per second at its head and increasing in area to a flow of 1 foot in 20 seconds at the discharge end, which should terminate in a broad, thin stream from the surface only, and so feed a filtering system if such is needed.

FILTERING SYSTEMS

The simplest forms of water-purifying devices for town supplies are what is known as filter wells, filter galleries, and filter cribs. A filter well is a curved hole in the sand or gravel on the banks of a river, drawing its supply by filtration from the river. A filter gallery is a series of drains of unglazed drain-pipe laid in gravelly soil on springy

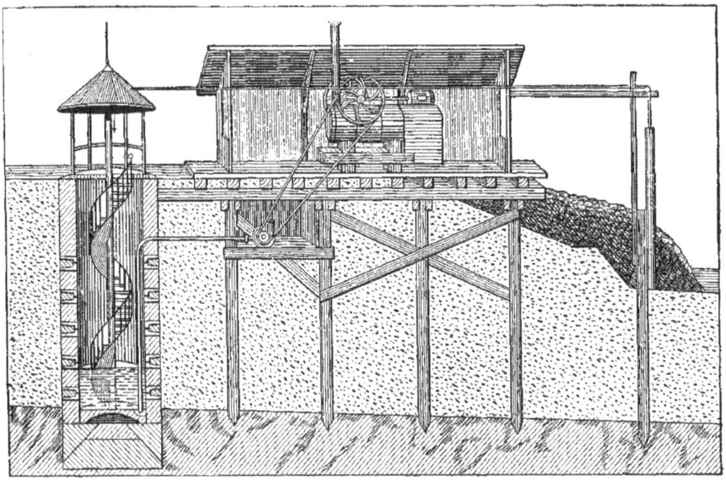

FIG. 141.—Filtering well, Nantes, France.

land or beneath the low-water level on the banks of rivers ior gathering the filtered water from larger areas than a well flow. Filter cribs are intakes in lakes or rivers, anchored in place and covered with gravel and sand.

An example of a filtering well built on a sand bank in the river at Nantes, France, is shown in Fig. 141.

During a whole year the water from this well showed no color in thin layers with a bluish color in deep strata; was absolutely limpid and of agreeable taste, with no odor, and equally pure at all stages of the stream.

A FILTER BASIN

The source of water-supply at Nantucket is a natural pond of about eight acres in extent and 14 feet in depth, having neither surface inlet nor outlet, and a very small area of shallow flowage, the shores being quite bold. The shores and vicinity of the pond are clean, and there is no possibility of sewage contamination, there being but one house within one-half mile of the pond, and the nearest houses of the town being more than a mile distant. The pond is fed mainly by springs, and the water has always been of very good quality, except in the autumn of certain years, when it has been infected with a growth of *Anabæna*.

The plan is a circular basin about 64 feet diameter at bottom and 6 feet deep, formed by a level clay-puddle bottom 1 foot thick, and

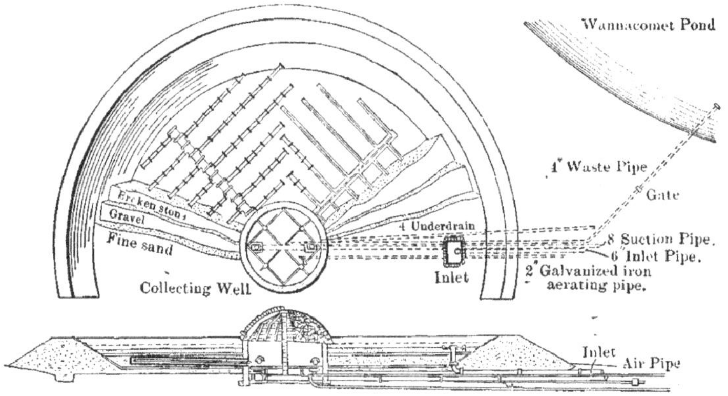

FIG. 142.—Filter of the Nantucket water-works.

an embankment of the same material having slopes of 2 to 1 and a top width of 3 feet. Clay puddle was made of 3 parts clay and 1 part sand. In the centre is built a circular brick collecting well, 15.25 feet diameter inside and 8 feet deep, having a dome-shaped roof built of wooden rafters covered with heavy laths, which were plastered on the outside, and small stone embedded in the cement plastering, to make an artistic finish. The bottom is concrete, water-tight.

On the bottom and inner slope of the basin was spread a layer of sand 1 inch thick, to prevent the moving water from coming in contact with the clay. Then on the bottom is a layer of round and broken stones about 2 feet thick, in which are embedded four lines of 12-inch vitrified pipe tees, radiating from the well, and from each branch tee lines of 4-inch vitrified pipes to receive and conduct the filtered water into the well. These pipes are all laid with a slight rising grade from the well, and the joints were packed with one turn of tarred yarn, loosely put in, and each joint was well covered with gravel.

Above the broken stone is a 6-inch layer of gravel, and above that the filtering sand, which was put in in three layers, each rolled with a stone roller, the depth when finished being from 12 to 16 inches. The surface of the sand is level and has an area of about 4,600 square feet.

Aeration pipes are laid from an air-compressor in the pump house for aerating the water in the collecting well.

Pond water is pumped through a 6-inch pipe to the surface of filter and kept at a depth of 12 to 18 inches above the sand. It is intended to filter at the rate of five gallons per square foot per hour. The filter, during the months it is in use, is always full of water, but filtration is intermittent, depending on the times of pumping from it. Filtered water is drawn from the well by an 8-inch pipe, the pumps being 300 feet distant.

Some trouble was experienced in August from odor, which was probably caused by the excessive temperature of the pond water (74° F.) and by exposure to sunlight in a 50,000-gallon tank nearer the town to which the water was pumped at intervals. The constant and slow operation of such a filter is faultless, if sunlight conditions can be avoided by covering both filter and tank.

FILTER REGULATION

The modern apparatus for regulating the velocity of filtration has been an evolution from the simple valves operated by hand, brought about by the increasing knowledge of the action of filters and of the conditions for most perfect operation. They may be divided into two classes—those operated entirely by hand, relying on the judgment and watchfulness of the attendant, and those which are automatic in their action.

Fig. 143 shows a section of an automatic filter regulator as in use in England and Japan.

The principle on which it acts is that by so regulating the height

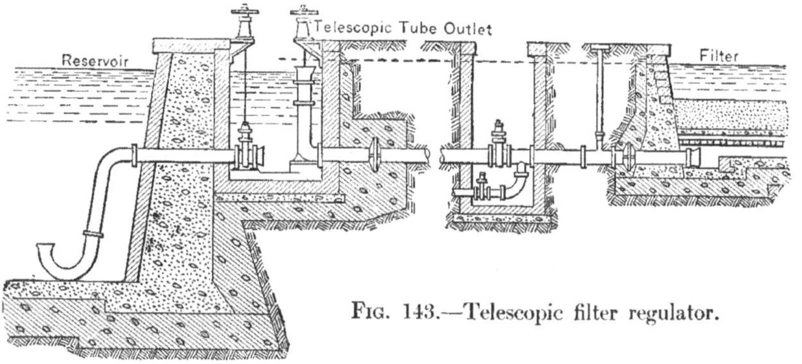

FIG. 143.—Telescopic filter regulator.

of the top of the telescopic pipe that a constant depth of water flows over its edge, a constant discharge is insured, and therefore a constant velocity of filtration. This can be secured by screwing down the top of the pipe as the resistance to filtration becomes greater with the increasing thickness of the sediment layer.

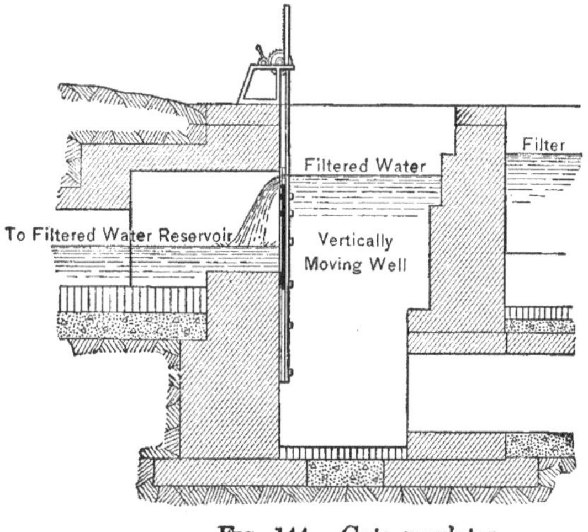

FIG. 144.—Gate regulator.

Another regulating device was long since proposed by Kirkwood, consisting of a weir which could be raised or lowered until the proper quantity would flow over it. This is shown in Fig. 144 and needs no comment further than to commend the keen insight of the distinguished engineer into the true principles of the efficient operation of filters long before the general engineering public was able to appreciate his work.

The floating-gate filter regulators are much in use in European water-works, of which a type is shown in section in Fig. 145. Of these the filter head is regulated by the height of the filtered-water reservoir within fixed limits over the assigned head fixed by screw or weights.

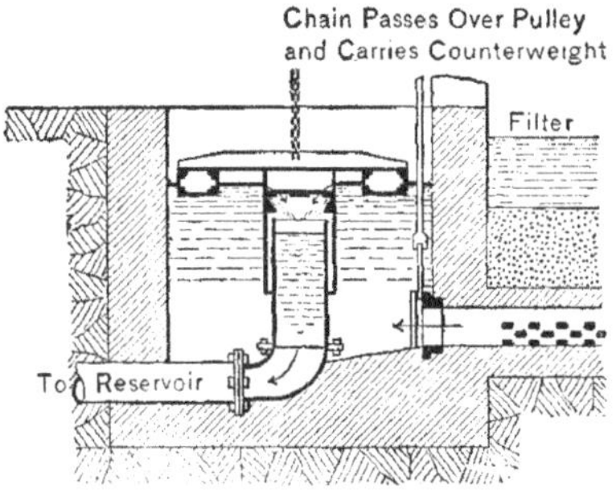

FIG. 145.—Floating gate.

In the weighted type a telescopic joint of pipe with vertical slits around the periphery at the top is suspended from a float. These floats swim on the surface of the filtered water as it comes into a closed chamber from the underdrains. The only outlet from this chamber is through this telescopic pipe, which connects with a conduit leading to the filtered-water reservoir. If the top of this pipe is kept at a constant depth below the surface of the filtered water a constant flow will be established.

In the apparatus, Fig. 145, the relative position of float and pipe is fixed, but the depth of immersion of the float, and therefore also of the pipe, can be altered by loading or unloading a counterweight attached to the chain shown, which passes over a pulley overhead. In either of these arrangements the discharge from the beds is free, and the difference in level between the water in the chamber and in the filter adjusts itself to the increasing resistance to filtration automatically.

DOMESTIC FILTRATION

The filtration of turbid water from many town water-works and private supplies from rivers and ponds often becomes a necessity for domestic use.

We therefore illustrate some of the devices applicable to hotel and household use.

Fig. 146 illustrates a hotel filter taking its supply from a town or private water-works. It is reversible, having a double diaphragm filled with filtering material so arranged that the water enters at the top and is discharged at the bottom, passing both ways through the trunnions. By reversing on the trunnions, the sediment may be washed downward and out through the opposite trunnion to the water-supply.

Fig. 147 represents a most desirable adjunct of household comfort in a large part of the United States and in other countries where the situation requires the use of rain water for domestic use.

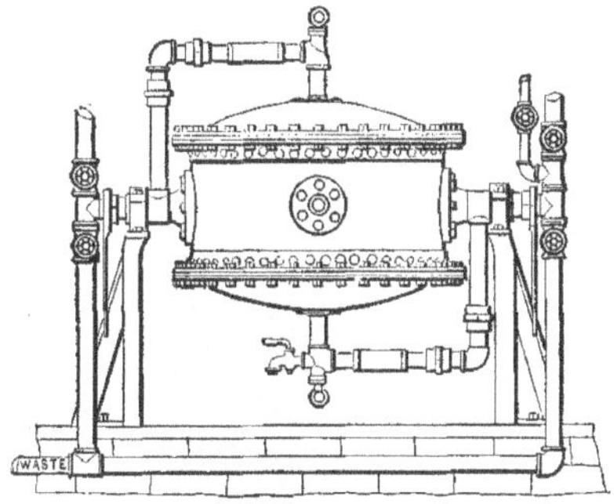

FIG. 146.—Reversible filter.

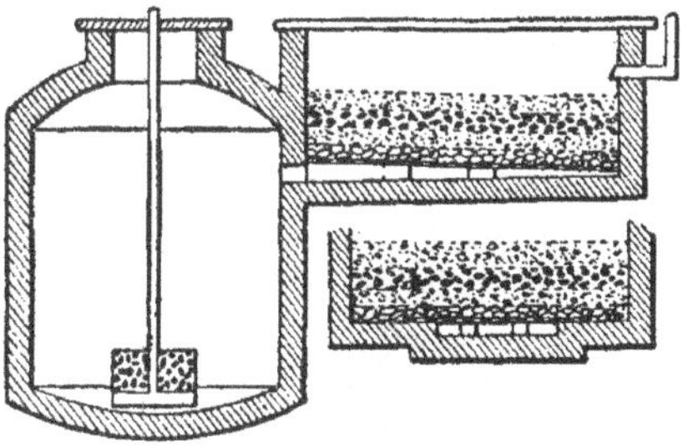

FIG. 147.—Filtering cistern.

The rain water is caught in a flat basin with gravel and sand spread on a perforated floor and drained into the cistern. The pump pipe is fixed to the perforated diaphragm of a two-chambered metal cylinder, the upper section of which may be filled with a bed of sand and charcoal in layers.

In Fig. 148 is illustrated the details of an upward-flow filter, which may be made large enough to supply the water for a steam plant of sufficient purity for boiler use. A perforated floor is made of any desired filtering capacity and charged with layers of gravel, coarse

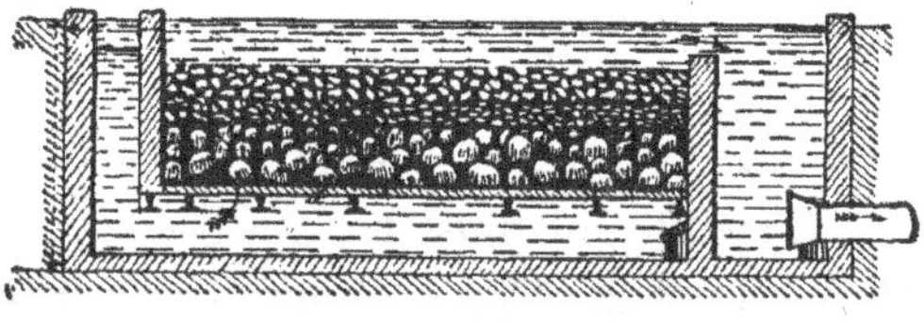

FIG. 148.—Upward flow filter.

and fine sand, with an inflow and overflow, as in the cut. A washout outlet should be made in the bottom of the lower compartment.

To make a filter with a wine barrel, procure a piece of fine brass wire cloth of a size sufficient to make a partition across the barrel. Support this wire cloth with a coarser wire cloth under it and also a light frame of oak, to keep the wire cloth from sagging. Fill in upon the wire cloth about three inches in depth of clear, sharp sand, then two inches of charcoal broken finely, but no dust. Then on the charcoal a layer of three inches of clear, sharp sand, rather finer than the

FIG. 149.

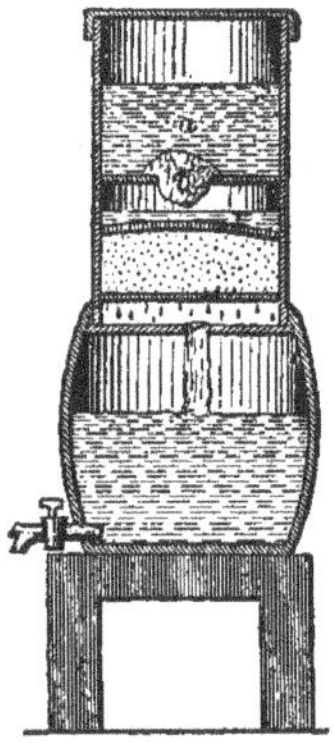

FIG. 150.

FIG. 151.

Domestic filters.

first layer. All the sand should be washed clean before charging the filter.

Fig. 150 shows a stoneware filter for household use. The lower jar is for storage of filtered water. The upper jar has a hole filled with sponge that filters the dirt out; beneath, a bed of charcoal on a porous stone or earthen plate.

Fig. 151 represents a home-made house filter composed of two stone pots or jars, the bottom one being a water-jar with side hole; if no faucet can be used, the top jar can be removed to enable the water to be dipped out. The top jar must have a hole drilled or broken in the bottom, and a small flower-pot saucer inverted over the hole. Then fill in a layer of sharp, clean sand, rather coarse. A layer of finer sand, a layer of pulverized charcoal with dust blown out, then a layer of sand, the whole occupying one-third of the jar.

CHAPTER IX

WELLS AND THEIR REËNFORCEMENT

SPRINGS AND UNDERGROUND WATER-WAYS

THE ordinary well, so much in use for domestic water-supply, has its failings in times of drought. Any means for restoring the water-supply, available by the village householder, the farmer, and the ranchman on the arid plains, are devices worthy of record.

No less is its value worthy of consideration for the relief of water-works supplied from large wells.

In Fig. 152 is illustrated a method of reënforcing a deficient well in a sandy or gravelly soil. Where a well bottom rests on rock, drilling to greater depth is probably the only recourse.

In ordinary situations a well may be deepened by driving a cylinder made of galvanized sheet iron with its sides punched by a thin chisel, as shown at the left in the cut. This can be pushed down in the centre of the well and the sand bored out with the sand augur shown at the second figure in the cut. The sand augur of 3 inches in diameter and the strainer of 6-inch diameter are sufficient for ordinary needs.

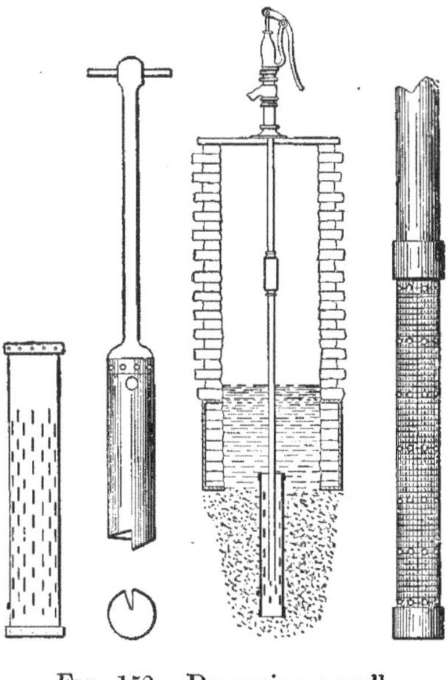

FIG. 152.—Deepening a well.

A drive strainer tube, as shown at the right, may be driven for greater depths and the sand drawn with an augur.

The great well of the Long Beach Improvement Co., at East Rockaway, L. I., 22 feet deep and 40 feet in diameter, on a bottom of quicksand, is a notable instance of the enlargement of flow of water into a well. The reënforcement was made without for a moment disturbing or interfering with the constant and necessary supply of water for the use of the great hotel at Long Beach, and at the height of the season, a time when a day's suspension of the water-supply would have been disastrous. In this well, as shown in Fig. 153, two pointed strainers, 6 inches in internal diameter, were sunk to a depth of 30 feet below the bottom of the well, over 50 feet from the surface of the ground, passing through a stratum of clay at a depth of 8 feet below the bottom of the well, and entering a substratum of water-bearing sand, which is supposed to be fed by the rainfall on the central part of Long Island, judging from the fact that the clay stratum crops out at Pearsall's and along the line of the Brooklyn water-works conduit.

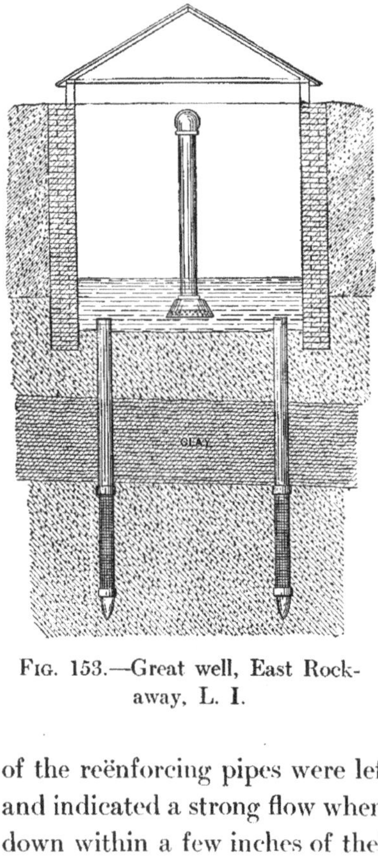

Fig. 153.—Great well, East Rockaway, L. I.

Upon trial, the pressure from the new source of supply raised the water in the pipes 4 feet above the water surface of the well. The tops of the reënforcing pipes were left 2 feet above the bottom of the well, and indicated a strong flow when the surface of the water was pumped down within a few inches of their open ends. The output of the well was doubled, the supply capacity rising to 130,000 gallons per day.

The large well of the Lawrence Cordage Co., in the Wallabout quicksand, Brooklyn, which was in progress of filling up by the movement of the quicksand from the excessive pumping of water for a condensing engine, was rescued by a single pipe strainer of 8-inch diameter driven to a depth of 18 feet below the bottom of the well, with its upper end about a foot above the quicksand bottom. The flow was largely increased, and became equal to the requirement for all purposes.

The great well in Prospect Park, Brooklyn, having a diameter of 50 feet and a depth of 54 feet to the water-curb, which is 10 feet deep by 35 feet diameter, as illustrated in Fig. 154, is probably the most unique well in existence in the peculiarity of its connection with the deep-water vein of Long Island. It is a model of what can be done in the reënforcement of a large well. It was at first reënforced with perforated tile pipe, driven horizontally around the water-curb, as shown in the cut. This was done to relieve the water pressure from beyond the walls, which had caused a flow over the floor of the curb bench, which is 7½ feet wide between the main curb and the water-curb. This gave a greater area over the water-way, but was not a sufficient reënforcement for the increasing requirement of the Park water-works, as the practical working of the well indicated; for as the surface of the water was pumped down to and below the open ends of the pipes, the flow from them gradually lessened, and finally ceased altogether, at a time when water was most needed. A further reënforcement was made for the purpose of obtaining water from a deeper source by driving four strainer pipes, 4½ inches internal diameter, to a depth of 20 feet below the bottom of the well, with open tops projecting 1 foot above the bottom. In addition to this, four strainer pipes, 6 inches internal diameter, were sunk through the bench floor between the curbs to a depth of 30 feet. These pipes were connected directly to the pumps, with valves so arranged that

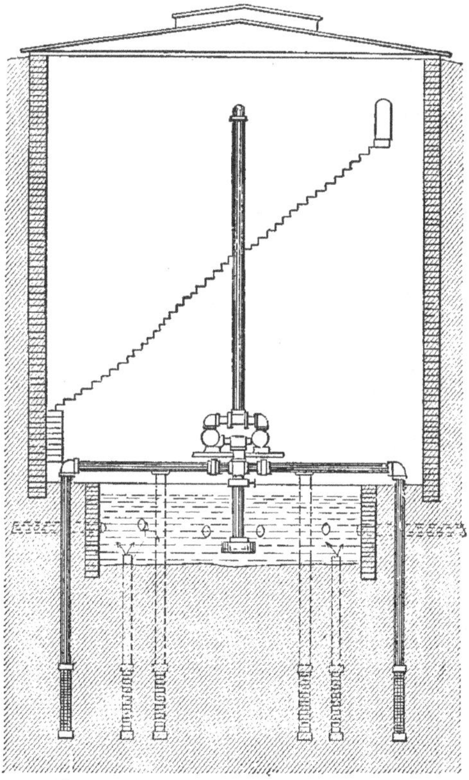

FIG. 154.—Great well of the Prospect Park water-works, Brooklyn, N. Y.

the suction could be taken directly from the well, or from the four large driven pipes. Thus, in a dry season, or when more water is needed than the well pump can supply, the direct pump connection with the deep-driven pipes will allow of the well being pumped entirely dry, and the water-line can be carried many feet below its bottom.

DRIVEN WELLS

The system of obtaining water by one or many driven wells is much in vogue, and undoubtedly furnishes a purer water than can be obtained from open wells in the same locality. The possibilities for a restricted depth for an increased volume are in favor of the multi-driven system or, for greater depth, by the wash system by which wells have been sunk to a depth of 700 feet.

The driving of a well pipe as ordinarily done is by a large mallet or ram striking an iron cap screwed to the top of the pipe, but an improved method is shown in Fig. 155, in which a clamp is strongly bolted to the well pipe on which the weight strikes to drive the tube. A clamp and two sheaves are bolted at the top of the tube with ropes rove through the sheave blocks and made fast to the weight for raising it. The weight is hollow, and rides loosely over the tube. The clamps are raised as additional pipes are screwed to the well pipe.

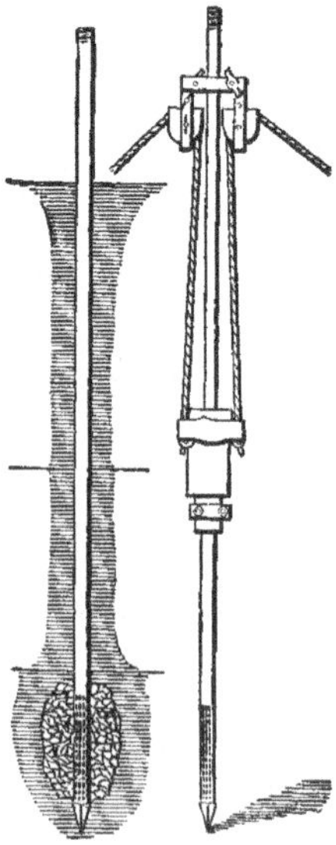

FIG. 155.—Driving device.

The bored pipe system for driven wells consists of an open bottom to the pipe below the strainer and the forcing of a strong stream of water down the pipe, which floods the outside of the pipe with water and the discharging sand. Small obstructions like bowlders will be washed aside, and by rotating the pipe slightly with a steady pressure, a well may be sunk at the rate of from 12 to 15 feet per hour. The bottom section should extend below the strainer to allow of plugging by dropping a heavy cylinder of iron or stone that easily closes the inside of the pipe.

The reënforcement of water-works by driving tube wells to a

greater depth and through an underlying clay strata has made a sensation in water-getting by examples at Jameco, L. I., and many other places in the United States. In such cases the air lift is resorted to for obtaining a large flow from great depths.

Wherever the water-table is within 20 feet below the ground surface, the direct draught from a pump with small clearance space in

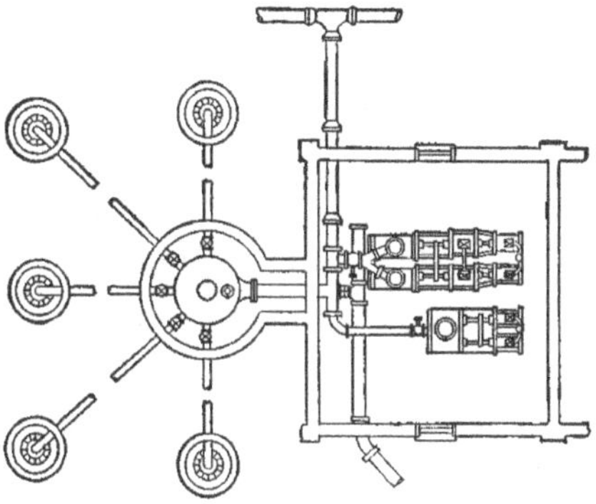

FIG. 156.—Plan of vacuum system.

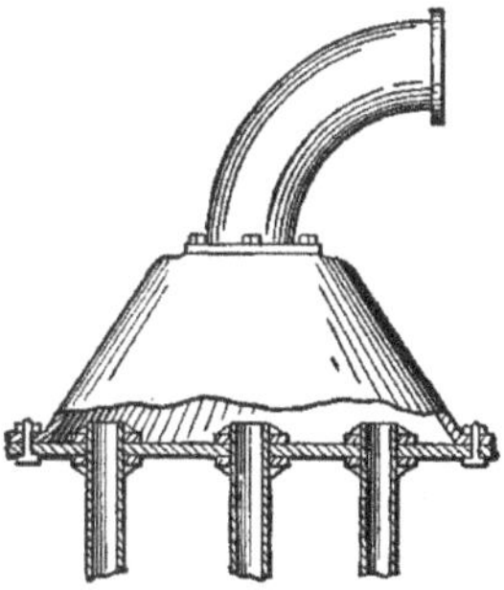

FIG. 157.—Gang chamber.

the cylinder is used for raising the water and forcing it to the stand-pipe or reservoir.

The independent vacuum system is also in use by which an air-pump exhausts the air from a receiver which is connected with the gang wells, thereby drawing the water into the receiver, from which it is forced to the stand-pipe, reservoir, or into the mains by a separate pump.

In Fig. 156 is shown a plan of this system, and in Fig. 157 the

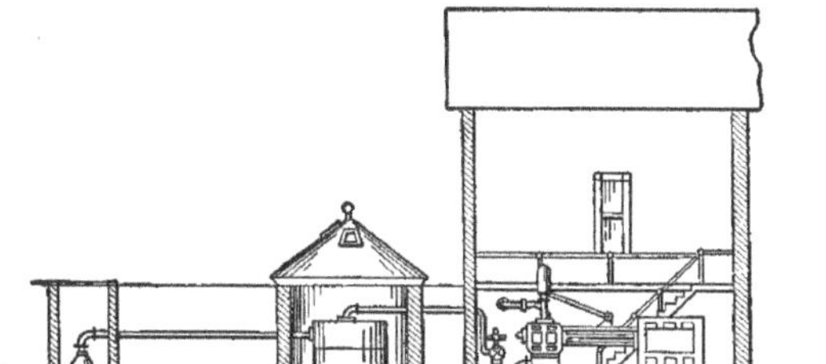

FIG. 158.—Elevation, pumping plant.

method of connecting the driven tubes with a common chamber, and in Fig. 158 an elevation of the tank and connections.

BORED WELLS

The boring of wells for water-supply by the drop-drill method has been in use from time immemorial in China and at depths more than 1,000 feet. This system appears to have been unknown in Egypt, Palestine, and in Europe in the early historic ages. In the province of Artois in France bored wells have been noted since the twelfth century and are probably the oldest in Europe; hence the name artosian or artesian was probably derived.

In choosing the location for a deep-bored well, the first consideration is that of the geology of the district and the nature of the underlying rocks as to their water-bearing quality, and for this purpose there is no better reference than the geological reports of the State or a communication with the United States Geological Survey at Washington. If the underlying rocks are observable it may be briefly stated that the light-colored sand rocks are generally free-flowing and the new red sandstone at great depths is charged with unpalatable minerals and iron.

The magnesian and other limestones yield scant supplies even at great depths with occasional exceptions. The gneiss and its overlying contorted dolomite are unsatisfactory and have given dry wells at 1,500 feet in depth. The Wheeling, W. Va., well is dry at 4,500 feet.

Wells located near the seashore have their water planes affected in sympathy with tidal action to a considerable percentage of the tidal range, according to their distance from the shore.

SPRINGS

Springs generally originate at the depressed edge of a water stratum lying upon an impervious bed of rock, clay, or hard pan, and are most

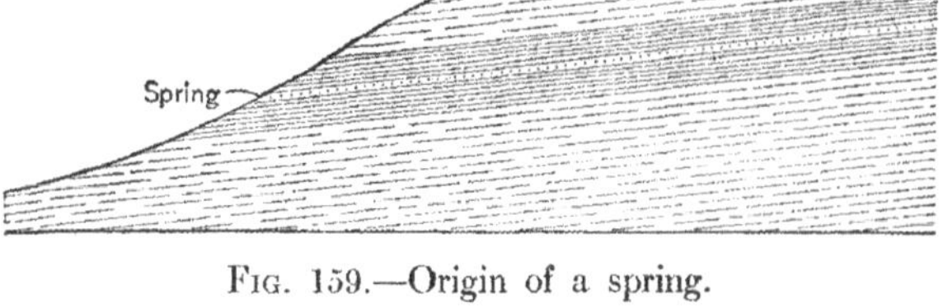

FIG. 159.—Origin of a spring.

in evidence along the banks of streams, to which they are the constant tributaries. Those that are located at an elevation to give a

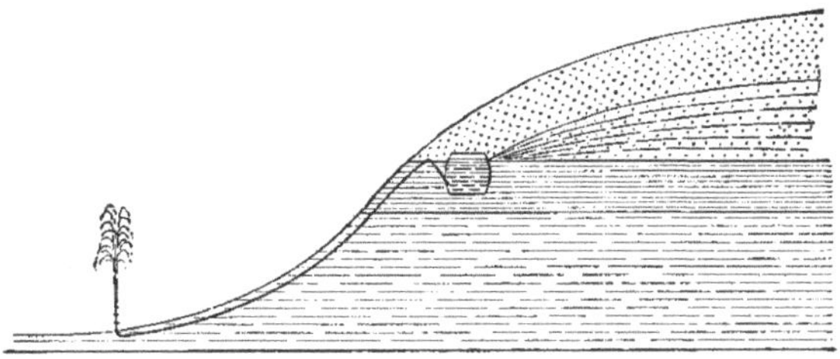

FIG. 160.—Fountain supply, geyser spring.

pressure head for household service should be appreciated and cared for in order to preserve their purity of flow, by walling and covering to prevent the possibility of contamination.

A typical example of a chambered hillside spring is shown in Fig. 161, which should be large enough for a person to enter through the trapdoor for inspection and cleansing.

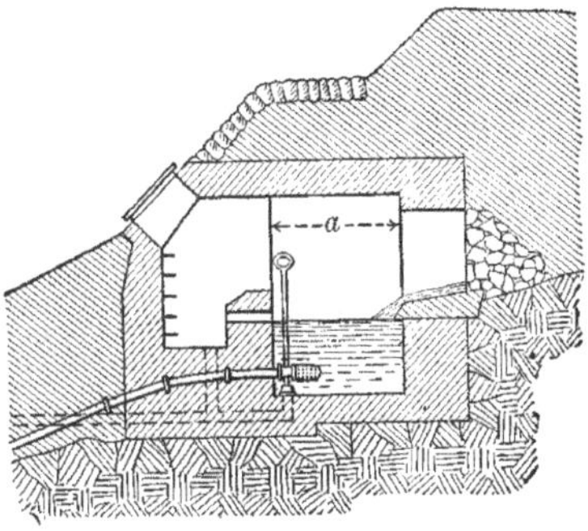

FIG. 161.—Chambered spring.

It is composed of a well of dressed stone or glazed brick with arched roof; an overflow leading to a drain-pipe; a plug connection with the drain-pipe for washing out the well, and the outlet pipe capped with a cylindrical screen.

THE DEPRESSION OF GROUND-WATER LEVEL

It has been determined experimentally that the depression curve of the ground-water surface made by pumping from open wells or driven wells of any depth may begin at from 200 to 400 feet from the well. Hence the natural fall of the ground water of a large area may be so changed by the excessive volume of water drawn that neighboring wells may be affected to even dryness, and that the water of a near-by pond may be drawn upon by filtration.

In this way small streams and rivers may be made the source of supply for water-works.

In Fig. 162 is shown a graphic ideal section of these conditions, in which an ordinary well may become dry and be reënforced by a deeper driven well when the depression has been caused by a deep-well pumping plant.

This effect is greatly increased in water strata of fine sand in which the margin of depression may be small, but in coarse, free-flowing gravelly water-ways it may be extended many hundred feet.

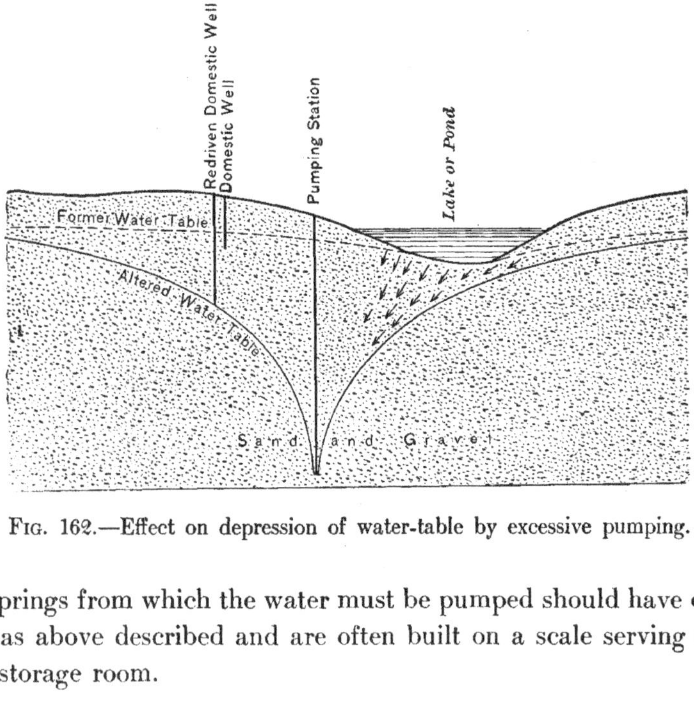

Fig. 162.—Effect on depression of water-table by excessive pumping.

Springs from which the water must be pumped should have equal care as above described and are often built on a scale serving for a cold-storage room.

SANITARY CONDITIONS

Too much care cannot be exercised in the location of wells. Not infrequently are wells located on the borders of sloughs, the waters of which are stagnant in the summer season and are heavily charged with organic matter. Fig. 163 shows a well so situated. While these sloughs are always lined with very fine sediment that is largely impervious, more or less water is able to pass through it into the adjacent sand. The drawing of water from a well situated as shown in Fig. 163, lowering the surface of the water in the well below that of the slough, invites the drainage from the slough toward the well.

In cases where, as a matter of convenience, it becomes desirable to locate a well near a slough, it should under no circumstances be an

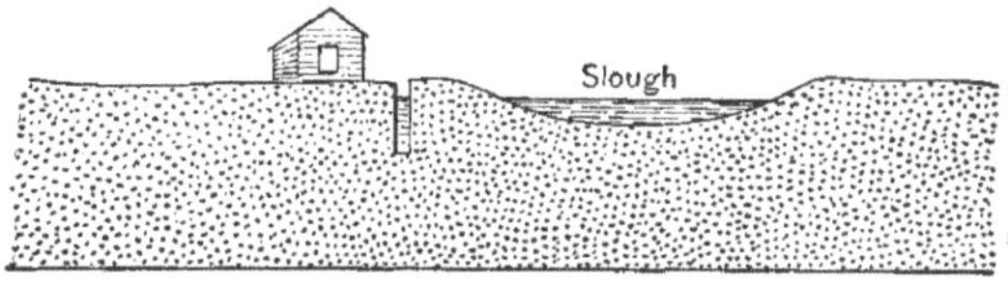

FIG. 163.—Well near a slough.

open one, and whether drilled or bored should go to such a depth below the bottom of the slough as to be out of reach of the downward-percolating, organic-charged waters.

When driven wells furnish water heavily charged with iron and organic matter the pipe should be pulled and driven elsewhere.

In no region should wells be located in barnyards or near other sources of surface pollution, except with intelligent advice. In a hilly region the surface drainage and stratigraphic conditions may be such as to permit a properly constructed well to be so situated; but in the lowlands, where there is but little, if any, surface drainage and the soil is porous, permitting the water freely to percolate downward, that practice cannot be too strongly condemned. The watering-places for barnyard fowls and the wallows for hogs are often immediately about wells from which water is used for drinking. Such disregard for sanitary conditions is excusable only among the most ignorant. The location of wells near cesspools and privies is one of the most dangerous conditions for generating typhoid fever and malaria.

The underground flow of water toward some lower level, and along the sea-coast toward the sea, has been made a subject of investigation by the United States Geological Survey with the result that the flow is constant with a velocity of from 1 to 100 feet per day, according to the slope and open condition of the sand or gravel in the water stratum.

BLOWING OR BREATHING WELLS

The phenomena known as blowing or breathing wells have been noticed in many parts of the United States, chiefly in the Central and Western States. The movement of air into and out of these wells is due to variations in the pressure of the atmosphere, as shown by the barometer, and its intensity is due to the freedom of air passage in the coarser sands above the water-table. The dry soil of the earth is a huge lung with a possible capacity of one-third of its volume in which the air moves into and out with every change in the atmospheric pressure. The coarser the subsoil, the stronger the breath. Caverns exhibit the same phenomenon to a marked degree.

ARTESIAN WELLS

Artesian wells generally include all deep-bored wells, either spouting, flowing, lapsed from overdraught, or those in which the water has never reached the surface. Others are dry wells such as the one at Wheeling, W. Va., nearly 5,000 feet deep, and one in Germany claimed to be 8,000 feet deep and terminating in a bed of salt 3,000 feet thick.

The artesian well districts most noted in the United States are as follows:

1. The wells of the Red River Valley in northeastern North Dakota.
2. The wells of the James River Valley in the two Dakotas (North and South).
3. The wells of the Yellowstone Valley at Miles City, Mont.
4. The shallow wells in the drift formation on the eastern side of the two Dakotas.
5. The wells of northern Nebraska.
6. Four groups of wells in southwestern Kansas.
7. The wells of the La Poudre, Denver, and Pueblo basins in Colorado.
8. The Fort Worth and Waco groups in Texas.
9. The wells of New Mexico.
10. The wells of Wyoming.

As might be expected, the wells in the various groups derive their flow from formations of very different geologic ages. From the latest to the oldest these are: the glacial drift, tertiary formations, Laramie, Dakota sandstone, and Triassic "red beds." The shallow wells in the eastern part of the two Dakotas are supplied from the glacial drift. In the Denver basin, and in and for some distance east of the city, the wells seem, generally, to draw water from the Arapahoe, a tertiary formation. There is one well in Denver, that at the court-house, which extends to the Laramie, a cretaceous formation. The whole of the James River group draws from the Dakota sandstone. It is thought that this formation extends south into northern Nebraska and gives rise to artesian conditions there. It is stated that on the western sides of the Black Hills, in Wyoming, the Dakota sandstones yield flowing wells of salt water, with oil and natural gas present. In southwestern Kansas and just over the line in Colorado a group of wells less than 300 feet deep are also supplied from the Dakota sandstones. This stone also crops out in middle Kansas and eastern Nebraska, giving evidence of artesian conditions, but it is thought their supply comes from local breaks rather than from the outcrops in the Rocky Mountains, as is the case in the James River group. In another part of southwestern Kansas, northeast from Meade Center, in the upper valley of Crooked Creek, a group of wells derives its supply from tertiary grit. At Larned, Great Bend, and in Morton County, near the Colorado line, the wells tap water in the Triassic "red beds."

The United States Geological Survey has made an attempt to enumerate the artesian wells of the Western States, but we fear that it falls far short of the present number. The reports cover the following enumeration:

One hundred and fifty high-pressure wells in the Dakota basin, including, with those in the Dakotas, the few in the Yellowstone Valley of Montana; several hundred flowing wells in South Dakota, evidently not connected with the great Dakota basin; over a thousand small flowing wells in the Red River group; in northern Wyoming 60 flowing wells; in the central section of the great plains, from the northern boundary of Nebraska to the southern boundary of Indian Territory, about 200 flowing wells, with several hundred more in which the water rises but does not reach the surface; in Colorado, at least 250 flowing wells in the Denver group and 100 non-flowing, together with 12 wells at Greeley and 8 at Pueblo; at Florence and Cañon City, Col., several heavy flows have been struck in boring for oil; in Texas nearly 700 flowing wells are referred to by Field Agent Roesler. We have here, enumerated by Mr. Hinton, in his review of the reports of the various agents, a total of 2,360 flowing wells, with "several hundred" additional flowing wells in South Dakota, 100 non-flowing wells in the Denver basin, and other non-flowing wells not enumerated. These figures would indicate that there are in the area investigated at least 2,500 flowing wells. The number of non-flowing wells cannot well be estimated.

Outside of the area under consideration, in notes and incidental references, wells are mentioned as follows:

Over 2,000 flowing wells in the San Luis Valley, west of central South Colorado, all sunk within the past few years, and mostly in the year 1889–90; about 2,000 flowing wells, of small bore, in Utah, in the region of Salt and Utah lakes, with water rising from 2 to 5 inches above the top of the casings; 67 flowing wells in Nevada; 2,000 flowing wells in the San Bernardino basin, California; 200 flowing wells at a high altitude, in Sierra County, Cal.; 100 flowing wells near Tulare Lake, Cal.; 40 large flowing wells in Kern County, Cal., just south of Tulare Lake, within a radius of 10 miles, and with an average daily flow of about 600,000 gallons. The number of wells just named is about 6,400, which, with those in the area to which the investigations were confined, makes a total of nearly 9,000 flowing artesian wells in the Western United States. Some of the figures given are probably estimates, but, on the other hand, it is likely that some groups are omitted. Were figures available for non-flowing artesian wells, we would have a much more surprising total, but one which it seems useless to estimate.

CHAPTER X

THE AIR-LIFT METHOD OF RAISING WATER

THE air-lift pump is said to have been invented in the eighteenth century and in use at Freiberg, Saxony. Siemens in England experimented with the air-lift in the middle of the nineteenth century, and it was patented as an air ejector by McKnight in 1864. The principle of its action became a theme with Dr. J. G. Pohlé, and to whom two patents were issued, Nos. 338,295 and 347,196, covering the system of elevating water by admixture of air under compression suitable for the height that the water was to be raised. This system, however, required a depth of water in the well more than equal to a height to which the water was to be lifted.

The original Pohlé system has been modified and improved with a number of patents on special points in the system with small gains in efficiency. Dr. Pohlé also introduced compounding or stage-lifting, which has been made available to such an extent that it is now possible to lift water to great heights from an ordinary sump in a mine or from ordinary wells.

We illustrate in Fig. 164 the receiver, air and lift pipe as usually operated in deep wells, in which the pressure in the air-pipe must be greater than the hydrostatic pressure of the water at the bottom of the pipe, and in quantities sufficient to make the ascending column of air and water in the flow-pipe lighter in its total height than the weight of an equal column of solid water of the depth of the well from the surface of the water to the bottom of the pipe, thus making this principle in pumping water essentially a differential gravity system.

The air-lift pump proper consists of only two plain open-ended pipes, the larger one with an enlarged end-piece constituting the discharge pipe, and the smaller one let into the enlarged end-piece of the discharge pipe constitutes the air inlet pipe, through which the compressed air is conveyed to the enlarged end-piece to the under side of the water to be raised. No valves, buckets, plungers, rods, or other moving parts are used within the pipes or well.

In pumping, compressed air is forced through the air-pipe into the enlarged end at the bottom of the water-pipe; thence by the inherent expansive force of the compressed air, layers or bubbles of air are formed in the water-pipe, which lift and discharge the water layers through the upper end of the water discharge pipe. At the beginning of the operation the water surface outside of the pipe and the water surface inside of the pipe are at the same level; hence the vertical pressures per square inch are equal at the submerged end of the pipe, outside and inside. As air is forced into the lower end of the water-pipe, it forms alternate layers with the water, so that the pressure per square inch of the column thus made up of air and water, as it rises inside of the water-pipe, is less than the pressure of water per square inch outside of the pipe.

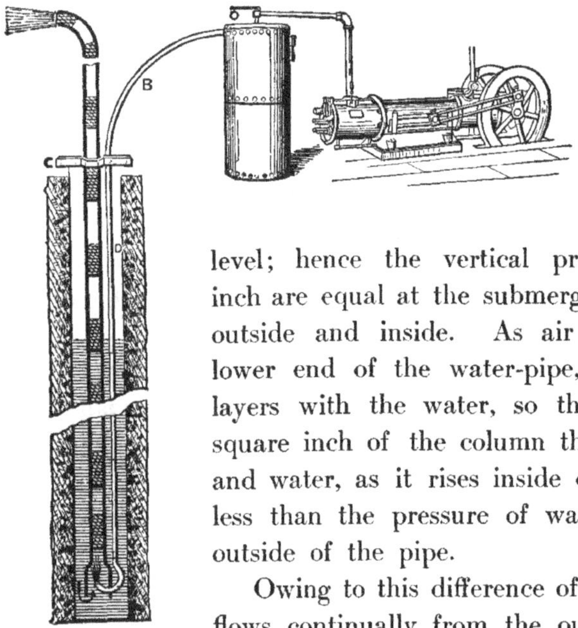

FIG. 164.—Air-lift pump.

Owing to this difference of pressure, the water flows continually from the outside to within the water-pipe by gravity force, and its ascent through the pipe is free from shock, jar, or noise of any kind.

These air sections or strata of compressed air form closed bodies, which, in their ascent in the act of pumping, permit no slipping or back flow of water. As each air stratum progresses upward to the spout, it expands on its way in proportion as the overlying weight of water is diminished by its discharge, so that the air section, which may have been, say, 50 pounds per square inch at first, will be only 1.74 pounds when it underlies a water layer of four feet in length at the spout; until finally this air section, when it lifts up and throws out this four feet of water, is of the same tension as the normal atmosphere; thus proving that the whole of its energy was used in work, and that this pump is a perfect expansion engine.

As the weight of the water outside of the discharge pipe (the head) is greater per square inch than the aggregate water sections within the pipe when in operation, it follows that the energy due to this greater weight is utilized in overcoming the resistance of entry into the pipe, and all the friction within it.

The Pohlé "air-lift" pump has been found to give above 80 per cent. of efficiency from the air receiver in water-pipes of large diameter, and, as a rule, above 70 per cent. in small-sized pipes. The efficiency from a steam-driven compressor is much lower, say from 50 to 25 per cent. of the horse-power expended for compressing the air. It retains this efficiency without repairs, or until the pipes rust through, whereas ordinary bucket-and-plunger pumps gradually lose efficiency from the first stroke they make, and lose it rapidly if the water contains sand or is acid in character.

The secret of the air-lift pump action is in the high velocity with which the air and water are discharged through the eduction pipe. Without this high velocity there would be no piston-like sections ex-

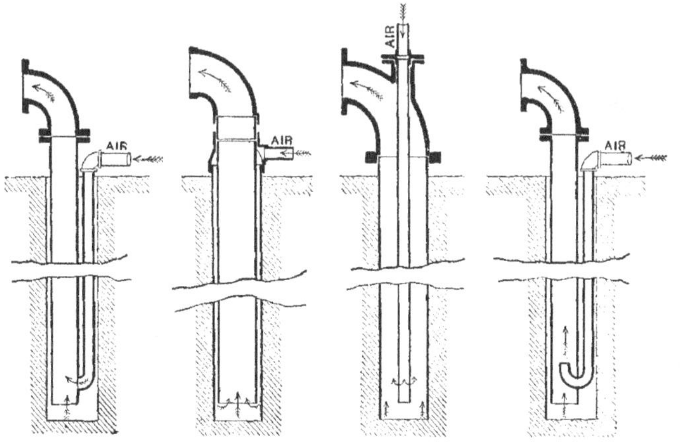

FIG. 165.—Different arrangements of air- and water-pipes.

cept perhaps in a small glass tube model where capillary attraction takes the place of velocity.

As the pump has no valves, no standing water remains in the pump column after the operation of pumping; it recedes into the well, and there is none left to freeze in cold weather. The capacity of the pump is unlimited, and, with the proper proportions of air to the water, will work efficiently in pipes several feet in diameter. Estimates have been made which indicate that a 30-inch pipe will deliver 16,660 gallons per minute, equal to 1,000,000 gallons per hour.

As sand, silt, gravel, and bowlders in water form no obstacles to interfere with the action of the pump, its adaptability for dredging is suggested as well as its utility for pumping sewage. Experience has proved that, by the use of this constant upward flow of water, artesian wells have been freed from their accumulated sedimentary deposits, as well as that lodged in the fissures and crevices of their wall rock, and have been thus made to yield greater quantities of water than they ever did before. For chemical uses, and for the liquids of the arts, there is no superior method than the "air-lift." It is used successfully for raising sulphuric acid of high specific gravities, and is well adapted for ore-leaching works, vinegar works, sugar refineries, dye works, paper-pulp works, etc.

As an irrigating pump for raising subterranean water in the arid regions of the West, its field of usefulness is very promising, for with one air-compressing plant at a central station, a number of wells, widely separated from one another, may be simultaneously pumped by branches of air-conveying pipes, taken from a main air-pipe from the air-compressor; for compressed air may be conveyed for miles without material loss of power.

It often happens that a single well does not yield the quantity of water desired, but that a number of wells would give the satisfactory result. By the old-fashioned deep-well pump, each well would require a separate "steam head," separate sets of rods, and the other paraphernalia, which, with the condensation of the steam, when conveyed to the several steam heads, would be very costly in the first outlay, and very wasteful of power in its maintenance, to say nothing of loss of time in repairs. By the Pohlé process, but one air-compressing plant is required, and this may be placed in the engine room or the boiler house, directly under the eyes of the engineer, from whence the air may be conveyed to the several wells, all of which may be pumped simultaneously and economically.

In the early trials for efficiency of the air-lift some curious comparisons were brought out relative to the ratio of the lift to the depth of submersion and the relative air pressure due to submersion.

Thus with 16 pounds air pressure with 41 feet water-lift and 10 feet submergence, 68 cubic feet of free air per minute lifted $\frac{7}{8}$ cubic foot of water 41 feet high, giving a computed efficiency of $3\frac{1}{2}$ per cent. of the steam-power. The efficiency was found to decrease with the increase of air pressure above what was necessary to do the work; for instance, with an equal submergence and lift of 26 feet and an air pressure of 20 pounds, 64 cubic feet of free air pumped 14 cubic feet of water 26 feet high per minute, showing an efficiency of 19 per cent. of the steam-power in the compressor. When the air pressure was reduced to $12\frac{1}{2}$ pounds, using 26 cubic feet of free air per minute and pumping $8\frac{1}{2}$ cubic feet of water 26 feet high per minute, the efficiency was raised to 42 per cent. It was found on trials that on a deeper submergence of 1 to 1.6 the efficiency rose to 53 per cent., and in all trials was greatest at the lowest pressure that the lift could be operated. It was found on a general average that the efficiencies that may be expected from the best conditions for air compression may be stated as follows:

$\frac{\text{Height}}{\text{Submergence}} =$	.5	efficiency	50	per cent.
	1.0	"	40	"
	1.5	"	30	"
	2.0	"	25	"

Mathematicians have formulated some complicated equations in relation to the action of the air in the ascending column of water; but as the air bubbles vary in size according to the form of the injecting nozle, and as their coalescence and expansion produce so many variable factors, reliable results can be obtained only from actual tests, and even these are merely approximate.

In a test of the Pohlé air-lift made at De Kalb, Ill., the air-pipe was placed inside of the well pipe with a water-lift of 133 feet, and the submerged nozle 123 feet below the surface, a nearly equal ratio. The well pipe was 6 inches diameter, air-pipe $2\frac{1}{2}$ inches, thus adding about 50 per cent. to the friction of the ascending water and giving to the whole length of 256 feet an irregular annular space for the passage of the water and air. With the expenditure of 42.7 horse-power indicated, there was raised 207 gallons of water 133 feet, with a volume of 310 cubic feet of free air per minute. The efficiency was found to be 17½ per cent. This shows very plainly that the friction of an internal air-pipe causes a loss of efficiency.

A series of trials with a gang-well system on the Pohlé plan was made at Rockville, Ill. In casings of 6¼ inches diameter inserted in four wells, 260 feet below the overflow, and air-pipes 1½ inches diameter, let down 250 feet, all in 8-inch drilled wells. After several trials with return bends and small nozles at the bottom of the air-pipes with unsatisfactory results as to water flow, the bottom of the air-pipe was closed and the sides slotted for 20 inches up from the bottom, giving a full and free opening for the air without any obstruction to the upflow of the water. In this manner the service was raised from 1,000 gallons to 1,400 gallons per minute, but still showing an efficiency of only 24 per cent.

Much doubt has existed from the early years of the air-lift system as to the possibilities in regard to conveying the water to a distance or direct to an elevation at a distance from the well. Lately there has been constructed at Point Pleasant, W. Va., on the bank of the Ohio River, a water-works employing the air-lift system to obtain water filtered into the gravelly soil beneath the river. The compressor was located in a power-house 500 feet distant from the location of the wells on the river bank. The receiving basin is situated at the top of the river bank, 67 feet above the top of the well pipes and 400 feet from the low-water bank of the river. In Fig. 166 is shown a profile of the situation. Well casings 10 inches in diameter were driven to the rock about 40 feet in depth.

After the 10-inch casings were in place 10-inch holes were drilled in the underlying rock 116 feet deep, and cased 8 inches inside diameter from bottom to top. This casing was also perforated similarly to the outer one, only the holes were larger—$\frac{5}{16}$ inch. The space between the two casings was tightly calked at the top to prevent water entering the wells at this point. Four-inch discharge pipes and 1¼-inch air-pipes were properly fitted and suspended in each of the wells, with their extremities 110 feet below the top of the 8-inch casing.

Both pipes were suspended from a water-tight cap, resting on the top of the 8-inch casing. It will be observed that no water can enter these wells except through the perforations in the casings, which are 10 feet to 20 feet below the flowing water in the river. None can enter at the bottom. It was the desire to allow the river water to enter the wells only through the perforations after having passed through the sand strata mentioned, which would serve as a filter; which has proved that, however muddy the river may be, the water taken from the wells is bright and sparkling at all times.

Just when the wells were completed and the pipes in place and extending up the sloping river bank a short distance, the river rose over the wells. For two months the wells stood unused. In the mean time the reservoir, receiving basin, and power-house were

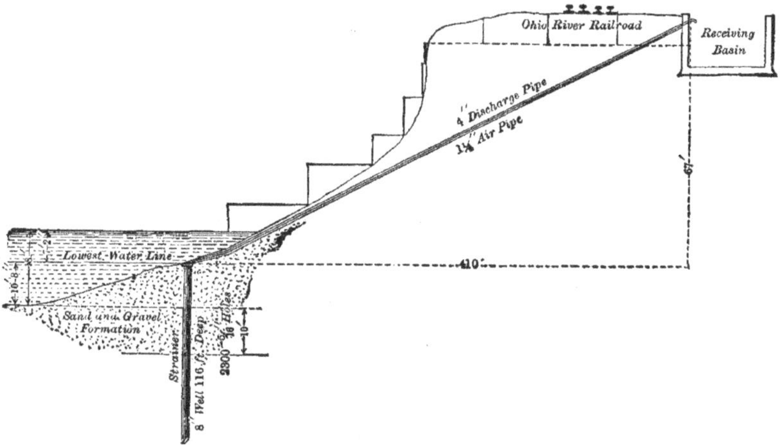

FIG. 166.—Profile of the Point Pleasant water-works.

completed, and the work advanced as fast as possible. Just as soon as the air-compressor was in place the air-pipes were connected up and the wells tested before the discharges were extended to the receiving basin. One well was found with a deposit of sand in the bottom reaching 5 feet above the foot of the discharge pipe. Several unsuccessful efforts were made to force air into this well. The river having receded, the air-pipe was disconnected at the top of the well and a $\frac{3}{8}$-inch gas-pipe coupled and lowered. It stopped 5 feet from the bottom. If was churned a few minutes and soon went down the remaining 5 feet. Again the air-pipe was coupled and the air pressure increased to 90 pounds per square inch. The effect was almost startling, but gratifying. The obstruction was cleared out very quickly. No other system of pumping could possibly have accomplished the clearing out of this well of the sand deposit.

The discharge and air-pipes to each well are independent. That is, each well has a separate discharge to the receiving basin and a separate air-pipe from the receiver. These are carefully graded and are not exposed at any point except where the discharges pass through the top of the walls of the receiving basin, and have open discharge.

The working pressure is from 45 to 50 pounds, varying with different river levels.

The discharge of water is not constant, however, but irregular or intermittent, as though the air and water formed alternate strata or volumes within the discharge pipes. It varies with the depth of water in the river, ranging from 1 volume of water to 8 volumes of free air, to 1 to 6. As the river is constantly rising and falling and is frequently 25 to 40 feet deep over the wells, the pressure on the sand surrounding the wells is constantly changing and affects the capacity of them as well as the necessary air pressure to pump them.

The reservoir is situated about 1½ miles distant and at 225 feet elevation. Water is taken from the receiving basin by belt-driven triplex outside-packed plunger pumps, 9 inches diameter by 12-inch stroke, operated at 37 revolutions per minute, delivering about 22,000 gallons per hour.

As there is no demand in the town for electric current during the day, the works are operated at night only. Usually the air-compressor is operated one night, and the following night the forcing pumps. The water received the previous night in the settling or receiving basin has about twelve hours to become cleared of any sand brought with it from the wells before going to the reservoir. This basin has a capacity of about 225,000 gallons; the reservoir about three times this quantity. The construction of the receiving basin is the same as the reservoir. The engine has ample power to operate all the machinery at the same time. Two men only are required to attend the combined plant. In addition to the public and private consumption of water, two busy railroads are consumers. All customers are served by meter, and therefore there is practically no waste.

There can be no doubt that water taken by air in this manner is purified to some extent, the admixture of air serving to oxidize and destroy organic matter. Samples of the water taken are bright and sparkling, have no odor, and remain apparently unchanged. There probably is not another town of 5,000 inhabitants in the country that has a better or more complete combined water and light works.

What has been accomplished at Point Pleasant can be done at hundreds of other small towns similarly situated where there is no water-works. Here it has been demonstrated that bright, sparkling water can be obtained from a muddy, filthy stream without the use of chemicals or mechanical filters.

Just use the filter nature has so abundantly supplied at the bottom of such streams, and by proper arrangement of the pumping system combined with an electric-lighting system, thus economizing the operating expenses to a minimum, establish first-class water and electric service on a paying basis when neither separately would pay operating expenses.

The air-lift system is undoubtedly the simplest as well as the best of all known methods of serving such towns with good water. Nor is the system less applicable to larger towns, as well as to factory and domestic supply.

Artesian wells, or wells supplied from land sources, generally yield hard water or water highly charged with mineral salts. The water at Point Pleasant is soft, pleasant, and wholesome. The railway companies using it speak very highly of it. It is simply Ohio River water freed of filth and all objectionable matter that render it so disgusting at many towns along the stream.

STARTING AN AIR-LIFT

The pressure of air necessary to start a pump is greater than that necessary to keep it in action when once started. That it must be so is clear when we reflect that the pressure in still water at a depth D is w D, but if the water be moving with a velocity V, the pressure is

$$w\left(D - \frac{V^2}{2g}\right)$$

Hence, the instant the discharge commences the pressure in the receiver will be reduced. This principle also accounts for the intermittent action which occurs under certain circumstances. Until a current is created, every bubble entering the pipe is under a pressure of w D; but the instant the discharge commences the pressure drops to

$$w\left(D-\frac{V^2}{2g}\right)$$

and all bubbles in the pipe expand suddenly, causing a mild degree of explosion which may nearly empty the discharge pipe, thereby bringing the pressure on the air-inlet orifice below

$$w\left(D-\frac{V^2}{2g}\right)$$

which is the pressure needed to maintain continued action, too much of the compressed air will escape, and, when the violent action is over, the store of compressed air will be practically exhausted and the water will have an opportunity to regain its full static head D against the escape of air. This completes one period of the action.

To prevent intermittent action the escape of air into the discharge pipe must be controlled. It should be throttled the instant the discharge of water commences. That intermittent action does not always occur is probably due to the effect of friction in the pipe conducting air to the point of admission. As this friction increases with the square of the velocity, it is evident that in long pipes of small cross-section it will serve to some extent as a governor, tending to control the discharge of air.

When the depth of submersion at starting is considerably greater than that previously referred to, viz., 85 per cent. of the total height of pump, the air-pipe should enter the water-pipe from 30 to 36 inches from the bottom of the water-pipe to facilitate starting.

The volume of air required to raise 1 cubic foot of water by means of the air-lift varies from 3.9 cubic feet as the minimum to 4.5 cubic feet as the maximum, giving a mean of 4.2 cubic feet of free air per cubic foot of water. This, however, is a general statement and only applies to air-lifts properly proportioned and working under favorable conditions. The volume of air required per minute to raise a given volume of water in the same length of time may be found by means of the formula:

$$\text{Cubic feet air} = \frac{L\ C}{16.824}$$

in which L = the lift in feet above surface of water, C, the number of cubic feet of water to be raised per minute.

Transposing the symbols in the foregoing formula we have for the capacity of the air-lift expressed in cubic feet per minute:

$$\text{Cubic feet water} = \frac{16.824\ A}{L}$$

and for the lift coresponding to a given discharge and the approximate volume of air we have:

$$\text{Lift in feet} = \frac{16.824\ A}{C}$$

A = the number of cubic feet of free air per minute.

The efficiency of the air-lift, as previously stated, varies with the ratio of depth of submersion to total lift, the efficiency generally increasing with increased submersion up to approximately 85 per cent. of the total lift, while the efficiency decreases slowly below 65 per cent. until the submersion reaches about 55 per cent. of the total lift. The capacity in cubic feet per minute for varying depths of submersion between these limits may be found by means of the formula:

$$\text{Cubic feet} = \frac{8.24\ A\ D}{L^2}$$

in which D = the depth of submersion in feet.

It is not practicable under ordinary conditions to attempt to raise water by means of the air-lift to heights exceeding 180 to 200 feet above the lowest water-level, nor to attempt to carry the discharge pipe horizontally to a greater distance than 700 or 800 feet. When greater horizontal distances must be covered by the discharge it is better to carry the pipe on an incline from the well or reservoir to the point of discharge.

AIR-LIFT WITH HORIZONTAL RUN

It is frequently necessary to deliver water to an elevated tank or reservoir at some considerable height above and distance from the well, in which event it has generally been considered inefficient to make the lift at the well force the water horizontally and end with a lift to a tank, but to just what extent the problem was affected by these conditions has always been a matter of guesswork. Confident, however, that serious disadvantage attached to forcing water from a well a long distance, the practice has been, wherever possible, to avoid doing so by substituting either:

1. Vertical lift at the well high enough to flow the water by gravity to the desired point, employing stand-pipes; or,

2. Raise the water to a surface reservoir or cistern near the well and deliver the water above ground by means of an ordinary pressure pump, or a pneumatic displacement pump, submerged directly in the receiving tank or reservoir, thus through a combination of the air-lift and pneumatic pump rounding out the whole operation as a complete compressed-air system of pumping.

The following tables have been prepared from actual tests made to show the decrease in efficiency under like conditions by extending the horizontal pipe from 200 to 600 feet:

Table XXIII.—Cubic Feet of Free Air and Gallons per Minute per Square Inch Area of Discharge Pipe for Lifts of 13 to 30 Feet at Well, Horizontal Run of 200 Feet, Subsequent Lift of 50 Feet.

Submergence—per cent.	Cubic feet free air per ft. gal.	Gallons per minute per sq. in.
40	.0231	3
45	.0176	5.2
50	.0138	6.4
55	.011	7.6
60	.0094	9

Table XXIV.—Cubic Feet of Free Air and Gallons per Minute per Square Inch Area of Discharge Pipe for Lifts of 13 to 30 Feet at Well, Horizontal Run of 600 Feet, Subsequent Lift of 50 Feet.

Submergence—per cent.	Cubic feet free air per ft. gal.	Gallons per minute per sq. in.
40	.0352	3.6
45	.0264	5
50	.021	6
55	.016	7
60	.0127	8

TWO-STAGE AND MULTISTAGE AIR-LIFT

The idea of compounding the air-lift was first proposed by Dr. Pohlé, and has since come into use for shallow sumps. Fig. 167 represents the conditions of a sump of about one-quarter of the total lift in depth, in which an auxiliary pipe is introduced to receive the water at about twice the depth of the sump to act as a pump-well for a higher lift. By this method the inconvenience and cost of a deep shaft or boring may be avoided and the compound system quickly applied in emergencies.

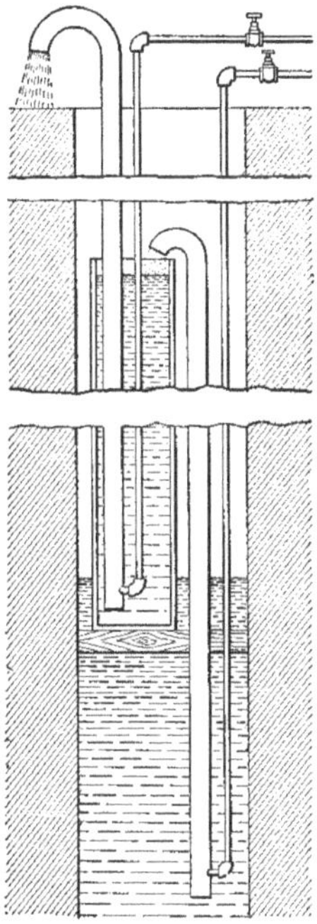

FIG. 167.—Two-stage air-lift.

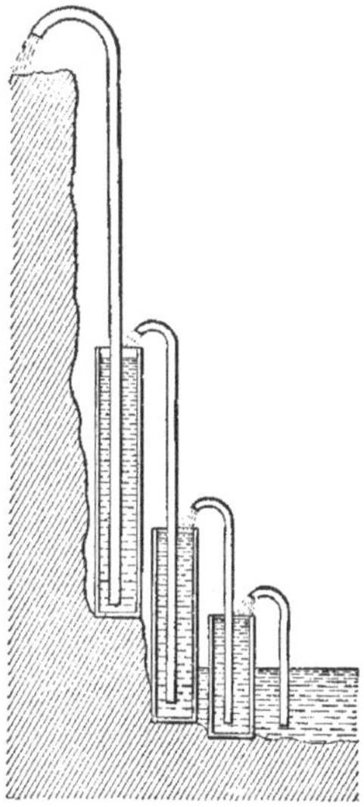

FIG. 168.—Multistage air-lift.

Its best efficiency is found in so equalizing the lifts that the air pressure may be taken from the same compressor for both stages.

MULTIPLE-STAGE AIR-LIFT PUMPING

In Fig. 168 we illustrate the possibilities in the work of compressed air in pumping water to great heights from shallow sumps by the Pohlé air-lift system. In order to show the detail of operations the illustration is spread out. In practice the several wells may be bunched together to occupy the smallest space in a mine shaft. It will be readily perceived that but one air pressure is needed, no more than sufficient to operate the highest lift in the multiple-stage system. The lesser lifts may be regulated by valves in the air branches to exactly meet the volume and pressure required for the lower lifts. Its air economy may balance the cost of a deep sump.

As yet we have no data as to its efficiency for permanent use, but there is no doubt that economy due to decreased air pressure will be found to warrant its adoption in mine and drainage work.

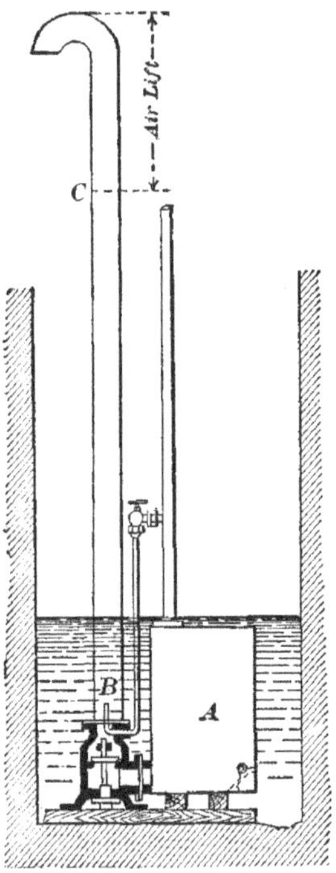

FIG. 169.—Combined air-lift pump.

A combination of the direct-acting tank system and the Pohlé expansion air-lift has been devised by Mr. Wheeler, by which the high-lift system may be utilized from a shallow sump by raising the water about one-half the height by direct pressure, then injecting air under the water column from the same air-pipe used for the direct lift, and thus doubling its elevation. In Fig. 169 is shown a sectional elevation of this system, in which A is the direct pressure or displacement chamber, from which the water is raised to a height at C; air is injected at B, and by its lifting and expanding action completes the lift; the pressure in the chamber A being equivalent to the deep immersion required in the Pohlé system. This system, as shown in the figure, is alternating, and evidently could not run constantly with one chamber; but by making a double-chambered direct lift, and connecting the air-pipe to the water column direct from the pressure side of the air-compressor, and using the air-switch only on the direct-lift pipes, a continuous flow would be obtained.

The efficiency of the Wheeler pneumatic pump compares very favorably with any of the other methods of pumping by air pressure. The computed efficiencies under varying conditions of air pressure of from 19 to 41 pounds per square inch for a lift of 105 feet from a shallow sump, as shown in Fig. 169, were from 24 to 48 per cent. of the least work needed from the compressor, or from 17 to 30 per cent. including the efficiency of the compressor.

VACUUM AND AIR PRESSURE FOR RAISING WATER

The illustration, Fig. 170, shows the details of the plant. The tank shown is of cylindrical form, is 6 feet in diameter by 18 feet long, rests in a horizontal position on a rise having an elevation of 17 feet from the river-level and holds 3,800 gallons. The water-pipes are 3 inches in diameter and the air-pipe $1\frac{1}{2}$ inches. The former enter the end of the tank through a tee, at either side of which is a check valve.

The air-compressor and engine are placed at the top of the river bank. The former is of the St. Louis Steam Engine Company make, 6×6-inch cylinder, and is used alternately as a vacuum pump and compressor. In order to secure a larger supply of water it will be replaced by an 8×8 compressor.

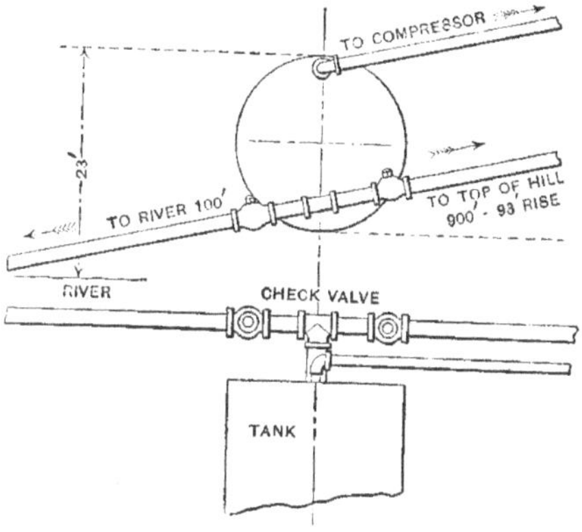

FIG. 170.—Vacuum and air-pressure system.

The operation of the plant may be reasonably well understood from the engraving. In exhausting the air from the tank the upper check valve closes and the lower opens, allowing the water to rush in and fill the vacuum. About 26 inches vacuum is required to bring in the water and fill the tank.

When the tank is full the operation is simply reversed. Air is driven back into the tank under pressure of 60 to 75 pounds to the square inch, and the lower check valve closing the water is forced up the hill to the reservoir. It takes a little less than two hours to deliver the contents of the tank in this way.

Compressed air possesses several advantages over steam, especially for transmitting power to considerable distances, and in many instances, including the pumping of water and other liquids, it is even cheaper than electricity for moderate distances, say from 1,000 to 3,000 feet. Direct air pressure can be used without the enormous loss from condensation incident to the use of steam, and by placing reheaters near the point of delivery, the efficiency of the compressed-air machines can be considerably increased. For pumping water, air has the advantage that it can be brought into direct contact with the water without loss by condensation, and in certain systems of pumping a large percentage of the heat required to compress and deliver the air can be reclaimed from the exhaust.

A simple construction of displacement pump is the Merrill, the principle of which is illustrated in Fig. 171. Two cylinders are provided side by side with an automatic air-valve situated immediately above them. Compressed air is admitted automatically first to one cylinder and then the other. When the air has expelled the water from one cylinder, the air is released and at the same time air is admitted to the second cylinder.

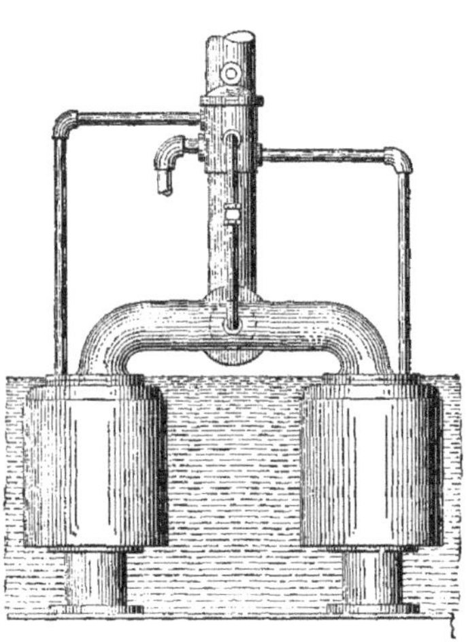
FIG. 171.—Duplex displacement pump.

One cylinder is filling with water while the other is discharging, so that an almost continuous stream issues from the delivery pipe as long as air is supplied. The pump is placed in the well or shaft and below the water-level, the cylinders filling, through large inlet valves in the bottom, by atmospheric pressure. The pressure of air on the surface of the water in the cylinder forces the water through the discharge valve, the height to which the water may be raised depending on the air pressure.

This system is not a new one, having been patented by Upham in 1809, and the system in its duplex form was patented in England in 1865.

The apparent difficulty in the use of this system lies in the loss of power when the compressed air, after driving the water out of the vessel, is allowed to escape into the atmosphere, thus losing all the power that was required to compress the air. The percentage of this loss increases with the head against which the water is pumped, and is about 50 per cent. when pumping to a height of 100 feet.

In the following system, the above difficulties are overcome to a degree that cannot be surpassed; for in it the air is not allowed to

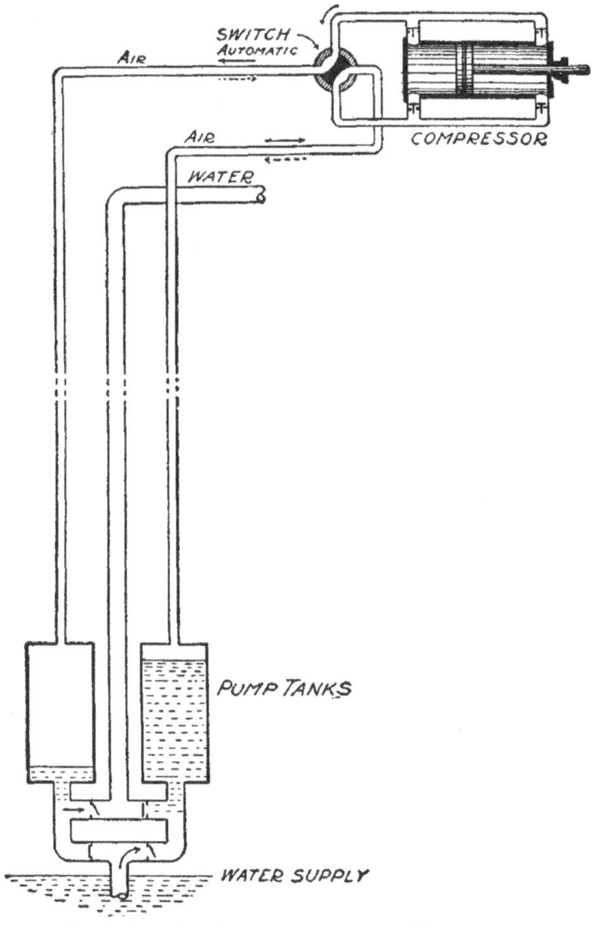

FIG. 172.—Duplex automatic water-lift.

escape, being used over and over so that none of the work done on it is directly lost.

Fig. 172 shows how the above conditions are attained. Suppose the compressor to be in operation with switch set as in the figure; the air will be drawn out of the right-hand tank and forced into the left-hand tank; and in so doing will draw water into the former and force it out of the latter. The charge of air in the system is so adjusted that when one is emptied the other is just filled. At that moment the switch will reverse the pipe conditions so that action in the tanks will be reversed.

The automatic control of the action of the pump is made by an air-switch at the compressor, which is thrown by the differential pressure in the air-pipes. The change in the pressure of these pipes alternating between the hydrostatic pressure in the air-force pipe and the absolute pressure in the air-suction pipe is equal to the head of water in the tank above the water-level in the well. At the moment of the greatest difference in pressure in the air-pipes, the automatic switch reverses the connections, and the compressor draws the air from the empty chamber and forces it into the full chamber. The compression and expansion nearly balance each other, and there is but little loss in power.

The same system as shown in Fig. 172 may be operated as a two-stage water-lift by placing one of the chambers in the sump and discharging the water into a sump at higher level.

They may be operated as before, and thus be made to raise one-half the volume to double the height, without increase of pressure.

Fig. 174 illustrates a multiple displacement pump. In this apparatus, instead of taking the air to the bottom and forcing the water the entire vertical distance at a single lift, a number of air-displacement tanks are placed at various intermediate points. The high-pressure air is conducted through the pipe A to a point close to the bottom of the lower tank. The discharge pipe from the lower tank runs to the next tank above, entering the bottom. The bottom tank is placed below the water-level, hence at the beginning of operations is full of water. When air is admitted through pipe A, the water in the lower tank is gradually displaced, being forced up into the next tank above, and is prevented from returning by the check valve.

The air rises to the surface of the water in the lower tank and forms a layer above the water. The discharge of water continues until the level reaches the perforated plate near the bottom of the tank and to which the water discharge pipe is attached. At this level is a second air-pipe, B, leading to the next tank above, as shown. When the water-level reaches the perforated plate or diaphragm, no more water can be expelled from the lower tank, and the air escapes through the pipe B into the bottom of the tank above and expels the water from this tank either into the atmosphere or into still another tank higher up, as the case may be. The pressure of the air after emptying the second tank is only one-half what it was after emptying the first, because the tanks are of the same capacity or volume, and the air therefore occupies twice the original volume.

Suppose the pressure of air in the pipe A is 100 pounds per square inch. This corresponds to a head of 230 feet in round numbers, at which height the second tank is located, the air pressure be-

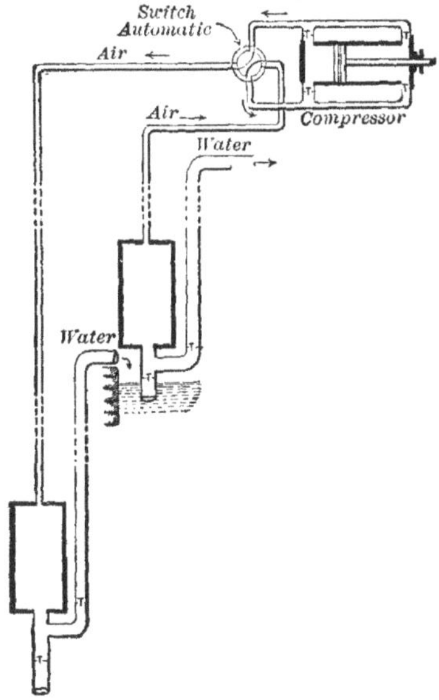

FIG. 173.—Arrangement of two-stage lift.

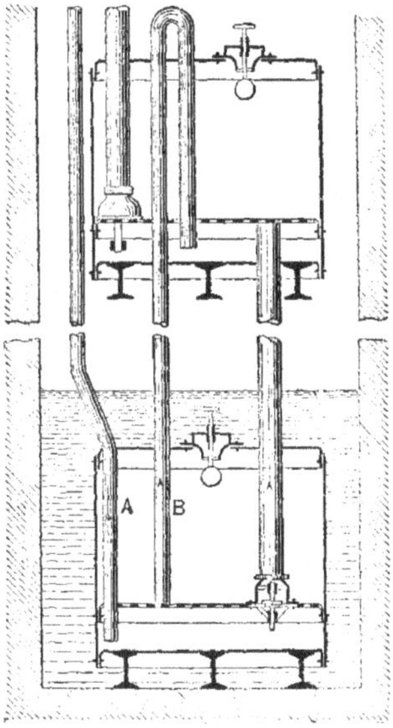

FIG. 174.—Multiple displacement pump.

ing merely sufficient to discharge the water into the second tank. The air, which has raised the water 230 feet, now fills the first tank from the bottom and the lower end of the air-pipe in this tank is exposed before the lower end of the water discharge pipe, and the air expands into the second tank above, and after emptying this tank has a volume equal to both or two tanks, which is twice the original volume. The pressure is now about $42\frac{1}{2}$ pounds by the gauge. The lift or head corresponding to this pressure is 98 feet, which is the height of the second lift, and the third tank will be placed approximately 98 feet above the second. At the end of the second lift the air will have a volume equal to three times the original volume, and the pressure will be one-third of the original pressure, or $100+15 \div 3=38.3$ pounds absolute, or 23.3 pounds by the gauge, so that the fourth tank will be placed $23.3 \times 2.3=53.59$ feet above the third, and so on to the top of the shaft or well.

By this method the air does about twice as much work with the same expenditure of work as the compressor, which increases the efficiency of the apparatus considerably.

In Fig. 175 is shown a Merrill water-pumping system for service where it is necessary that the valve mechanism or working parts be

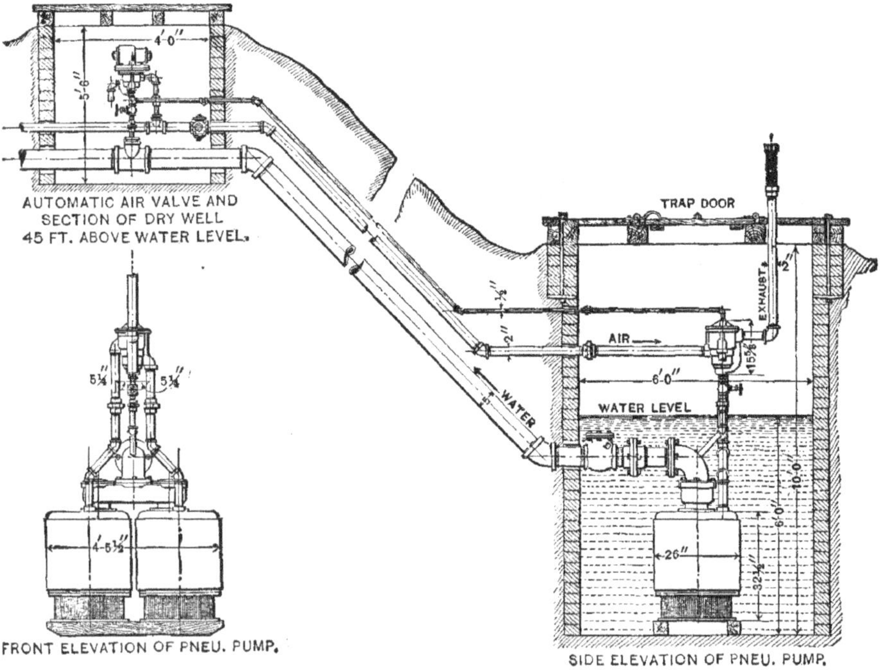

FIG. 175.—The Merrill pneumatic pump.
Direct-acting, with elevated air-valves.

placed some distance from and above the water chambers, as in the case of rivers where the rise and fall of water are great, and where it is desired to have the controlling valve above high water, and accessible at all times.

By this system of arranging the location of the air-valve above and at a distance from the location of the well or intake, and thus facilitating a pure water-supply for public and private use by locating the wells in the filter sands of streams and water-courses with the air-valves on the bank and an air-compressing station at any convenient distance, a valuable water-supply service may be made available at all times and under any condition of flood that would otherwise derange the old systems of water-supply from rivers. The only precaution necessary would be to build the well curb above the flood line, or cover the well with sand and carry the exhaust pipe up the bank, or to a safe place out of flood-water range. In this manner the neglected and scanty water-supply of towns and factories may be re-enforced with the pure and filtered element so essential to life and prosperity.

HYDRAULIC AIR-COMPRESSORS

One of the earliest compressed-air devices was the trompe or hydraulic air-blast for forges. Its capacity was sufficient for the wants of the times, which made it the principal means for furnishing a steady blast for the Catalan forges of the early years of the iron age. It could produce a pressure from an ounce to one pound or more, according to the height of the water-shaft and the depth of the water-seal. In the trompes of the best construction the water-seal was a sliding gate which could be operated to produce any desired pressure within the range of the apparatus. Its operation was as follows (Fig. 176): the falling column of water draws in air through the small inclined orifices, at the contracted vein, carrying it into the reservoir, where it separates, and is discharged through the tuyère pipe. The outlet discharges the water through an inverted siphon, carried high enough to balance the air pressure.

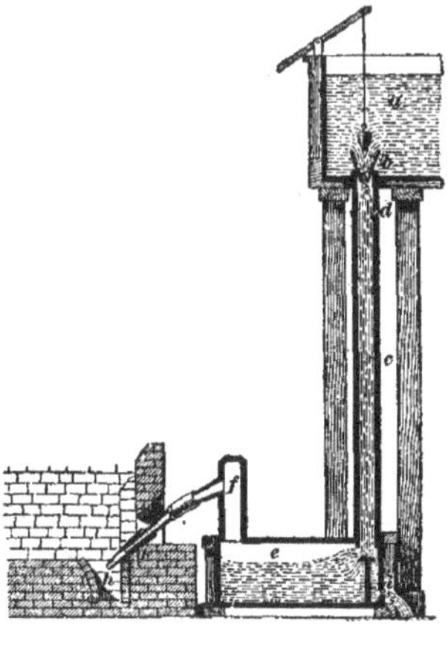

FIG. 176.—The trompe.

In the principles of the trompe is found a correspondence and suggestion of the experiments made by J. P. Frizell in 1877, and since carried out on a larger scale by C. H. Taylor in the practical hydraulic air-compressors at Magog, Quebec, and at Ainsworth, B. C.

Many experiments have been made to compress air by the direct and injector system for small quantities, by the use of water under pressure from city water-supply.

By direct pressure it requires an equal quantity of water to the volume of free air compressed to nearly the same pressure as the water. By the injector system, the only available experiments are those of M. Romally, in France, who found that with 35 feet head only 46 per cent. of the volume of the water used was equal to the volume of free air at a pressure of 21 pounds per square inch; thus realizing an air pressure of 138 per cent. of the hydraulic head and less than one-half the volume, an efficiency of about 63 per cent.

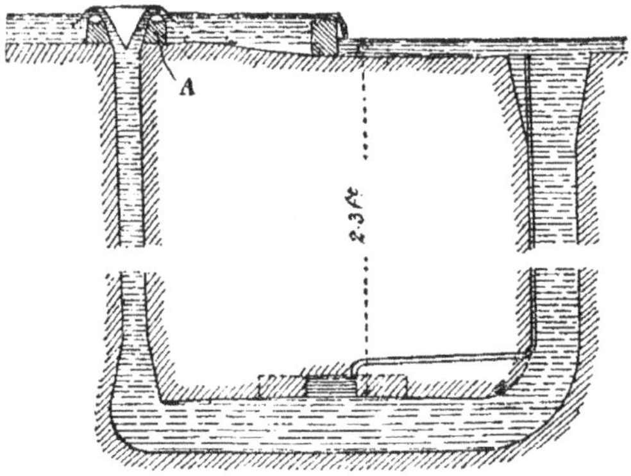

FIG. 177.—The Frizell system.

Mr. Frizell's experiments involved a large outlay in cost of plant, and where there is a moderate water-fall and plenty of water this is no doubt the cheapest working method of compressing air. The general idea of Mr. Frizell was to utilize a high water-fall with built-up shafts and air-chamber, or with a low water-fall to sink shafts with an air-gathering chamber at the bottom and air-pipe leading to the surface, as shown in Fig. 177. The entrance at A in the cut was a circular hollow dam with a conical inlet. The annular chamber under the dam communicated with the outer air and was perforated, so that the falling water drew down the air and by its velocity carried the air to the receiving chamber below. This suggestion and experiments lay in abeyance under the Frizell patent for many years, and was supplemented by a similar patent to Mr. George Waring. The efficiency in Frizell's early experiments was 26 per cent. of the fall of water used in the apparatus. Later improvements by him raised the efficiency to 52 per cent. with a head of 5 feet.

The hydraulic compressor system of Mr. Taylor is illustrated in

Figs. 178 and 179, in which a large number of small air-tubes are distributed around an annular water inlet to the down-flow pipe. One of its several forms of construction is shown in Fig. 178, and more fully illustrated in Figs. 179 and 180.

A number of air-tubes, *c*, *c*, terminate at the conical entrance of the down-flow pipe, B, at *a*, *a*, Fig. 178. A supply of water to the chamber A, A, and its flow down the pipe, draws air through the small pipes, carrying it down to the separating tank, *e*, *e*, where it is liberated at the pressure due to the hydrostatic head. The air is delivered through a pipe, as shown in the cut, and the water rises through a pipe or open shaft to the tail race.

The compressor as erected at Magog, Quebec, gives in air-power 62 per cent. of the water-power used and delivers 155 horse-power in compressed air at 52 pounds gauge pressure.

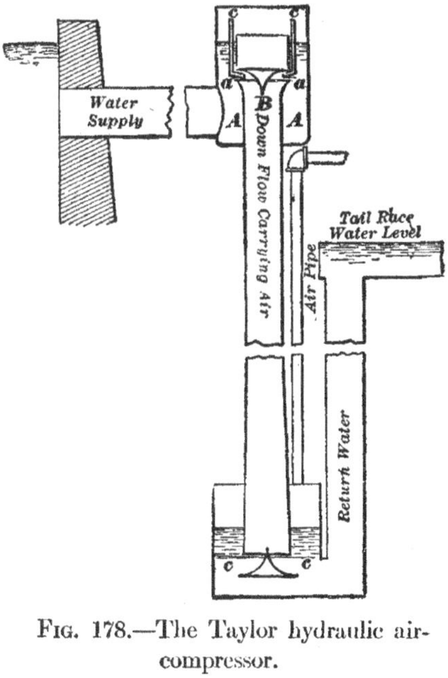

FIG. 178.—The Taylor hydraulic air-compressor.

A most remarkable feature of this system is that, notwithstanding that the air is compressed by the weight of the water and in actual contact with it, the air so compressed is delivered in the receiver and thence to the transmission pipe drier than when drawn in from the atmosphere.

At first sight this would seem impossible, but it is well known that in a high temperature moisture is held longer in air than in a lower temperature, hence the contact of the air globules with the cold water keeps down the temperature usually caused by the compression of air, and the atmospheric moisture held in the globules condenses, as it were, on the walls of these globules, and at the point of separation the air and water are absolutely separated, leaving the air all ready for distribution at the same temperature as the water it has just left, and drier than when first taken in through the small air-pipes.

Another feature is that the power of the water can be converted into compressed air at any pressure per square inch, giving the same efficiency at either high or low pressure with a far less loss of energy

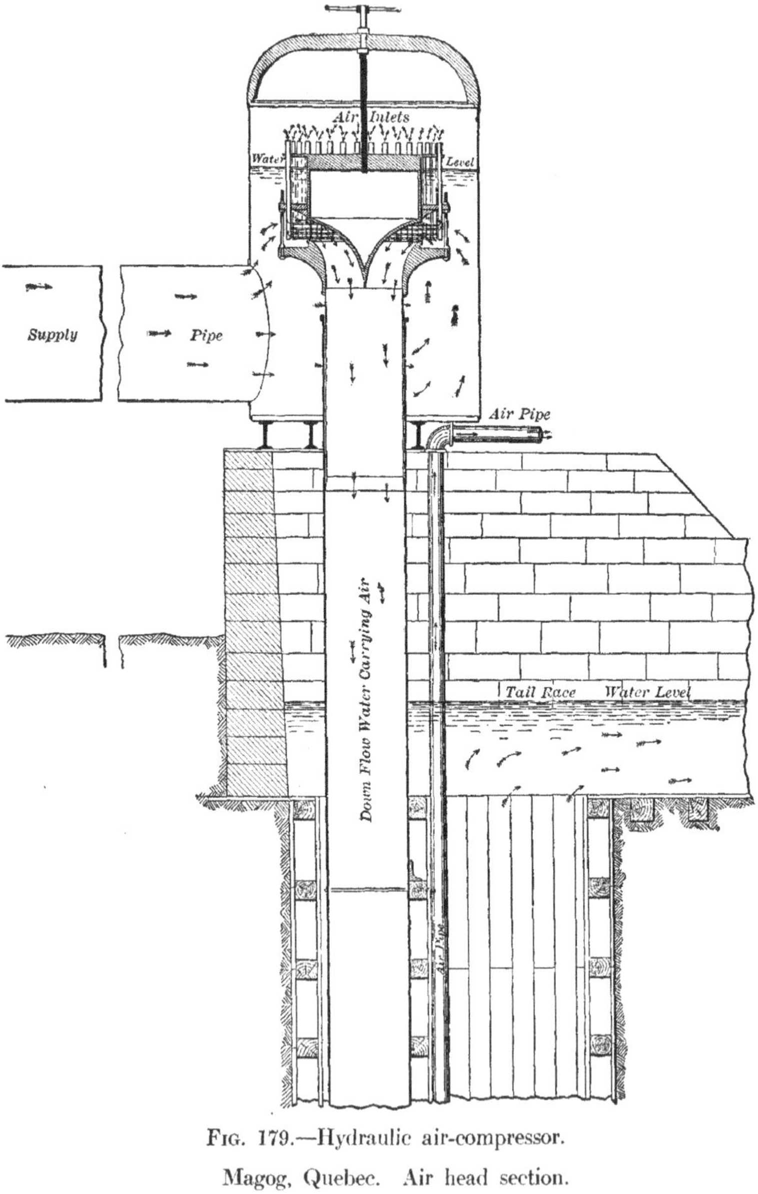

FIG. 179.—Hydraulic air-compressor.
Magog, Quebec. Air head section.

than by any other process of transforming a water-power into transmittible force, and with unvarying pressure.

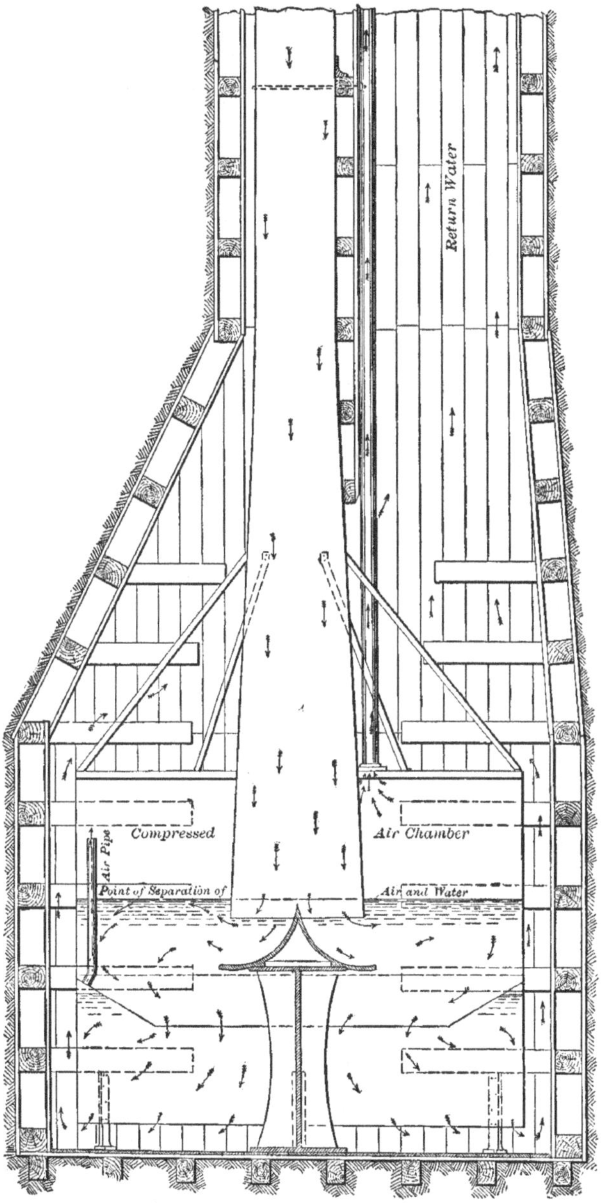

Sectional View

FIG. 180.—Hydraulic air-compressor.
Magog, Quebec. Air-chamber section.

Should the volume of air taken down be greater than that being used, it accumulates in the receiver until it forces the water below the lower end of the receiver, and the surplus passes up with the return water, thereby forming a perfectly automatic safety-valve, without requiring any attendance whatever. It will be observed that the material used in the construction of the down-flow pipe need only be of sufficient strength to carry the weight of water and pressure generated in the working head of the water-power, as once it reaches the tail-race level the internal pressure is gradually neutralized from that point down by the pressure in the return water surrounding the down-flow pipe; so that any pressure almost may be reached without increasing the strength of the down-flow pipe. The material for the down-flow pipe may be of iron, or wood hooped with iron, and the shaft may be constructed of the cheapest of timber; and as it is preserved by being constantly in the water, there is practically no limit to its durability.

By this system low falls, otherwise useless, may be utilized, and the same pressure obtained as from high falls, the horse-power being determined by the diameter of the down-flow pipe, and the height and volume of water in the fall, while the pressure depends solely upon the depth of the well or shaft; therefore any desired pressure can be obtained.

In the apparatus at Magog, Quebec, the receiver is sufficiently large in diameter to allow the air to rise to the surface of the water therein, from whence it is taken through the air-pipe for transmission to be utilized as power or for other purposes. The water, being kept down by the pressure of the air, is forced out through the open bottom of the receiver and up the shaft around the down-flow pipe to the tail-race level.

The compressor is so constructed as to permit of its being regulated to furnish any proportion—from one-third of its capacity—using water proportionately with a like efficiency.

By reference to the head section (Fig. 179) it will be noticed that the head-piece is telescoped into the down-flow pipe, and raised or lowered by means of a hand-wheel on top, thus permitting the flow of water to be regulated, or to lift it above the water-level and stop entirely the flow of air, the water being regulated by the head-gate.

Briefly stated, the air is compressed by the direct pressure of falling water without the aid of any moving machinery, and practically without expense for maintenance or attendance after installation.

By this system any fall of water varying in working head may be utilized, and any pressure required can be produced and uniformly maintained up to the capacity of the water-power, delivering the compressed air at the temperature of the water, and in a drier state than is possible by any known means of compression, thereby avoiding all loss by condensation or shrinkage by cooling of the air after compression.

The water may be conveyed to the compressor by means of an open flume; or, as shown in the diagram, through a pipe supplying a

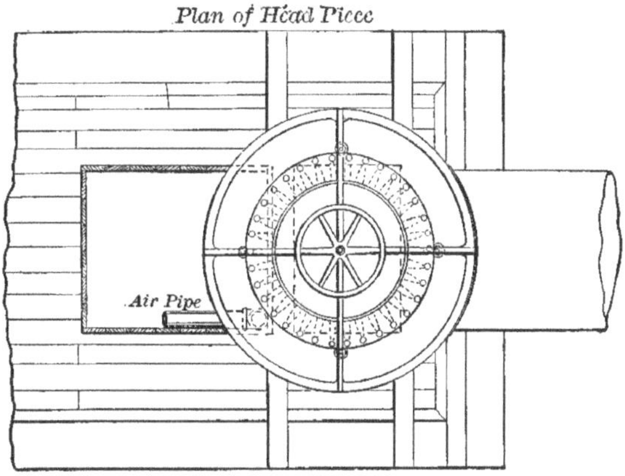

FIG. 181.—Plan of air-tubes.

tank or stand-pipe around the head-piece of the compressor, where it can attain the same level as the water in the dam or source of supply.

Around the head-piece are placed a large number of small, horizontal air-pipes, drawing their supply of air through larger vertical pipes, which extend above the surface of the water and open to the atmosphere.

As the water enters the down-flow pipe and passes the ends of these small air-pipes, it draws in the air in the form of small uniform globules, which, becoming entangled in the descending water, are carried down to the receiver at the bottom of the pipe, compressing the air by the pressure of the water surrounding these globules until they reach the point of separation. This pressure is maintained so long as there remains any air in the receiver chamber.

The enlargement of the down-flow pipe at the bottom section was made to lessen the velocity of the water and air at that point, which was found to facilitate the separation of the air from the water by coalescing the small globules of air and the better separation at the deflecting plate below. The deflecting plate prevents the plunge of the down-flowing water into the separating part of the tank and by its deflections gives the air a more ready separation from the water. By this arrangement no air was found in the water discharge pipe.

In tests of efficiency it has been found that the gross power of the water passing through the compressor due to its natural fall was 158 horse-power, of which 111 horse-power was utilized in the work of air compression, giving an efficiency of 70 per cent. of the gross power used.

Later experiments indicate that an efficiency of 75 per cent. may be obtained by a modification of the air-inlet pipes and water head.

In Fig. 182 is illustrated the Taylor hydraulic air-compressing plant at Ainsworth, B. C., which was established in a trussed tower in order to carry up the air head to a level with the flume, of which Fig. 179 represents the elevation and arrangement of the head. The available working head from the water-level in the head stock to the tail race is 102 feet; the depth of the shaft is 210 feet, and the depth of the air-chamber at the bottom of the shaft is 17 feet, from which the water closure of the down-flow tube leaves 200 feet as the available hydrostatic pressure, which gives an air pressure of 87 pounds per square inch. The flume supplying water from Coffee Creek, 1,350 feet distant, is 5 feet in diameter, of stave-barrel construction. The tower head is also of wood staves, is 12 feet in diameter and 20 feet high. The down-flow pipe is of the same construction, 2 feet 9 inches in diameter, widening slightly at the bottom to retard the velocity of the descending water and allow it to impinge upon a whorling cone that produces a circling current in the air-chamber that facilitates the separation of the compressed air from the water. The air rises to the top of the separating chamber and is delivered through a 9-inch pipe to the various branches for air distribution at the ground surface. A secondary pipe is carried from midway in the separating chamber to the surface above the tail race that seals the air-space with water when the air is being used in excess of compression, and allows the air to escape when it accumulates and pushes the water surface below the mouth of the air-pipe; thus making an air-pressure regulator within the limit of one-pound air pressure.

The regulation of the air-inlet pipes, of which there are about three thousand tubes, $\frac{3}{4}$-inch diameter and the conical ajutage, is made by raising or lowering the air-pipes and cone by a screw and wheel, as shown in Fig. 182. The velocity of the water in the down-

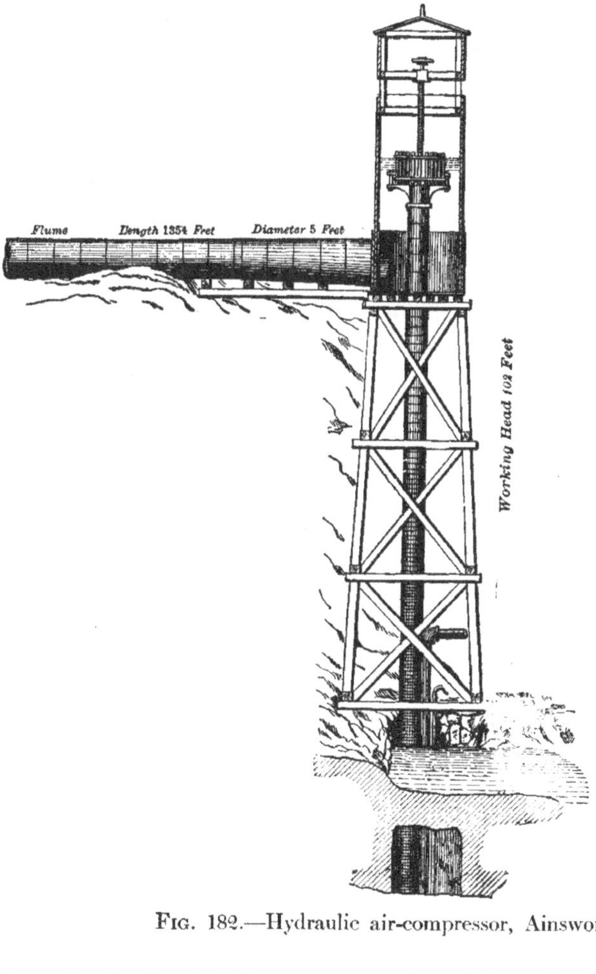

FIG. 182.—Hydraulic air-compressor, Ainsworth, B. C.

flow pipe is about 34 feet per second, and the velocity of the indraught of air is nearly the same. The air is received by the water in millions of globules, which in a great measure retain their individuality, gradually becoming smaller by the increasing water pressure until they are liberated in the air-chamber below.

CHAPTER XI

ARTESIAN WELLS, PRINCIPLES OF FLOW AND WATER-SUPPLY FOR IRRIGATION

THE meaning of "artesian" has probably been derived from Artois in France, where the earliest flowing wells were bored, but is now applicable to any deep-bored well.

The arid regions of the Dakotas, Kansas, Nebraska, and the States farther West have received great benefits to agriculture by the success of their artesian-well flow. Flowing and pumped wells of great depth have given a most useful service in all countries and are largely increasing in number for domestic and irrigation purposes, and in many localities furnish water for power.

ESSENTIAL FEATURES OF ARTESIAN WELLS

The artesian stream has its source, its underground water-way, its ascent through the well, and its final descent in the rill that runs away. It is peculiar mainly in its underground conditions. Upon these, chiefly, the ascending flow depends.

To fashion a simple idea of the common class of flowing wells, picture to the mind a pervious stratum through which water can readily pass. Below this let there be a water-tight bed, and let a similar one lie upon it, so that it is securely embraced between impervious layers. Suppose the edges of these layers to come to the surface in some elevated region, while in the opposite direction they pitch down to considerable depths, and either come up again to the surface at some distance, thus forming a basin, or else terminate in such a way or take on such a nature that water cannot escape in that direction. Now, let rainfall and surface waters penetrate the elevated edge of the porous bed, and fill it to the brim. That such beds are so filled is shown by ordinary wells, which commonly find a constant supply in them at no great depth. Now, it is manifest that if such a water bed be tapped by a boring at some point lower than its outcrop, the water will rise and flow at the surface because of the higher head in the upper edge of the bed. If the surface water continually supplies the upper edge as fast as the water is drawn off below, the flow will be constant.

The leading conditions upon which artesian flows depend are involved in this simple conception drawn out as follows:

1. A pervious stratum to permit the entrance and the passage of the water.

2. A water-tight bed below to prevent the escape of the water downward.

3. A like impervious bed above to prevent escape upward; for the water, being under pressure from the fountain-head, would otherwise find relief in that direction.

4. An inclination of these beds, so that the edge at which the waters enter will be higher than the surface at the well.

5. A suitable exposure of the edge of the porous stratum, so that it may take in a sufficient supply of water.

6. An adequate rainfall to furnish this supply.

7. An absence of any escape for the water at a lower level than the surface at the well.

These may be considered in detail, and then attention directed to some special practical questions.

In Fig. 183 is illustrated an ideal section of an artesian basin in

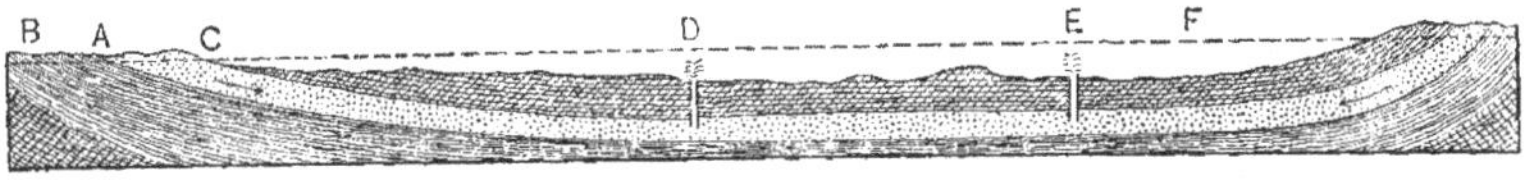

FIG. 183.—Ideal section illustrating the chief requisite conditions of artesian wells.

A, a porous stratum; B and C, impervious beds below and above A, acting as confining strata; F, the height of the water-level in the porous bed A, or, in other words, the height of the reservoir or fountain-head; D and E, flowing wells springing from the porous water-filled bed A.

which the water bed is spread between layers of non-porous rock or clay.

It is the proper function of geological investigation to ascertain the stratigraphic conditions which determine success or failure. As the outcome of completed investigations, it is possible to map off the face of the country into (1) areas in which success may reasonably be anticipated, (2) areas in which the conditions are nearly balanced, and in which local and indeterminate conditions will decide success or failure, and (3) areas in which the conditions are adverse, if not altogether prohibitory. The service which such maps are competent to render is in encouraging the utilization of an important resource, and are now in preparation by the Geological Survey for the United States.

From the foregoing considerations it is manifest that the areas of probable success must be the relatively low tracts, that the areas of adverse probabilities are the relatively high regions, and that the doubtful belts lie between.

The greatest sources of indefensible expenditure are (1) the selection of a location of too great relative altitude, in defiance of the simple fundamental principle of artesian flowage, and (2) the penetration of unproductive strata, after all favorable chances have been exhausted, in disregard of the fundamental laws that govern the distribution of subterranean waters.

In the face of many possibilities of failure, the importance of a special consideration of the assemblage of conditions that surround

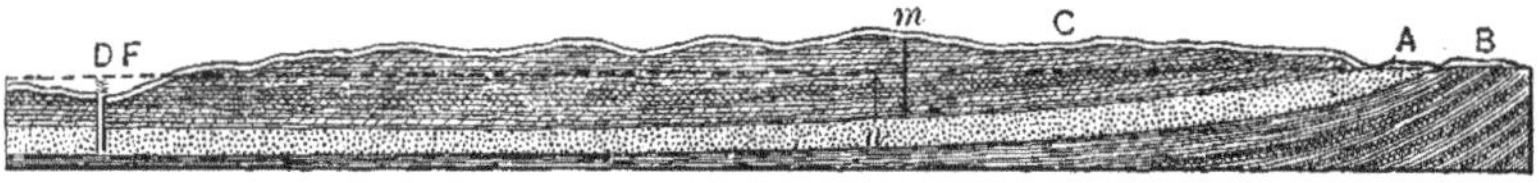

FIG. 184.—Subwater bed and saturated upper strata.

Section intended to illustrate the aid afforded by a high-water surface between the fountain-head and the well: A, a porous bed; B, a confining bed below; and C, a confining bed above. The dark line immediately below the surface represents the underground water surface. Its pressure downward is represented by the arrow *m*. The pressure upward due to the elevation of the fountain-head is represented by the arrow *n*. The line F represents the level of the fountain-head. There can be no leakage upward through the bed C except near the well D. There may be some penetration from the bed C into A, which would aid the flow.

a proposed undertaking, in a region not previously proved, or in a situation in any important respect dissimilar from those already tested, need not be pressed. One of the most favorable conditions for securing a fountain is found when thick semiporous beds, constantly saturated with water to a greater height than the fountain-head, lie upon the porous stratum, and occupy the whole country between the well and its source, as illustrated by Fig. 184. This is not only a good, but an advantageous, substitute for a strictly impervious confining bed. Under these hydrostatic conditions, limestone strata reposing on sandstone furnish an excellent combination.

If, on the other hand, the underground water surface between the proposed well and the source of supply is much lower than the fountain-head, there will be considerable leakage, unless the confining beds are very close textured and free from fissures. For example, if it be 100 feet lower, there will be a theoretical pressure of nearly three atmospheres, or about 45 pounds to the square inch, upward greater than that of the underground water downward, disregarding the in-

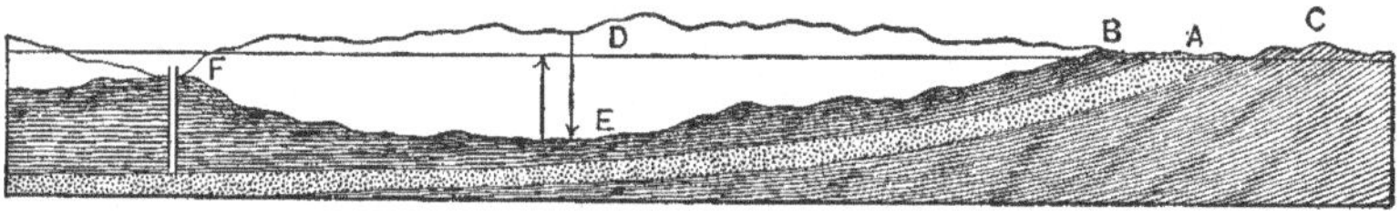

Fig. 185.—Double section illustrating the effects of high- and low-water surface in the cover area.

fluence of capillarity, and this will be competent to cause more or less penetration of the water upward through the pores and crevices of the rocks, and consequent loss of head and forcing power.

Both of the above points may be illustrated by profiles, Figs. 185 and 186, in which A represents a porous stratum enclosed between the impervious beds, B and C. The source of water-supply is at A, and the proposed well at F. Let E be supposed to represent the surface of the ground (and for convenience, also, the surface of the common ground water) in one of the two supposed cases, and D the surface in the other. The arrow springing from the surface, E, rep-

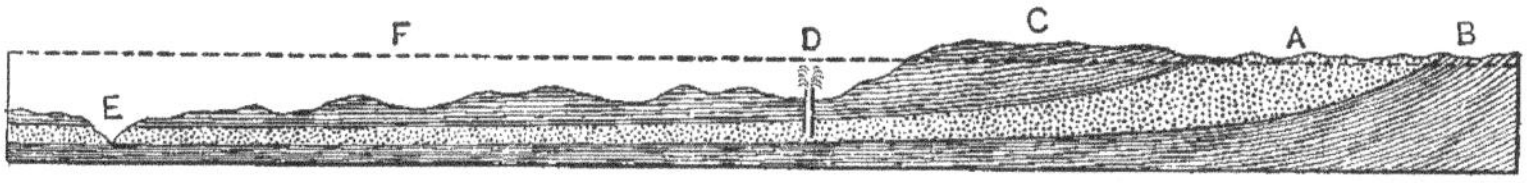

Fig. 186.—Section illustrating the possibility of a flow from a bed even when exposed at a lower level.

A, a sandstone bed, thick and coarse at the right, its shore edge, and thinner and finer at the left. B and C, confining impervious beds. F, the water-level in A. D, a well which may flow notwithstanding the lower exposure at E.

resents the upward tendency of the water in the porous bed, owing to pressure in the fountain-head, while the arrow depending from the line D represents the downward pressure of the ground water, whose surface is represented by D, and is, it will be observed, more than equivalent to the upward tendency due to pressure from the fountain-head. A flow at F could very safely be predicted if the surface were as represented by D, while it might be doubtful whether one could be secured if the surface were as represented by E.

It is often convenient to speak of the source of supply as the reservoir. Erroneous impressions, however, are likely to arise from the use of the term. Two such are quite current, and need to be dismissed. The one is the assumption that the reservoir is a surface lake, the other that it is an underground pool, occupying a cavernous cistern, as it were. A surface lake is an extremely improbable source of an artesian flow. It has already been indicated that the water must have a ready entrance and flow through the porous stratum to give an efficient fountain. But most lakes owe their existence to the fact that they have impervious bottoms, otherwise the water would pass into the earth beneath. This fact stands in the way of their serving as sources of artesian wells. Far from being looked upon as special fountain-heads, lakes are to be regarded in precisely the opposite sense. They show that the rainfall, instead of going into the strata to feed the fountain, is held at the surface and exposed to loss from evaporation and overflow.

The rainfall of a region is discharged in three ways: (1) by evaporation; (2) by surface drainage; (3) by underground percolation. Artesian wells can avail themselves only of the last. Whatever increases the first two decreases the last. In so far, therefore, as impervious surface basins aid evaporation and surface discharge, they detract from the copiousness of the underground supply.

It is a compensating fact, however, that surface drainage is usually imperfect in lake regions, as the existence of the lakes themselves testifies. Possibly all, or more than the loss from evaporation may be gained by this reduced surface flow. This, however, does not destroy the force of the general observation, that a lake is not to be regarded as the special reservoir of an artesian fountain.

The notion of a subterranean pool has little more to support it. Tubular channels and cavernous spaces undoubtedly exist, and are occasional sources of flow, or means of passage, and so are, in a sense, reservoirs, but not in the import of the term as used in connection with artificial fountains, i. e., in the sense of a fountain-head.

The reservoir or fountain-head of most artesian wells is simply the water contained in the water-bearing stratum above the level of the point of flow, or, in other words, the water in the elevated margin of the water-filled stratum.

No stratum is entirely impervious. It is scarcely too strong to assert that no rock is absolutely impenetrable to water. Minute

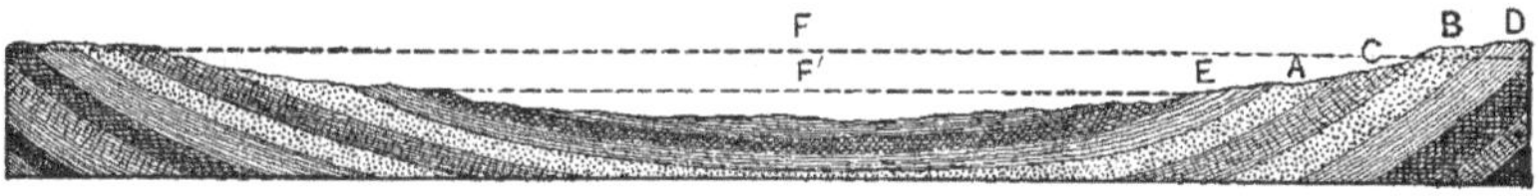

FIG. 187.—Section illustrating the usual order in which the strata of a basin come to the surface.

A and B, porous beds; D and E, impervious beds; C, a half-impervious bed; F′ and F, the water-levels of A and B, respectively.

pores are well-nigh all-pervading. To these are added microscopic seams, and to these again larger cracks and crevices. Consolidated strata are almost universally fissured. Even clay beds are not entirely free from partings.

But in the study of artesian wells we are not dealing with absolutes, but with availables. A stratum that successfully restrains the most of the water, and thus aids in yielding a flow, is serviceably impervious. It may be penetrated by considerable quantities of water, so

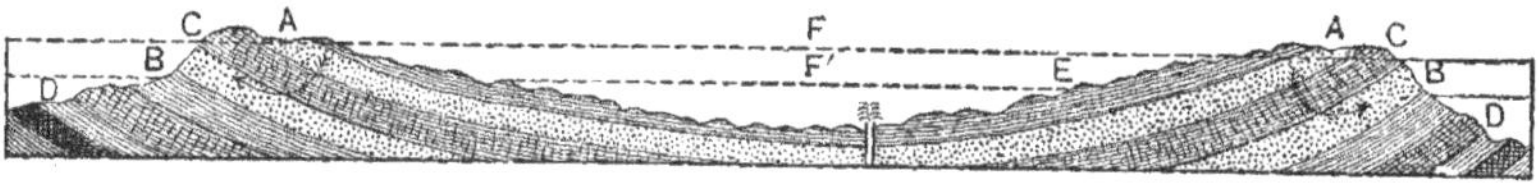

FIG. 188.—Section illustrating the possible effect of erosion upon strata originally like those in Fig. 183.

A and B, porous beds; D and E, impervious beds; C, a half-pervious bed; F and F′, the water-levels of A and B, respectively. If the stratum C is not practically a confining layer, the water from A will pass through it and escape at the edge of B, so that a flow cannot be obtained at a higher level than it, but may be had below the line F′.

that the leakage is quite appreciable, and yet be an available confining stratum. The nearest approach to an entirely impervious bed is furnished by a thick layer of fine, unhardened clay. In this case solidifying permits the formation of fissures, and the clay rocks are less impervious than the original clay beds. The clayey shales rank next as confining strata, after which follow in uncertain order shaly limestones, shaly sandstones, the various crystalline rocks, and even compact sandstones.

The limestones are likewise much traversed by crevices near the surface, and are, besides, subject to the solvent action of the waters passing through them. These often form extensive underground channels, mammoth examples of which are found in the great elongated caves of Indiana and Kentucky. But, like the above, these prevail mainly in the superficial portion of the beds, and are chiefly confined to regions where the limestone stratum is not overlain by other rocks, and hence not available as a source of fountains. The reason of this is manifest when it is considered that the solvent action is mainly accomplished by surface waters. These exhaust their solvent power before penetrating far. When the limestone is overlain by impervious beds, these surface waters are cut off, and hence sol-

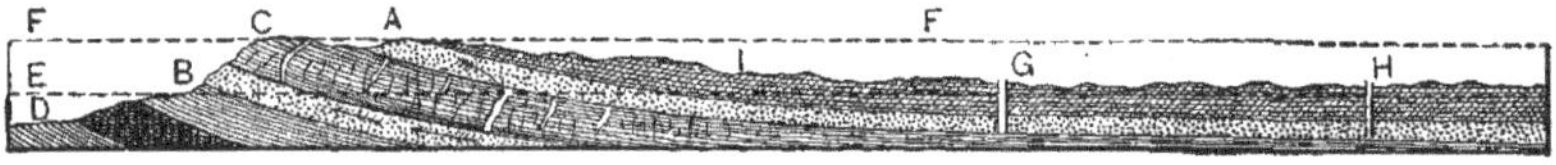

FIG. 189.—Section illustrating the failure of an artesian well, because of defects in a confining bed below.

A and B, porous beds; D and I, impervious beds; C, a defective confining bed; E, the water-level of the stratum B; G and H, wells that do not flow. The bed A might give a flow at G and H but for the defect in C, which permits the water to descend into B and escape through its outcrop, which lies below the surface of G and H.

vent action is limited to such waters as entered at the distant outcropping edge. The cracks and cavities of deep-seated limestones are often found to be healed up with calcite, an index that the waters there are depositing, rather than solvent.

The grounds, therefore, for anticipating success in penetrating limestone for fountains are not very flattering, though less adverse than in the crystalline rocks. However, limestones that have once been exposed to surface action, and thereby fissured and channelled, and subsequently buried beneath a thick mantle of drift clay, are not altogether unpromising. Not a few important flowing wells have been derived from them. But when the beds have always lain deeply buried beneath impervious strata, they have rarely been found productive, so far as our knowledge extends.

In the employment of several wells, their distribution is a matter of some consequence. The normal direction of flow when it is once set up by virtue of the opening of an avenue of discharge is along a line drawn from the outcropping edge of the bed down its slope to the wells. Now, it is clear that if several wells are arranged along this line, the first one will be better placed than those which stand in its lee. These will be, indeed, measurably supplied by lateral flowage under the law of equal pressures, but less directly and freely. If the wells are disposed in a cluster, those on the exterior will partially cut off the supply of the interior wells. A more fortunate disposition than either of these would be an arrangement in a line at right angles to the direction of flow.

A still more advantageous arrangement, subject to local modification, would be to dispose the wells in a curved line, convex toward

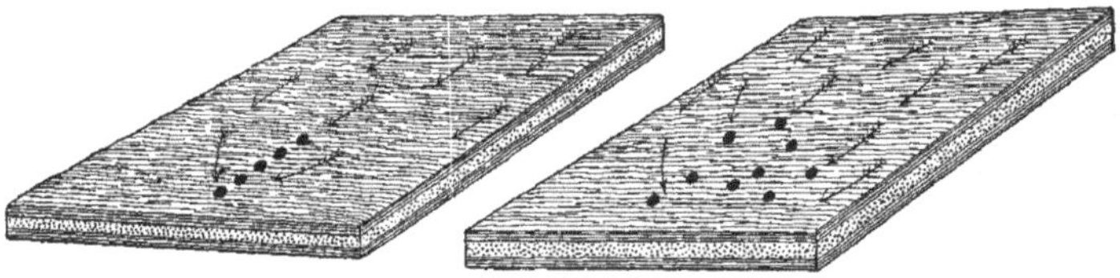

FIGS. 190 and 191.—Tabular sections of strata, showing disadvantageous arrangements of wells.

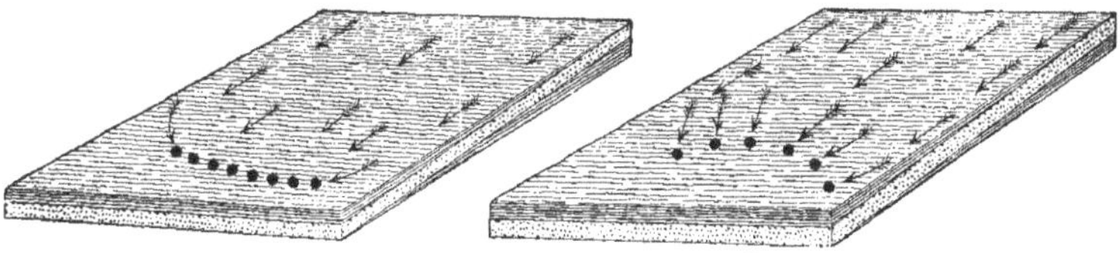

FIGS. 192 and 193.—Tabular sections of strata, showing advantageous arrangements of wells.

the collecting tract; for when the draught of the wells has made itself felt upon the sheet of water flowing most directly from the collecting belt to them, the higher pressure which the flanking portion still suffers will cause a lateral inflow, and the curved disposal of the wells will be more favorable for receiving the ingathering currents than a rectilinear arrangement, being more nearly normal to the resultant pressure and flowage. The positions of the wells in regard to the underground water flow is shown in Figs. 190, 191, 192, 193.

In respect to the degree of separation of the wells, it is obvious that so far as the mere question of the greatest reception is concerned, the farther they are apart the better, for they will affect each other less; but, of course, practical considerations put a limit to their dispersion.

A second condition of delivery relates to the well itself. It is clear that, if the bore merely touches the upper surface of the water-bearing bed, only a small space is afforded for the entrance of the stream. If, on the other hand, the well penetrates the formation deeply, the water can run in all along the sides, and, though the inflow at any one point may be moderate, the total amount from the large surface presented by the sides of the bore may be great.

Quite in contrast with the close-textured beds, that are water-carriers only by virtue of fissures and channels, are the open-textured strata that constitute continuous water-filled sheets underspreading wide areas, and which can therefore almost certainly be tapped at the proper depth. Speaking in general terms, these are the only reliable sources of artesian wells.

To this class belong beds of sand, gravel, sandstone, conglomerate, and other less common rocks of loose granular texture. Some of the more porous chalks and granular limestones may be classed here. The common feature of the class lies in the construction of the rock from separate particles, loosely put together, leaving small open spaces between them. The porosity is of an interstitial, not vesicular, kind.

ESCAPE OF WATER AT LOWER LEVELS THAN THE WELL

It is manifest that if the confining beds are pierced either naturally or artificially at a point lower than the surface of the well, the water may find relief from pressure by escaping there, and fail to flow from the well. This is not often a source of failure from natural causes, where the overlying strata are thick, since the tendency in the deeper beds is rather toward the closing of openings and the healing of fissures than to the opening of a free passageway. However, in those regions in which profound fracture and displacement are common, failure from leakage through fissures is a source of apprehension.

The artificial defects consist mainly of wells previously sunk. It is a well-known fact that, where several are located near each other, those which are lower than the proposed well may already have consumed the full delivering capacity of the water-bed. The reverse may also happen. The new well, if lower than the previous ones, may draw off their flow. The remedy in these cases is simple. Either the flow of the lower wells may be reduced until the upper ones discharge, or else all stopped.

When the rocks have been broken and displaced or faulted, the conditions are not favorable for obtaining artesian water. Even if an artesian basin is present, the impervious beds, which would otherwise confine water under pressure, may be broken and permit the water to escape. In this connection it is to be noticed that hot springs are frequently situated on fault lines. In the case of the Stein Mountain fault, a spring with several surface openings having temperatures ranging from 168° to 177° F. occurs on the southwest side of Alvord Desert, Oregon.

It is principally from the study of the structure of the rocks in any region that the presence of artesian basins is to be determined. The reason why structure may be considered as the controlling condition in this connection is because the other necessary conditions, such as the presence of pervious and impervious layers in a series of stratified rocks, the presence of water in pervious beds, etc., are of common occurrence; while the requisite conditions for storing water beneath the earth's surface and under pressure are much less frequent, and it is these conditions which are dependent principally on the positions which the rocks occupy—that is, the geological structure. When a saucer-shaped basin is present, the chances are that the other necessary conditions will also exist.

The sedimentary beds are in many instances open-textured and of such a nature that water will readily percolate through them, while in other instances they are of the consistency of clay and tend to retain water in the porous beds, if any is present, between them. These conditions indicate that wherever the sedimentary beds have been bent or displaced from their original horizontal position so as to form basins, there is a probability that flowing water may be obtained by drilling wells.

When the topographical and geological conditions have been determined, after careful consideration, to be favorable for the construction of a deep or artesian well, it is necessary to ascertain how many wells in the vicinity reach to the stratum it is proposed to tap, and how they affect one another. If the starting of a well causes much diminution in the discharge of others in the neighborhood it is an indication that about all the available water in the porous stratum is already being drawn from it. Another well may so diminish the flow from those previously constructed that serious inconvenience

FIG. 194.—South Dakota irrigation well.

may arise. If, however, the starting of one well has little or no influence on those about it, one more may be sunk with fair assurance that it will not be a source of vexation.

It is, of course, desirable to sink the well some distance into the porous stratum so that, by perforations in the lower portion of the pipe, the water may enter the well through a much larger surface than is offered by the cross-section of the pipe itself. In case the flow remains inadequate after the well has been sunk far into the water-bearing stratum, various expedients may be tried to increase it. One of these is the explosion of a torpedo at the bottom of the well. This rends and shatters the rock, and opens fissures through which the water may pass more freely to the well.

In Fig. 194 is illustrated a view of a spouting well near Aberdeen, South Dakota. It is more than 1,000 feet in depth, of 6-inch bore and piped; when closed it shows a water pressure of 150 pounds per square inch and supplies water sufficient for irrigating a farm of 800 acres.

A few artesian flowing wells have been obtained in Illinois and Wisconsin from the deep beds of the Potsdam and St. Peter's sandstone. Along the eastern coast of the United States low-pressure flowing wells have been obtained, and also at a few locations in the Middle States and the borders of the Gulf of Mexico.

The oil gushers are an anomaly and belong to the department of the oil industry.

CHAPTER XII

IRRIGATION OF ARID DISTRICTS

IRRIGATION, which is of unknown antiquity in all parts of the Old World, Europe, Asia, and Africa, and which has been long in practice by the prehistoric inhabitants of the New World, as the remains of their canals and reservoirs attest, is of the utmost importance to the success of agriculture and its sustaining life in all the arid districts of our country.

In its modern methods it is comparatively new and limited in the United States; but enough has been done individually to encourage the undertaking of extensive irrigation works by the Government and by large incorporated water companies; from which vast tracts of arid land have been and are in course of being converted into valuable and productive land for agricultural, pomological, and horticultural use.

It has been estimated that there are about 2,000,000,000 acres of arid land in the United States, of which a large percentage is reclaimable by irrigation. Probably 100,000,000 acres can be redeemed to agriculture by the judicious use of the perennial streams and subsurface waters.

The first systematic application of irrigation in the arid West by English-speaking people was made by the Mormons. The soil was so barren that crops could not be raised by ordinary means, and, forced through fear and privation to adopt new and extraordinary devices, they turned the waters of the little cañon streams upon the ground where Salt Lake City now stands. After many years of meagre success or disheartening failure, they succeeded in mastering the art of irrigation.

At about the time when the Mormons were building up the State of Deseret, now Utah, the gold miners in California were building ditches for placer washing, and were using water from these ditches for irrigation. The results obtained through the use of these ditches, and also by those of the old missions, attracted public attention, and irrigation slowly developed, at first as an adjunct to mining. With the stoppage of hydraulic mining in many places, consequent upon the passage of the antidébris law, the ditches built for mining were either abandoned or used exclusively for irrigation. Many of them have been enlarged and have now even greater value than in the old days of mining excitement.

A notable epoch in the development of the West was the founding of the colony in northern Colorado, named after Horace Greeley, its chief promoter. Although unfortunate in its early years, the colony succeeded in learning how to control and utilize the waters of Cache la Poudre River for irrigation. The success ultimately attained by the Greeley colony, and the wonderful results shown by the Mormon communities, which have spread from Utah north into Idaho and Wyoming and south into Arizona, have attracted public attention and have greatly stimulated the colony idea. As a consequence, many organizations have been formed for the purpose of bringing people in large bodies from the Eastern States, and even from Europe, and placing them upon small farms located near each other and supplied with water from a common ditch. Individual settlers also have sought opportunities for bringing land under cultivation by artificial watering, and thus, at many widely scattered points, irrigation has been introduced. The Union colony settled at Greeley, Col., in 1870, twenty-three years after the Mormons had begun irrigating. There are no statistics concerning the area irrigated in 1870, but it is probable that in that year there were not over 20,000 acres under irrigation in the whole United States. From 1870 to 1880 was an era of rapid development of small ditches, constructed by individuals and associations of farmers. At the end of that period there were probably 1,000,000 acres under irrigation.

In the decade 1880 to 1890 occurred the "boom" of speculative enterprise in irrigation canals. Large sums of money were obtained for irrigation works by the sale of stocks and bonds, and great enterprises were projected, canals of upward of 100 miles in length being planned and in some cases built. Nearly all of these failed of financial success, and although they have aided in the extension of irrigation, they have not enriched the investors.

The Eleventh Census was the first to devote attention to irrigation, and the statistics obtained show that in 1889 there were 3,631,381 acres irrigated on 54,136 farms, with an average irrigation area of 67 acres. During the following decade, the irrigated acreage doubled in extent. This has been due rather to the extension and enlargement of the many canals existing in 1889 and to the more complete practice of irrigation on the lands then under ditch than to the construction of new and large systems of irrigation.

A relatively small part of the West is under irrigation. It must be remembered, however, that only a small portion of the Western States is as yet in private ownership or included in farms.

There is great need of systematic study concerning the actual effect which the water has upon the soil and upon the plants. Where the supply is abundant the quantity of water used is generally far in excess of that theoretically demanded by, or actually beneficial to, the crops.

The means for raising water for irrigation and farm use from subsurface water strata and sluggish streams is a serious matter in all the arid regions; but is amenable to the efficiency of modern methods. The windmill stands preëminent for individual farm use and is the cheapest and most effective power for constant work.

For the purpose of irrigation alone the windmill is of the greatest advantage to the agricultural interests of the United States, and even in our Eastern States, where irrigation has been heretofore almost totally neglected, it has been found by trials that by the use of a windmill with a small storage capacity for water to meet contingencies the increase in a garden or small-fruit crop alone will amply pay the interest on the plant, and in seasons of severe drought the saving will pay for the plant. These are serious matters for consideration and for the success of our gardeners and small-fruit raisers.

The average velocity of the wind in a large portion of the United States, and for the lowest force that will do effective work with a windmill, is 8 miles per hour for from 5,000 to 6,000 hours in a year, and an average of 16 miles per hour may be expected for 3,000 hours per year; so that for a power that does not require daily attention and can be utilized for twenty-four hours of the day, it is the cheapest for all uses within its sphere of action. For pumping water for storage for all uses, there is no more economical prime mover. In the larger sizes, of 50 and 60 feet diameter, wind-power is doing excellent work in our Western States for milling, and in all sizes is largely extending its usefulness in irrigation. The following table gives the sizes of windmills in common use, their power and capacity for pumping water with an average of a 16-mile wind for eight hours per day:

TABLE XXV.—THE WINDMILL AND ITS WORK.

Diameter of mill, feet.	Horse-power from shaft.	Horse-power in water pumped.	Gallons of water 15 feet high per hour.	Irrigation in acres, column 4.	Gallons of water 25 feet high per hour.	Irrigation in acres, column 6.
8½	0.09	0.04	616	0.18	370	0.10
10	0.16	0.12	1,918	0.57	1,151	0.339
12	0.25	0.21	3,420	1.02	2,036	0.60
14	0.40	0.28	4,530	1.37	2,708	0.798
16	0.50	0.41	6,460	1.84	3,876	1.142
18	0.70	0.61	9,768	2.83	5,861	1.727
20	1.	0.79	12,465	3.65	7,479	2.20
25	1.50	1.34	21,233	6.27	12,743	3.75
30	3.	2.25	31,660	12.88	19,000	7.61

For gardens and small-fruit orchards, the smaller sized windmills are much in use, and a 30-foot mill of the model shown in Fig. 195 will lift enough water 25 feet high to irrigate 7½ acres every day with a 16-mile wind for 8 hours, or nearly ¾ inch in depth for each day, and if a reservoir can be made available, a much larger acreage

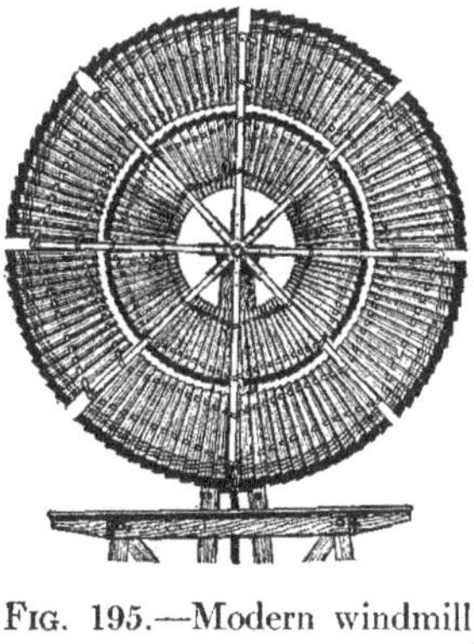

FIG. 195.—Modern windmill.

FIG. 196.—Prairie windmill.

may be covered, and the duplication of windmills may be made to any extent.

Fig. 195 shows the wheel plan of the most efficient windmill now in use; it consists of two series of concentric blades fastened to the purlines of a braced radial frame. The blades are fixed at an angle of about 35° to the plane of the wheel. A peculiarly constructed mechanism turns the wheel edgewise to the wind to stop it, or to regulate its position in a high wind. Fig. 196 shows the prairie windmill, called in Kansas the "Jumbo Mill," generally made with 6 arms on an axle placed in a north and south position and lower half covered with a box to shield the lower paddles from the wind. A simple crank connection to a pump supplies sufficient water to irrigate 6 acres of garden land.

In many of the arid districts the home-made windmills of various designs are in view and doing good work in providing water storage for the dry season. Where large areas need irrigation, there is no more prompt and satisfactory power than the gasoline- or oil-engine, which is now made with pumping attachment for any volume required.

Along the river bottoms where a current of 4 feet per second can be made available, the current wheel can and is found a cheap and useful means for a continual supply of water for storage and irrigation; see Chapter I.

Where dams or barrages can be made, if not more than 4 feet in height, a most efficient service can be obtained from the turbine wheel and pump. They are made to operate well with 2 feet head, and in Europe are running submerged.

In Fig. 198 is illustrated a view of an irrigation works with a low head turbine wheel at Columbus, Neb. The water-ram is also much in use and its first instalment costs much less than other devices for irrigation wherever a fall can be obtained of from 3 to 10 feet; see Chapter VI on Water-Rams.

AMOUNT OF WATER REQUIRED FOR IRRIGATION

The quantity of water required for raising crops varies according to the character of the soil. The plants themselves need a certain minimum supply, but a far larger quantity is required to saturate the surrounding soil to such a degree that the vitalizing processes can continue. The soil is constantly losing water by evaporation and by

FIG. 197.—Current irrigation wheel, Elgin, Utah.

FIG. 198.—Irrigation by a turbine wheel, Columbus, Neb.

seepage, so that the amount which the plant takes from it is relatively small. Among the most important investigations on this subject are those by Prof. F. H. King, of Madison, Wis., who has found by direct measurement that from 300 to 500 pounds of water are required for each pound of dry matter produced. In other words, for each ton of hay raised upon an acre, from 300 to 500 tons of water must be furnished either by rainfall or by artificial means.

Water, covering an acre of land to the depth of 1 inch, weighs about 113 tons, and to produce 1 ton of hay the depth of water required is approximately from 3 to 5 inches. It is necessary to furnish at least this amount, and sometimes several times as much, in order to produce a crop. The actual amount required to produce 5 tons of barley hay is about 20 inches in depth.

When the ground is first irrigated an enormous quantity of water is sometimes required to saturate the subsoil. The quantity of water turned upon the surface during the first year or two has frequently been sufficient to cover the ground to a depth of 10 to 20 feet, and in some cases an amount equal to a depth of 5 feet or more per annum has been thus employed for several years. Gradually, however, the dry soil is filled and the water table is raised nearer the surface.

The pioneers in irrigation sometimes use excessive quantities of water, often to their disadvantage. They are actuated in part by the consideration that, having paid for the water, they are entitled to a certain quantity, and if they do not take it their claim to a perpetual right may be disputed. Equity demands that flowing water shall be considered as a common stock or fund, the right to the use of which shall be regulated, and beneficial use shall be the measure and the limit of such right.

The quantity of water used in irrigation is usually stated in one of two ways. First, in terms of depth of water on the surface, and, second, in quantities of flowing water through the irrigating season. The former method is preferable, since it is susceptible of more definite examination and is also more convenient for comparison with figures for rainfall which are given in inches of depth. In the humid regions the rainfall is usually from 3 to 4 inches per month during the crop season. In the arid region, where the sunlight is more continuous, and the evaporation greater, there should be for the ordinary crops at least enough water during the growing season to cover the ground from 4 to 6 inches in depth each month. Carefully tilled orchards have been maintained on far less.

The second method of stating the quantities necessary for irrigation is of convenience when considering a stream upon which there is no storage. It is estimated that 1 cubic foot per second, or 1 second-foot, flowing through an irrigating season of ninety days, will irrigate 100 acres. One second-foot will cover an acre nearly 2 feet deep during twenty-four hours, and in ninety days it will cover 180 acres 1 foot deep, or 100 acres to a depth of 1.8 feet, or 21.6 inches. This is equivalent to a depth of water of a little over 7 inches per month, during the season of ninety days.

It is instructive, in this connection, to know what is the least amount of water which has been used with success. To learn this, it is necessary to refer to southern California. Successive years of deficient rainfall in that State from 1897 to 1900 served to prove that with careful cultivation, crops, orchards, and vineyards can be maintained by using very small quantities of water. In some cases an amount not exceeding 6 inches in depth was applied during the year, this being conducted directly to the plants, and the ground kept carefully tilled and free from weeds.

In the measurement of water for irrigation in this country there has never been a unit common to all sections. The "miner's inch," or the quantity discharged through an opening 1 inch square under a given head, has generally been used, but with different heads so that the quantity of water discharged has been variable. According to the figures given below, the miner's inch in California is taken to equal $\frac{1}{50}$ cubic foot per second, or 1,728 cubic feet per day, and in Colorado a little less than $\frac{1}{40}$ cubic foot per second, or 2,160 cubic feet per day. Recently the "acre-foot" seems to be coming into favor, especially for stating the capacities of storage reservoirs. This, as is evident from the name, is the amount of water necessary to cover an acre of ground to a depth of 1 foot or 43,560 cubic feet.

As estimated by various water companies in southern California, 1 miner's inch of water will irrigate from 5 to 10 acres. The miner's inch in this connection is defined as a quantity equaling 12,960 gallons in twenty-four hours, or almost exactly 0.02 second-foot, this being the amount delivered under a 4-inch head, measured from the centre of the opening. Under this assumption 1 second-foot should irrigate from 250 to 500 acres. This is on the basis of delivering the water in pipes or cemented channels in the immediate vicinity of the trees or vines to be irrigated. If it be assumed that 1 miner's inch is allowed for 10 acres, or 1 second-foot for 500 acres, this quantity of water flowing from May to October, inclusive, will cover the ground to a depth of a little over $\frac{7}{10}$ of a foot, or 8.8 inches. This quantity, with the care and cultivation usually bestowed, has been found to be sufficient.

1 Cubic Foot per Second Equals:

2 acre-feet in 24 hours.	7.5 gallons per second.
60 acre-feet in 30 days.	449 gallons per minute.
180 acre-feet in 3 months.	50 California inches.
720 acre-feet in 1 year.	38.4 Colorado inches.

100 California Inches Equaes:

4 acre-feet in 24 hours.	15 gallons per second.
1 acre-foot in 6 hours.	900 gallons per minute.
120 acre-feet in 30 days.	77 Colorado inches.
360 acre-feet in 3 months.	2 cubic feet per second.
1,440 acre-feet in 1 year.	

100 Colorado Inches Equals:

$5\frac{1}{6}$ acre-feet in 24 hours.	19.5 gallons per second.
1 acre-foot in 4.2 hours.	1,170 gallons per minute.
155 acre-feet in 1 month.	2.6 cubic feet per second.
465 acre-feet in 3 months.	130 California inches.
1,860 acre-feet in 1 year.	

All figures where miner's inches are used must be regarded as of doubtful value for comparative purposes unless a full definition of the miner's inch used is given. The character of the orifice, as well as the head in inches, needs to be known.

The method of applying water largely governs the amount used. In the case of alfalfa, flooding is practised; with small grains the water is run in furrows; while in the case of orchards the water is sometimes applied directly to each tree. In this case a little earth basin about 6 feet or more across and 6 inches deep is formed around each tree and partially filled with water. The better way, however, is that of running water in furrows, four or five of these being ploughed between each two rows of trees. The water is applied very slowly, several days being spent in the process, and when dry the ground is thoroughly cultivated.

CANALS AND DITCHES

The shape of the cross-section of a canal depends largely upon the character of the surface soil. In light or sandy soil, where the earth is easily eroded, very gentle side slopes are given, while in harder materials the side slopes can be steeper.

When the fall of the canal is so great that it is impracticable to allow the water to flow freely down the slope, devices known as drops are introduced. These consist of an arrangement whereby the water can drop to a lower level without injury to the canal. The force of falling water is very great and rapidly digs out earth and gravel, so that wherever possible the drops are made with the fall upon solid rock. Such cases are rare, and to take up the force of the water it is usual to provide what is known as water cushions. The stream falls into a pool of sufficient size and depth for the eddying or boiling of the water to take up and dissipate the erosive effect.

These drops are usually built of planks, with a sharp overfall edge, and a low dam or obstruction below the fall in order to maintain the pool. Occasionally they are made in the form of an incline, with a

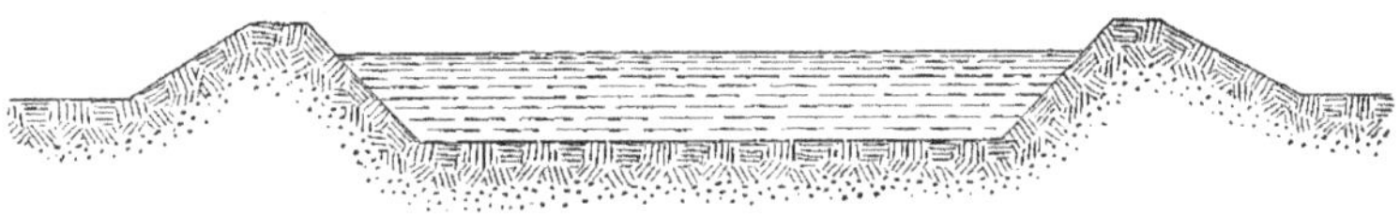

FIG. 190.—The broad ditch on the plain.

pocket at the bottom to break the force of the falling water. These drops are expensive to build and difficult to maintain, because of the rapidity with which the timbers decay and the wearing action of the water, which constantly tends to cut exposed portions.

The earlier canals, being usually of small size, were built with heavy grades. When the canals have been enlarged, the increased volume has attained excessive velocity, and thus it has been necessary to introduce many of these drops.

In localities where frosts do not occur to any considerable extent, and where water has greatest value, experience has shown that it is desirable to line the ditches and canals with concrete or cement, reducing the loss by percolation, and making the channel so smooth that the water moves rapidly even on slight grades. Often it is possible to trim the banks of the ditches to a uniform surface, and this is found to be sufficiently firm to serve as a foundation upon which to put a layer of cement mixed with sand and having a thickness of from $\frac{3}{4}$ of an inch to $1\frac{1}{2}$ inches. Where the bed and banks are not firm, it is necessary to pave or revet them with small stone, and then place upon this a coat of concrete made of small gravel and sand. The economy of water resulting from this careful construction has been found to be sufficiently large to justify a considerable outlay.

The pioneer irrigators in planning a ditch use a straight-edge or board a rod long (16.5 feet), on one end of which is a block projecting half an inch or an inch. When this board is placed horizontally, the lower projecting point will thus indicate a fall of half an inch or an inch to the rod. By this means points are determined at intervals of a rod where stakes may be driven into the ground, marking out the course of the ditch upon a slightly ascending or descending grade, according as the work is begun from the lower or upper end.

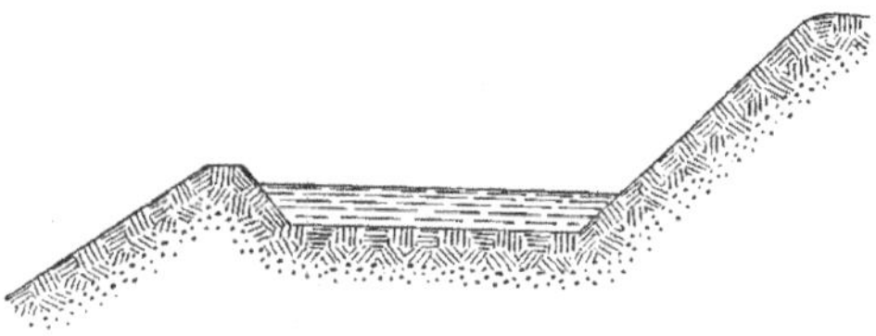

FIG. 200.—Ditch on the hill-side.

The ditch having been staked out in this manner or by means of surveying instruments, a furrow is ploughed along the course and the earth thrown out by shovels or scrapers. Rocks may be blasted away and depressions filled, or crossed by means of wooden flumes. As far as practicable, however, ditches are carried into and around gulches or depressions in the surface of the ground in order to avoid building these wooden structures, since they decay rapidly and are sources of considerable expense.

A considerable slope can be used for small ditches, since the volume of water is not sufficiently great to move the large particles of sand and gravel. For example, on the farm lateral, carrying 1 or 2 second-feet, a fall of 50 feet or more to the mile may not be excessive, the velocity being retarded by the relatively great friction. On the other extreme, a large irrigation canal carrying 1,000 second-feet may be in danger of injury if a grade of much over 6 inches to the mile is given it.

As a general rule, it may be said that conduits of this character built in common earth should be so proportioned as to have an average velocity of a little less than 3 feet per second, or 2 miles per hour, when carrying their full capacity. It is necessary, therefore, to take into consideration the amount of water to be carried and from this deduce the size and shape of the cross-section of the canal or ditch in order to obtain the desired velocity. Many of the older irrigation works have been given an excessive grade through fear on the part of the builders of getting too little fall. Some of these grades are as much as 50 feet to the mile, giving a velocity to the water of 5 feet per

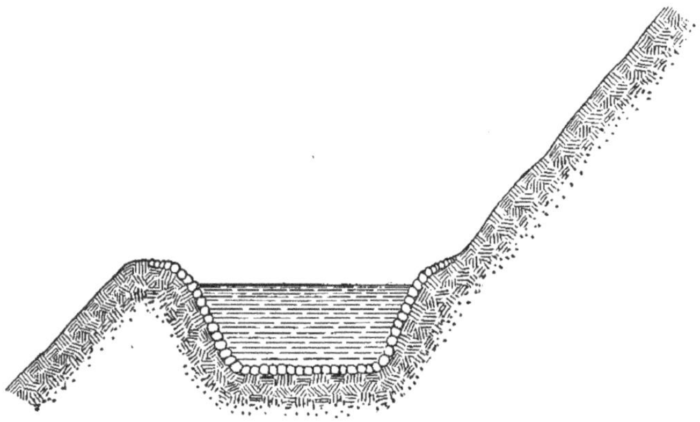

FIG. 201.—Stone-lined ditch on mountain-side.

second, washing the bed of the channel and leaving only a mass of cobbles. The seepage through this material, even though the water is flowing rapidly, has been known in one instance to be over 20 per cent. of the total flow in a course of 4 miles.

FLUMES AND WOODEN PIPES

It is necessary in the construction of nearly every ditch or canal to take water across a depression at some point in its course. This is usually done by means of a flume or long box, usually rectangular and supported above the ground by a frame or trestle of timber or iron. Such flumes are often used across rocky ground where it is im-

Fig. 202.—Wooden flume at Little Rock Creek, Antelope Valley, Cal.

Fig. 204.—Old flume and new stave-pipe replacing it, Redlands, Cal.

practicable to dig a ditch. This is particularly the case near the head, where the water, after being taken from the river, is often carried through a narrow, steep-walled canyon. Here the foundation for a flume is prepared along the rocky cliffs, supports being devised to suit the inequalities of the ground.

A better, though more expensive, type of flume is that having a semicircular section. These flumes are built of narrow planks or staves laid side by side and held in place by iron bands run around the flume, joined by nuts and threads by which the bands can be drawn up and the staves brought together. In crossing very deep depressions it is necessary to have a correspondingly high trestle in order to carry the flume across on grade. Such high trestles are expensive and liable to destruction from storms. In their place there have been built inverted siphons, or wooden stave-pipes. These pipes are similar in construction to the semicircular frame of narrow plank, carefully planed to a given dimension, and held in place by circular iron bands or hoops.

TABLE XXVI.—SIZES OF TRAPEZOIDAL DITCHES WITH SIDE SLOPES 1 TO 1; MEAN VELOCITIES, GRADES, AND DISCHARGE BY KUTTER'S FORMULA.

Bed, width in feet.	Depth in feet.	Side slopes.	Grade per mile.	Computed mean velocity in feet per second.	Computed discharge in cubic feet per second.
100	6.5	1 to 1	10 inches	2.50	1,730.6
80	5.5	1 to 1	13 inches	2.50	1,175.6
60	5.0	1 to 1	15 inches	2.48	806.0
40	4.5	1 to 1	19 inches	2.52	504.6
20	4.0	1 to 1	2 feet	2.43	233.3
10	3.0	1 to 1	4 feet	2.64	103.0
6	2.5	1 to 1	5 feet	2.45	52.0
4.5	4.75	1 to 1	14.5 inches	2.25	100.0
2.5	4.0	1 to 1	15 inches	2.00	50.0

PLANNING AN IRRIGATION SYSTEM

In laying out an irrigation system it is usual to begin at the highest point to be irrigated and to run a trial line to the water surface across lands where ditch construction is most practicable, and on a slightly ascending grade, a foot, more or less, to a mile. Rocky obstructions and depressions are avoided as far as possible by detours in the route.

In planning the larger canals and irrigating systems bringing water by gravity, it is necessary to consider carefully the slopes to be given the conduits. This is especially true where a broad valley is to be irrigated from a stream whose upper course is only a few feet above the general level of the land. If the grade is steep it will be necessary either to lengthen the canal or to take water only to the lower land, leaving the higher portions of the valley dry. If, on the other hand, a very gentle grade is given, the water will flow slowly and a very large canal must be built to carry the necessary volume. Of importance equal to the relative height of the source of the water

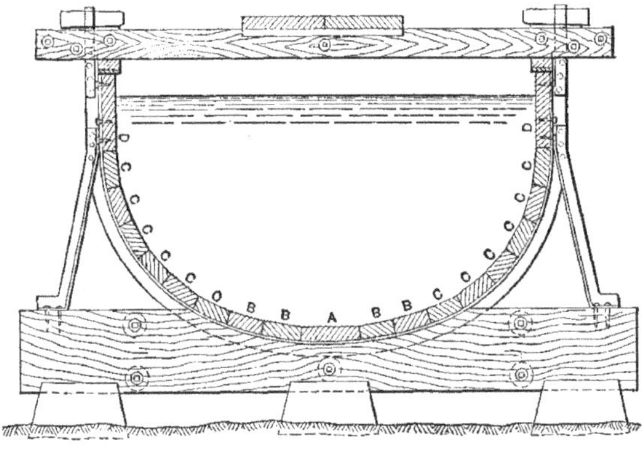

FIG. 203.—Section of U-shaped wooden flume.

and the land to be irrigated, are the effects of the slope of the canal upon the velocity of the water and the consequent cutting or filling of its channel. With a steep grade, the water moves with such rapidity as to pick up and carry along fine particles, and with increasing velocity larger and larger grains of sand or pebbles are moved, eroding the bottom and producing bars or obstructions at points of its greatest width.

There is in irrigation a benefit of great value beyond the mere moisture. The fine silt brought in suspension by the water is deposited upon the surface of the soil, making a light, rich covering, which protects the young plant and supplies it with food. This sediment is of greatest value when the source of the stream which carries it is from forested areas with decayed vegetation; on the other hand,

FIG. 205.—Diversion dam and headgate, Carson River.

FIG. 206.—Concrete drop weir for canal.

a stream from the bare mountain slopes may bring fine quartz grains of little value to the land. The artificial control of silt is a reproduction on a small scale of the method by which nature has made rich farm lands. The rivers have brought the fine material from the higher slopes and spread it out, filling the old lake bottoms or covering the flood plains of the rivers.

Thus it is, when sufficient water to saturate the soil is regularly applied to the land, its productive qualities are maintained in many cases without impairment for an indefinite period, and soil, apparently poor and worthless, produces crops which compare favorably with those grown on the richest soil of the humid region.

THE WORK OF THE UNITED STATES RECLAMATION SERVICE—THE NEVADA PROJECT

The territory included in the Truckee-Carson Irrigation Project is situated in western Nevada. The major portion of the water for this project is derived from the drainage basin of the Truckee River, while the lands to be reclaimed are mostly within the basin of the Carson River, so that the project involves a union of the two. The principal lands to be reclaimed lie in Churchill and Lyon counties. The territory is traversed by the Ogden route of the Southern Pacific Railroad, and also by a branch running from Hazen to Tonopah.

The average annual rainfall over most of this territory is 5 to 6 inches. The climate is mild, and the proportion of sunshiny days is high. Conditions are favorable for the raising by irrigation of the ordinary crops of the temperate zone.

This project, in its ultimate plan, will involve the reclamation of about 350,000 acres of land, at an estimated cost of $9,000,000. The more immediate object is the reclaiming of about 200,000 acres in the Carson Sink Valley, reaching from the low divide between Wadsworth and Hazen, southeasterly to Carson Lake and easterly to Stillwater. The town of Fallon is near the centre of the area that is being reclaimed.

The irrigation system so far constructed consists essentially of the Truckee Canal, for diverting the waters of the Truckee River from its lower canyon over the low divide to the Carson Basin and thence to the Carson River; the diversion of the united waters of the

Truckee and Carson from the Carson River by the main distributing canals on to the principal bodies of land in Carson Sink Valley; the lateral distributing system, for the further distributing of water to the lands to be reclaimed.

The canal is 30.9 miles long, divided into three divisions at the 6- and 13-mile points. The divisions are numbered 1, 2, and 3 from the head. An important part of Division 1 is the diversion dam and headworks. There are several expensive structures on Divisions 1 and 2, and also three tunnels on the latter. Division 3 is entirely in open country.

Following the principle adopted by the Reclamation Service, by which constructed works must be of the most permanent character, all the structures are of Portland cement, concrete, and iron or steel, no wood being used except for minor parts, as for flashboards.

The headworks consist essentially of a diversion dam across the river, and of headgates at the entrance to the canal. The diversion point is in the canyon about 10 miles westerly and upstream from Wadsworth. The elevation is about 4,200 feet.

The canyon has sides more or less rocky and steep, but the bottom is of some width. It is traversed by the Southern Pacific Railroad (Ogden route), the reconstructed line of which runs quite close to the canal in places.

The operation of the gates is as follows: The lower gate is raised directly by the screw stem. It slides up the front of the upper gate until it reaches its limit ($4\frac{3}{4}$ feet), when it strikes the horizontal flange projecting from the top of the upper gate, when the further movement raises both gates. The flashboards can be pulled out at the top, one by one, as the double gate is raised. The total height of each opening is 15 feet, and the 16 openings give a total water-way of 1,200 square feet. The greatest observed discharge of the river is about 9,000 second-feet.

The 16 gate openings are each 5 feet in the clear between the piers. Each gate is virtually composed of three sections: (1) the lower gate, used alone for the ordinary regulation of water; (2) the upper gate, called into action only when a large stream is required to pass; (3) the flashboards, which may be removed for passage of extreme high water. The lower gate is of cast iron $1\frac{1}{4}$ inches thick, strengthened by radial ribs, and has lugs cast on the back for receiving the screw

Fig. 207.—Headgates, dam, and canal sluices, Truckee River, Nev.

Fig. 208.—Top of the dam, gate-lifting screws.

stem. The upper gate is of cast iron 1 inch thick, reënforced by rectangular ribs. The two gates together give a total clear opening 10 feet in height.

An interesting work now in progress is the Minidoka irrigation project of the United States Reclamation Service, in the south central part of Idaho. It embraces all the details of a general distribution from a dam across the Snake River with a long spillway, controlling and headworks. Two long irrigation canals, one from each end of the dam.

METHODS OF IRRIGATION

The methods of irrigation practised in various parts of the United States differ with the climatic conditions and soil, and especially with the early habits or training of the irrigators. The methods of conserving and applying water have been improved under the stimulus of modern invention, although there has been little if any scientific or well-considered information available.

Water is applied to the irrigated field in three ways: by flooding, by furrows, and by subirrigation.

Flooding.—Flooding is done by the check system and by wild flooding. By the latter process the irrigator turns the water from a ditch over a level field and completely submerges it. Perfectly level fields are, however, comparatively rare, and the first step in primitive agriculture by irrigation has been to build a low ridge around two or three sides of a slightly sloping field, so that the water is held in ponds. These low banks are commonly known as levees or checks. In construction they are frequently laid out at right angles, dividing the land into a number of compartments. Water is turned from a ditch into the highest of these compartments, and when the ground is flooded the bank of the lower side is cut or a small sluiceway opened, and the water passes into the next field, and so on until each in turn is watered.

This flooding in rectangular checks is practised most largely by the Chinese gardeners and by the Mexicans living along the Rio Grande. The banks are thrown up by spade or shovel, and the ground between the banks is tilled with a heavy spade or mattock.

Water, when had in abundance, is turned into these checks, and the quantities used are often extraordinarily large.

Many of the early settlers in the Southwest imitated the Mexicans, or employed them as laborers, building checks upon the same general plan, but usually enclosing more ground. Fields from 1 up to 20 acres or more in area have been levelled and surrounded by low levees of from 1 to 2 feet in height and 5 to 10 feet in width. These are made relatively wide at the bottom, in order that the slopes may be gentle, so that mowing-machines can be driven over them. Rectangular fields are connected by gates set in the levees, so that water can be turned from one field into the other without cutting the banks

A better method of procedure with these beds is to let the water flow through the upper one until the lowest is covered to a depth of

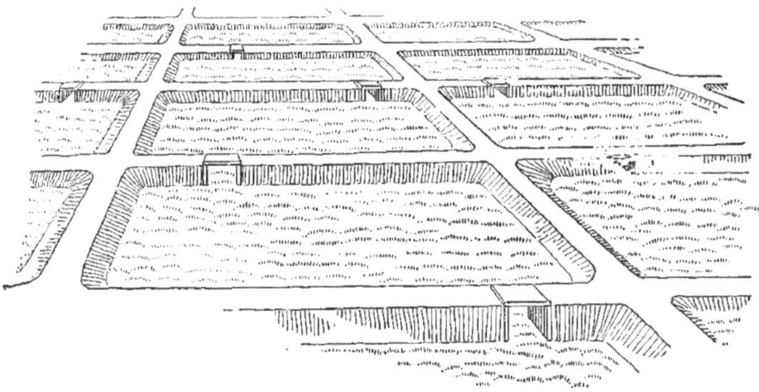

FIG. 211.—Flooding or block system of irrigation.

about 3 inches, then obstruct the opening to this bed and permit the water to accumulate in the next square above, and so on, filling each in succession from the lowest to the highest and allowing the water to soak away. It is claimed that by following this course the land receives water more uniformly.

For crops such as tomatoes, sweet potatoes, and chili, which are the most important foods of the Mexicans, ridges are made in the beds in such a form that the water is compelled to flow around and along these until the bed is filled nearly to the top of the ridges. Then it is let into the next bed and the operation is repeated. Instead of turning the water from one bed into another, it is customary to provide lateral ditches in such form that the water can flow into each compartment without passing through the other. In this way wash-

FIG. 209.—Deep cut, Truckee-Carson Canal.

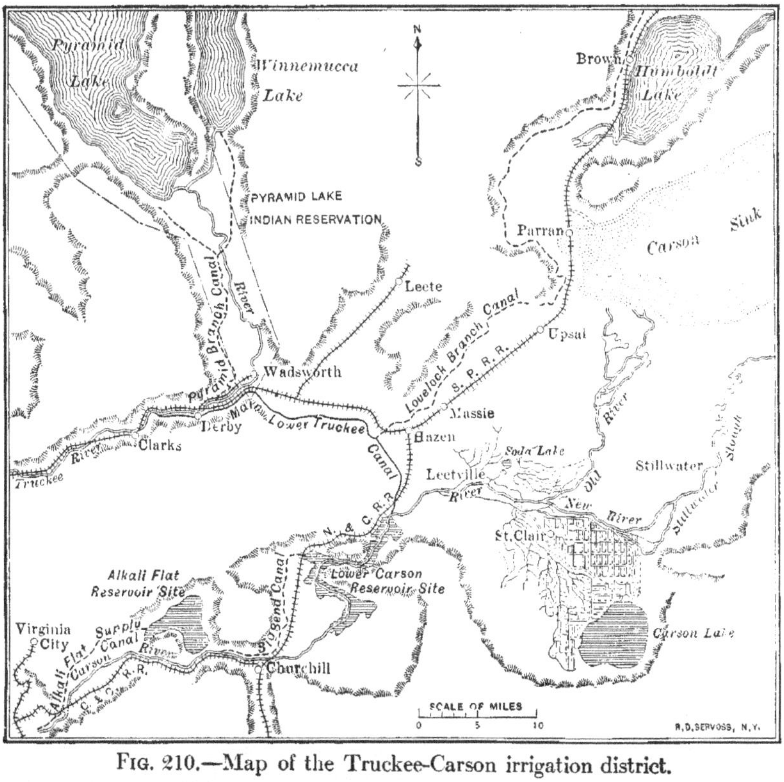

FIG. 210.—Map of the Truckee-Carson irrigation district.

ing of the soil is prevented and the amount of water can be regulated to suit the needs of each variety of crop.

On land that is nearly level, small inequalities are smoothed off by plough and scraper, or by dragging a heavy iron beam across the field, pulling the hummocks into the hollows.

WATERING BY FURROWS

The furrows are ploughed in such a direction that the water when turned into them from the lateral ditches will flow freely down them without washing away the soil.

When the water has completely filled the furrows and has reached the lowest points, the little streams are cut off and turned into another

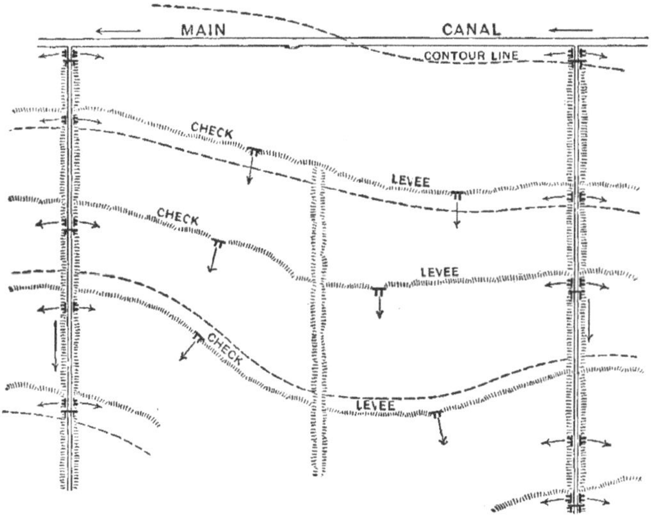

FIG. 212.—Furrow system by check levees.

set of furrows. The methods of doing this differ in various parts of the country. Sometimes the irrigator simply cuts the bank of the distributing ditch with a shovel, and then closes the opening after sufficient water has escaped. A more systematic method commonly employed in California is as follows: Water is carried to the furrows in a small box flume with openings in the side. These openings are closed by little shutters and a number can be opened at once, permitting a certain quantity of water to escape into each furrow.

The slope given the furrows determines to a certain extent the amount of water received by the soil. If the fall is very gentle, the water moves slowly and a large portion is absorbed while the furrow is being filled. If steep, the water quickly passes to a lower end and the ground does not absorb so much.

When the entire field has been watered and the surface has become sufficiently dry for cultivation, the furrows are usually ploughed out and a thin layer of the top soil stirred to make an open, porous covering or mulch, preventing excessive evaporation and allowing the air to enter the ground. Without such cultivation a hard crust may be formed, which, although retarding circulation, permits continual evaporation. The loosening of this crust breaks the capillary connection with the moisture beneath and thus lessens the loss of water.

For irrigating small grain, such as wheat, the ordinary plough furrows are not used, excepting in the construction of the distributing laterals. The fields, brought to a uniform surface, are thoroughly cultivated, and after the grain has been sown, small parallel lines are made similar to furrows, but smaller and nearer together. These tiny channels are made either by a peculiar drag or by a roller upon which are projections so arranged as to make small grooves in the soil. These are made in the direction of the desired slope, so that the water can flow down the marks through the grain as it would in furrows through a cornfield. The rapidly growing grain shades the surface and prevents the formation of crust, rendering subsequent cultivation unnecessary.

In order to cause the water to spread from the lateral ditches into the furrows through the ground, the tappoon has been applied. This consists usually of a small sheet of metal of such shape as to fit across the ditch. This can be forced into the soft earth, making a small dam and causing the water to back up and overflow the field of grain. Sometimes a canvas dam is used. This consists of a piece of stout cloth, one edge of which is tacked to a stick long enough to extend across the lateral ditch or furrow. The canvas falling into the furrow fits the sides and bottom, and is held in place by throwing a clod of dirt upon it. Water meeting the obstruction still further forces the canvas down, making a fairly tight dam, against which it accumulates and overflows into the field. After sufficient water has

FIG. 213.—Irrigation on terraced slope.

been turned out the canvas dam is pulled up and carried farther down the ditch, again placed in it, and another section of the field is irrigated.

Furrow irrigation is usually employed in watering trees and vines. In some localities, however, basin or pool irrigation is practised. It is frequently the case that where the best orchards and vineyards are situated water is procured with the greatest difficulty and its value is such that large expenditures are incurred in guarding it carefully from loss. The supply is conducted often in cement-lined ditches and by wooden flumes as near as possible to the trees and vines, and is then turned out into the furrows ploughed around or near the trees. The water issuing from small apertures in the side of the wooden box falls into the furrows and is immediately conducted to the vicinity of the trees.

Care is usually taken that the water shall not actually touch the tree trunks, and it is extended far enough to wet the ground about the extremities of the roots to encourage these to spread outward as far as possible. After the water has traversed the furrows to the lower end of the orchard the supply is cut off and the ground is tilled as soon as the surface dries sufficiently.

The cost of levelling is usually very great, and it is only for the most valuable crops and orchards that this is done. Where the undulations are of such extent that they cannot be removed by this method, it is necessary, in order to practise check flooding, to adjust the shape of the banks or levees to suit these conditions. Instead of making them rectangular, the levees are built along the slopes to fit the contour of the surface. The canal brings water to the upper side of the field and follows along on gentle grade. Below this, at such distance that a bank which is a foot or two in height will pond the water back to the side of the canal, a ridge is built. The distance of this ridge from the canal will depend, of course, upon the slope of the ground. If very gentle, the bank or levee can be 100 feet or more away, while with steeper slopes it must be nearer.

In the irrigation of grass land, clover, alfalfa, and similar forage plants it is not feasible to level the ground and build checks. Water must be applied by some form of flooding. It is conducted to the upper part of the field and there turned loose in such a way as to cover the surface with a thin layer. Much care is required to do this, far more than when checks or furrows have been made. To get the water to the right places it is usual to provide through the fields shallow depressions which serve to guide the water. From these it spreads out in thin sheets.

The irrigator takes advantage of all of the smaller ridges or inequalities, running the water out upon these and not allowing it to escape into the depressions until it has thoroughly wet the surface. Not all of the water will soak into the ground, and the amount in excess which collects in the depressions is again conducted along contours to the next lower series of ridges. The streams of water are distributed, gradually vanishing into the grass land or cultivated field. A portion of the stream reappears in the low places; these streams, when they attain considerable size, are gradually conducted out and used in lower portions of the field.

SUBSURFACE IRRIGATION

Attempts have been made to conduct the water beneath the surface immediately to the roots of the trees, thus preventing waste by evaporation from the surface of the ground. Few devices have been successful, owing to the fact that the roots of the trees rapidly seek and enter the openings from which the water issues, or, surrounding the pipe by a dense network, cut off the supply. Porous clay tiling has been laid through orchards, and also iron pipes perforated so as to furnish a supply of water along their length. A machine has been invented and successfully used for making cement pipe in place. Small trenches are dug through the orchard between the trees and the pipe-making machine deposits the material in the trenches, which are filled with earth as soon as the cement is set. Water is thus distributed underground where needed.

In a number of orchards where subsurface irrigation has been unsuccessful because of roots stopping up minute openings beneath the surface, the system has been reconstructed and water has been brought to the surface at or near each tree by means of small hydrants. Vertical pipes are placed at short intervals leading to the level of the ground, and in these are small gates so arranged that the flow can be cut off in the buried pipe. Pushing down one of these gates the

FIG. 214.—An irrigated farm.

water rises and overflows the surface until a sufficient amount has been obtained. This gate is then raised and the next pushed down, and so on, until water has been caused to overflow at each point in succession down the slope of the ground.

For annual or root crops subirrigation has been successfully practised by the use of small iron pipes partly open at the bottom, allowing a small amount of water to escape. These pipes are laid at 12 inches or more beneath the surface, and are connected with lines of tile or clay pipes leading from the reservoir or source of supply. As the crops are removed each year, and the ground cultivated, the roots have no opportunity to stop up the pipes.

The term subirrigation is occasionally applied to conditions occurring in nature where water percolates freely beneath the ground for a considerable distance sufficiently near the surface to supply the need of crops. The ground is not actually saturated, but moisture is transmitted in sufficient quantity to nourish plants without drowning or waterlogging the soil. These subirrigated areas, so called, are often located in broad valleys along a stream from which the water finds it way outward beneath the surface. They are occasionally found also upon gentle slopes where the moisture tends to form springs near the edge of the valley.

Where the subsoil transmits water freely, irrigation ditches may subirrigate large tracts of country without rendering them marshy. Thus farms may obtain an ample supply of water from ditches a half mile or more away without the necessity of distributing small streams over the surface. In the San Joaquin Valley of California, vineyards in certain localities are thus maintained in good condition, although water has not been visibly applied for many years. The closing of the ditches would, however, result in drying up the ground, and this obliges the farmers who are benefited by subirrigation to pay their share of the cost of maintaining the ditches, although they do not receive water directly.

It occasionally happens that the lower part of a subirrigated field must be drained to remove the excess of water. This can be done either by gravity ditches or by pumping devices.

CHAPTER XIII

WATER-POWER—WATER-WHEELS

THE conservation of water-power by the storage of a 24-hour run of a stream for a day's industrial use has become one of the most economical features in the use of nature's power. The reservation of the night flow may be made to increase the gross power that may be obtainable from the day flow alone by two and a half times, which not only applies to small streams, where the whole power is used by an individual industry, but is also applicable to large streams serving a multitude of individual industries.

There are vast areas in the United States, subject to a scant and seasonal rainfall, where the construction of great dams in the mountain gorges and across the rivers of the great plains for storage reservoirs of power and for water-supply of towns and for irrigation are of vast value in the development of agriculture and the industries of our arid territory.

The dry canyons may be made the catch basins of the cloudbursts and thus contribute to the necessities of the mining communities by a prolonged supply of their most essential element.

The potential power of a reservoir is equal to the ratio of the hours of filling divided by the hours of running a mill-wheel, in addition to the power of the stream flow. Thus a mill day of 10 hours may have a storage flow of 14 hours, which may be made to give a total power of $1+\frac{14}{10}=2\frac{4}{10}$ of the power of the flowing stream, In this manner the water-power of small streams may be made valuable properties, when their natural flow is not sufficient for mill requirement.

The loss of head by drawing from the reservoir will depend upon the ratio of the head to the drop in the water level in the reservoir, and this again will be in proportion to the area of its surface.

For example, a small stream flowing at the rate of $15\frac{1}{2}$ cubic feet per second with a possible 10 feet wheel-head, is equal to $12\frac{1}{2}$

gross horse-power, and $12\frac{1}{2} \times 2.4 = 30$ horse-power; then, as the flow during 10 hours is 400,000 cubic feet, the storage for 14 hours will require 560,000 cubic feet capacity in the reservoir within the available limit of depth. If the limit may be made at 2 feet, then $\frac{560000}{2} =$ 280,000 square feet will be the surface area, and if 100 feet in width can be obtained, $\frac{280000}{100} = 2{,}800$ feet in length, or any width and length that will give the required area.

For the utilization of water-power, one of the earliest and cheapest as well as the least efficient of the many devices, is the current wheel illustrated in Fig. 215. The most efficient velocity of the wheel

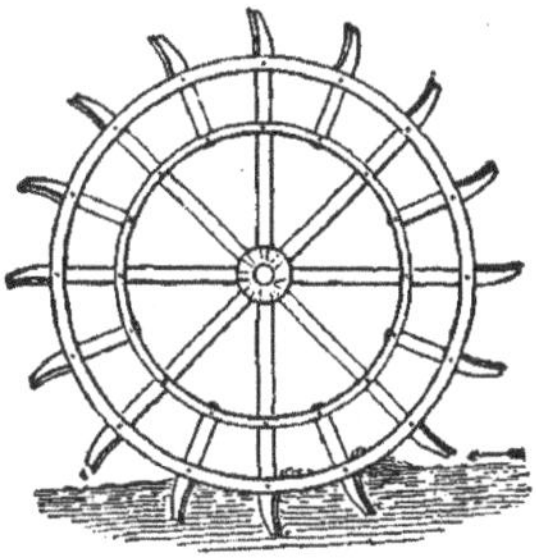

FIG. 215.—Current wheel.

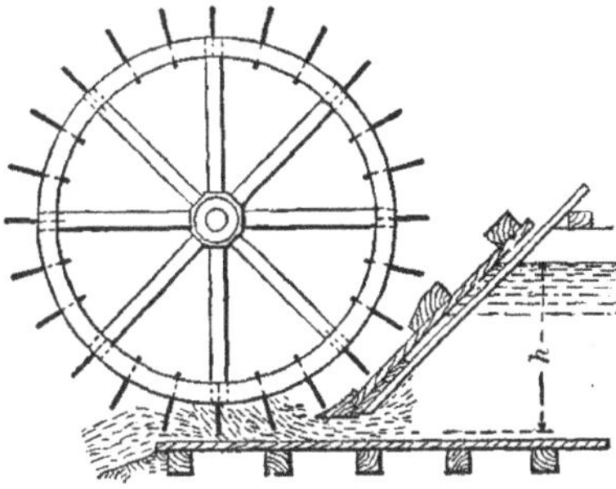

FIG. 216.—Undershot wheel.

periphery is 40 per cent. of the current velocity. The stream velocity may be increased under the wheel by a wing dam and in this way some very useful power plants have been made for milling and pumping. Their horse-power is:

$$\frac{\text{Area of immersion of blades}}{150} \times (V - S)^2$$

in which V = velocity of the stream; S = velocity of the periphery of wheel, both in feet per second; area of blades in square feet. The undershot wheel, Fig. 209, is mostly used where small water-heads are available from wing dams or races from a local rapid. In this way a head may be obtained from 1 to 4 feet, that can be cheaply turned into power.

The flood nature of the stream should be taken into account in fixing upon the size of the wheel. If the flood rise does not exceed

3 or 4 feet, an 8- to 12-foot wheel may be used. If floods reach 6 to 10 feet, wheels should be made from 15 to 20 feet diameter—their widths according to power desired, but not much wider than their diameter. Their best efficiency is when the speed of the periphery is about one-half that of the wheel-race stream, and will vary, according to their lightness and perfection of design and construction, from 30 to 45 per cent. of the value of the water passing the gate.

As the power realized depends upon the velocity of the issuing stream from the gate, and which is variable with the variable head of water behind the gate, the formula for the velocity of the stream in the wheel-race will be the square root of the height of the water surface behind the gate above the centre of the gate slot in feet multiplied by the square root of twice gravity, $\sqrt{2g \times h}$, from which 30 per cent. should be deducted for friction on the soleboard and curb. For example, with a head of 4 feet above the centre of the gate opening $\sqrt{4 \times 2g} = 16.04$ velocity of the issuing stream, less 30 per cent. = 11.22 feet per second.

For best effect, the periphery of the wheel should run at one-half the velocity of the stream, or 5.61 feet per second.

For the horse-power of a wheel with a gate 3 inches deep by 36 inches wide we have $\frac{3 \times 36 \times 11.22}{144} = 8.41$ cubic feet per second. Then $8.41 \times 62\frac{1}{2} = 525$ pounds of water falling 4 feet, and $\frac{525 \times 4}{550} =$ 3.8 horse-power of the water passing the gate; which with a wheel well constructed, true, and running close to the soleboard and curb, having a diameter of 12 feet by 3 feet in width, with 30 blades, each 9 inches wide, should have an efficiency of 40 per cent., or a delivery of $1\frac{1}{2}$ horse-power, and in proportion for any other width.

The Poncelot wheel, Fig. 217, may be made of the same sizes and construction as the undershot wheel, with the exception that the buckets or blades are curved forward and the soleboard also curved or inclined to fit the periphery of the wheel, it having been found that this form with thin metallic blades caught the water to a better advantage, showing an available effect or efficiency of from 50 to 60 per cent.

The computation for the horse-power of this wheel is precisely the same as for the undershot wheel, as above; only that the efficiency is higher and is due to the peculiar curve of the buckets, which for best effect should be made of iron plates, and, as shown in Fig. 217, should give an efficiency of 60 per cent. of the value of the gate opening as

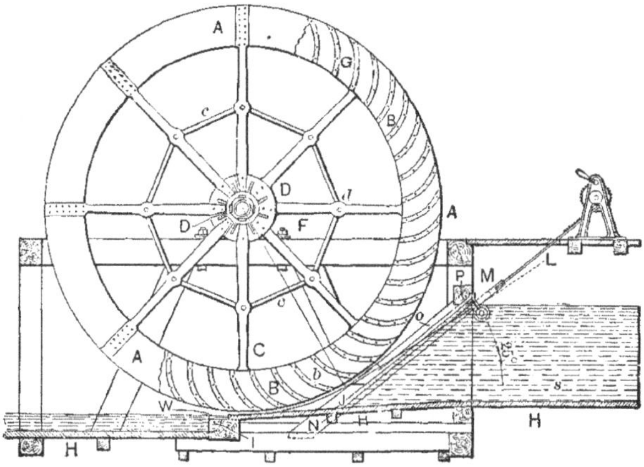

FIG. 217.—Poncelot wheel.

computed for the undershot wheel. In size the wheel may be three times the head of water, and in width up to twice its diameter.

The breast wheel is a desirable form where a head of water can be made equal to one-half the diameter of the proposed wheel, or from 5 to 12 feet; or a 10-foot wheel for a 5-foot head and a 24-foot wheel for a 12-foot head, for best effect.

Fig. 218 represents a low breast wheel in its ordinary form with plain blades running in a curb to prevent waste at the ends of the buckets.

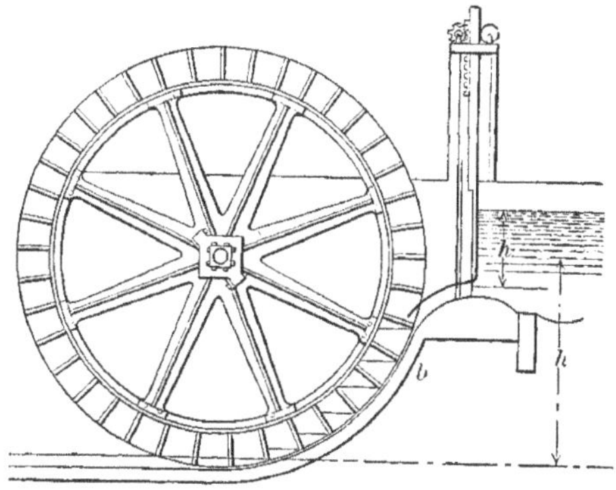
FIG. 218.—Breast wheel.

The volume of water flowing to a breast wheel may be approximately assumed as six-tenths of the velocity due to head, multiplied by the area of the gate.

For the theoretical velocity, $8.02 \times \sqrt{\text{height}}$ in feet from the centre of the gate opening to the surface of the water in the race; and the actual volume in cubic feet per second by multiplying the area of the gate in square feet or decimals of a foot by six-tenths of the theoretical velocity, or $\sqrt{2g \times h'} = V$ and $V \times 0.6 \times A =$ cubic feet per second, $A =$ area of the gate in square feet.

When the lip of the gate is depressed so as to give the water an impact upon the blades in the direction of their motion, one-half of the height of the water above the gate may be the point of measurement from the bottom of the wheel for the height in the formula for the horse-power of the water issuing from the gate.

Then the $\frac{\text{cubic feet} \times 62\frac{1}{2} \times \text{height}}{550} =$ horse-power of the issuing water.

This wheel as ordinarily constructed has an efficiency of about 50 per cent. of the value of the water-power. In the latter expression the height may be equal to h, as shown in the cut.

The Sagebein wheel is a low breast wheel of peculiar form and illustrated in its best form in Fig. 219. It is well adapted for heads of about one-third the diameter of the wheel, and for a large volume of water. The buckets are slightly curved and placed leaning backward and tangentially to a circle one-half the diameter of the wheel, open at the back, and of a width to prevent the spill on the inside of the wheel at the level of the water in the race.

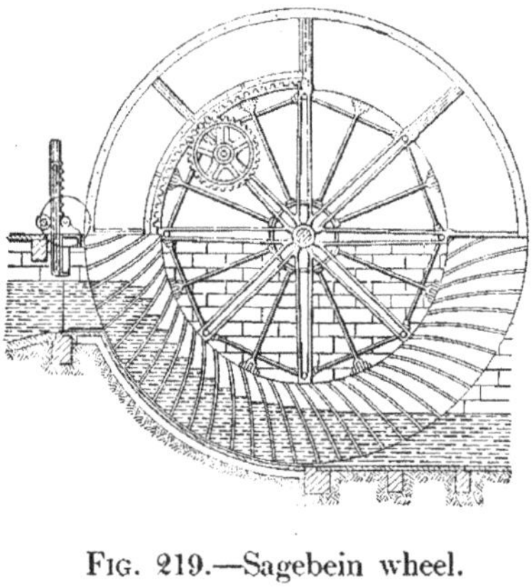

FIG. 219.—Sagebein wheel.

The wheel should run close to a well-fitted sole plate, which may be curbed, or the buckets shrouded. It is made to wallow in the tail-race and to carry such a volume of water by its wide buckets and freedom from spill as to raise its efficiency to 60 per cent. of the value of the water-power due the area of the gate and the whole head.

The power of this wheel is equal to 60 per cent. of the weir flow of the gate in pounds of water, multiplied by the height of the surface of the flood-race above the tail-race.

The weir flow for any depth up to 24 inches for 1 inch in width per minute may be taken from the table, and by multiplying the quantity in the column opposite the given depth by the width of the gate in inches, will give the total flow of the weir in cubic feet per minute. Then the $\frac{\text{cubic feet} \times 62\frac{1}{2} \times \text{height in feet}}{33,000}$ = horse-power of water flow, and this power $\times 0.60$ = the horse-power of the wheel.

In this case the gate should be clear from the surface of the water, and is only used for regulating the speed or stopping the wheel.

The overshot wheel has its peculiar province in the greater heights and smaller quantities of water flow, but cannot compete with the turbine and Pelton wheels in efficiency, nor well utilize the high heads for which these wheels are adapted. Its efficiency is about 70 per cent. of the value of the water flow. Within the range of its size, or from say 12 to 30 feet, a light overshot wheel may be made and set up in a home-made way that will do good work at reasonable cost. Nor is their size confined to the above figures, as wheels of this type are in use from 4 to 50 feet in diameter.

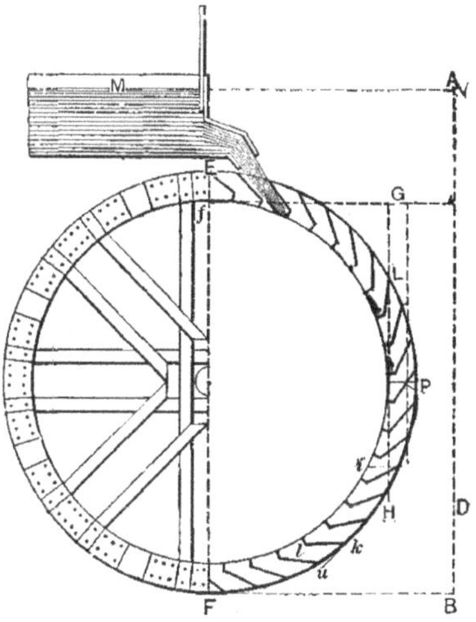

FIG. 220.—Overshot wheel.

The power of the overshot wheel is derived principally from the weight of water flowing into and held by the buckets. The velocity of the stream from the gate by its low head should only be a little more than the velocity of the wheel periphery, which for best effect should be about 5 feet per second. The bucket should be so proportioned as to hold more than the required volume of water that is intended to be utilized at the best speed of the periphery, so that in computing areas, if the gate issues theoretically 3 cubic feet per second, and the wheel intended to run at 5 feet per second, then 5 feet of the periphery should hold the 3 cubic feet of water with sufficient surplus space to prevent the spill taking place until the buckets have passed two-thirds of the half periphery.

The horse-power of an overshot wheel is computed by multiplying the volume of water passing the gate in pounds per second by the height from the centre of the gate to the bottom of the spill from the buckets in feet, and dividing the product by 550 and multiplying the quotient by the coefficient 0.70 for the horse-power of the wheel. The volume of flow from the gate may be obtained by the formula for the current wheel.

The sawmill wheel, so much in use in the lumber districts where water-power is available, is a quick running wheel deriving its power more from the impact of water under the velocity due to higher head than other wheels of this class.

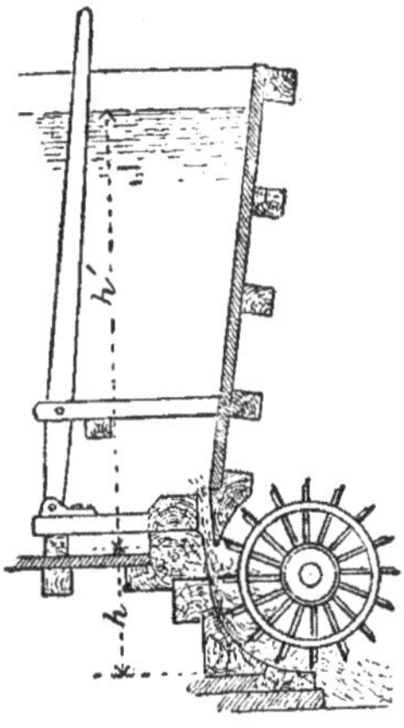

FIG. 221.—Sawmill wheel.

Its best effect, like other wheels, is at one-half the peripheral velocity of the water from the gate, the varying work of the saw affecting the velocity to a great extent.

The velocity of the water at the mouth of the gate opening, if it have a slightly taper form, will be, by the formula for theoretical velocity,

$$\sqrt{2g \times h'} \text{ or } 8.02 \times \sqrt{h'}$$

and this product multiplied by the coefficient 0.8 for this class of nozle. Thus for a 10-foot head in the flume, the velocity would be

$$8.02 \times \sqrt{10} \times 0.8 = 20.28 \text{ feet per second.}$$

For speed of wheel, 3 feet in diameter, the variation in work would allow of from 50 to 75 revolutions per minute.

For the horse-power we may assume a gate opening of 2 inches wide by 8 feet in length, or an area of 1.33 square feet. Then 20.28 $\times$ 1.33 = 26.97 cubic feet discharge per second. Adding two-thirds of the diameter of the wheel, or h, Fig. 221, will make the total height of effective power 12 feet, and

$$\frac{26.97 \times 62\frac{1}{2} \times 12'}{550} = 36.7 \text{ horse-power.}$$

These wheels when well made with the outer edge of the blades bevelled back and slightly curved on the face should have a coefficient of 60 per cent. of the gross power, or for the above wheel 22 horse-power.

On streams where wing dams or other barrages cannot be made available, a current wheel may be mounted on a float with the blades attached to a chain belt as shown in Fig. 222. The float may be

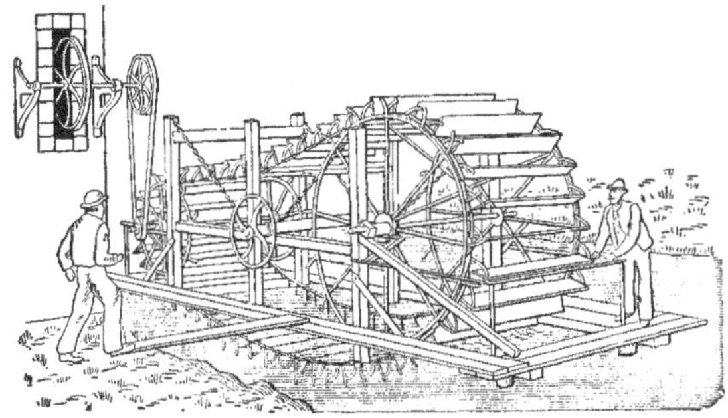

FIG. 222.—Chain-belt motor.

guided by piles or anchored to the shore and its power conveyed to a mill on shore by belts or rope drive. Another method of obtaining power from deep-water streams with low velocity may be obtained by hanging a multiblade propeller in a sliding frame in a well on a flat-bottom barge, with its power transmitted by chain and gear. This plan, as shown in the two sections, Fig. 223, is a useful power for

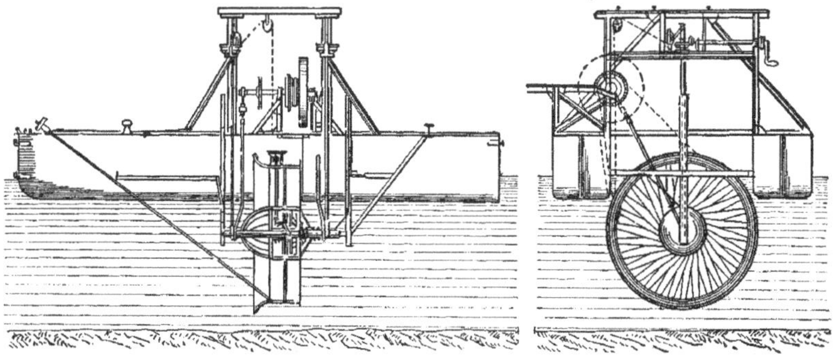

FIG. 223.—Propeller wheel on float.

pumping the water from the larger streams for irrigation. The facility for lowering the wheel to the mid-depth of a stream, the section of its greatest velocity, and for raising the wheel for mooring the float in safe shallow water, when not in use, are advantages claimed for this method of obtaining temporary power.

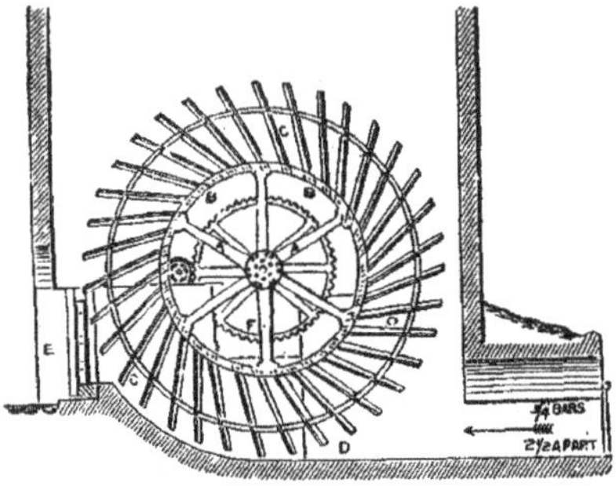

FIG. 224.—Drainage wheel.

In Fig. 224 is a section of a drainage wheel, a modification of the type of wheel used in lifting the water of the south branch into the old drainage canal at Chicago. It is much in use in Europe for draining fens and lowlands.

Power-driven drainage-wheels with broad buckets of back or tangential slope, and revolving in a shield with proper speed, will lift a large volume of water to a height of nearly half their diameter.

IMPACT WATER-WHEELS

The force of a jet of water as applied to the buckets of a revolving wheel has a theoretical value, of which about 87 per cent. has been realized in practice. The dynamic force of a jet is the weight of

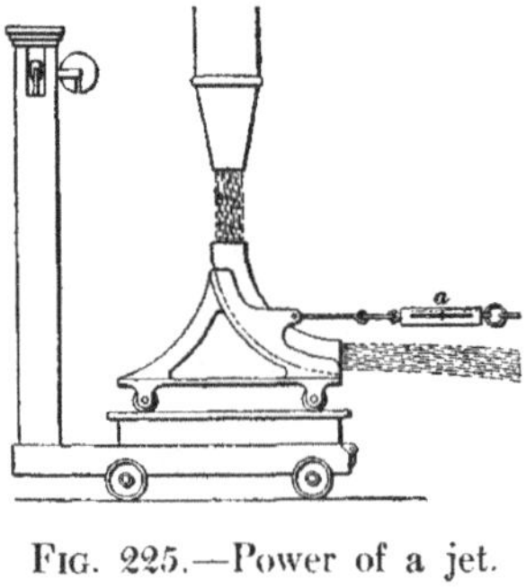

FIG. 225.—Power of a jet.

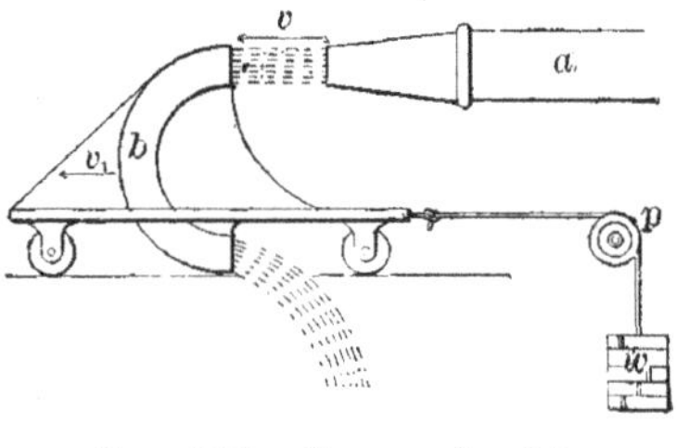

FIG. 226.—Power of a jet.

water flowing per second, multiplied by its velocity in feet per second, and the product divided by gravity—

$$\text{Pressure} = \frac{W\,V}{g} = \text{foot-pounds second.}$$

When a jet impinges upon a moving blade the formula becomes $P = \frac{W}{g}(V - U)$ and when diverted by a curved blade becomes $\frac{2WV}{g} \times$ sine $\frac{\text{Angle}}{2}$. For the best effect of a jet upon the blades of an impact wheel, the nozle should be of the best form, as shown in Fig. 73, Chapter IV.

In Figs. 225 and 226 are shown the method of measuring the actual force of a jet when it is made to impinge upon curved troughs of 90° and 180°. The first giving the contact and exit pressures and the second the two pressures combined.

In Fig. 227 is shown the division of the jet as used on the Pelton wheel and its exit at a slight angle to clear the following buckets,

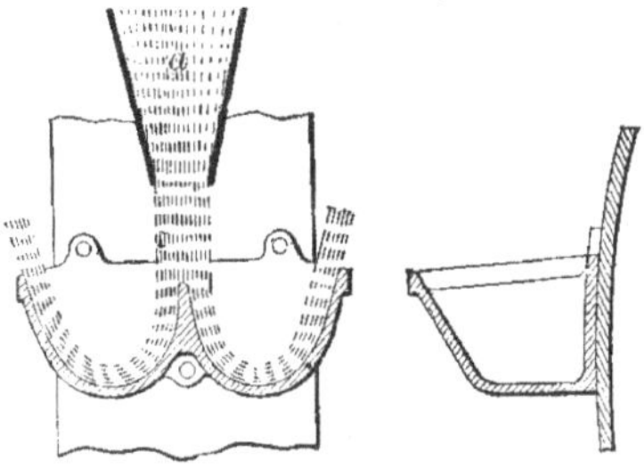

FIG. 227.—Jet dividing bucket.

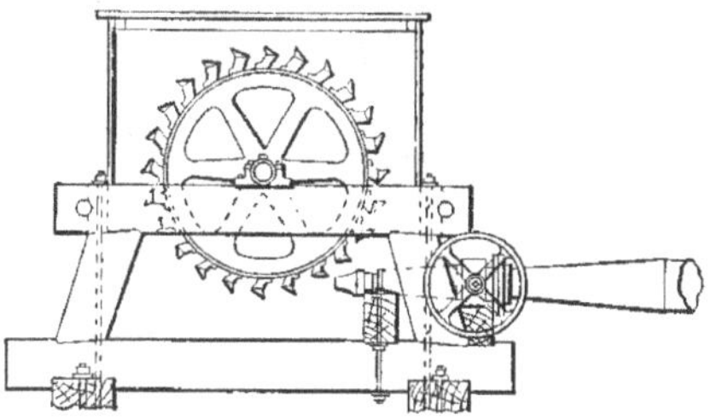

FIG. 228.—Pelton wheel.

which with other points in construction had contributed to the high efficiency of this wheel. An excerpt from the record of an experimental test gives the following as desirable points of construction:

"1. Given a certain stream of water at a given spouting velocity, it is advisable that this be taken upon the bucket surfaces of just enough buckets to catch every particle of water on the dividing wedges, turn it all on the curved surfaces, and discharge it at just enough velocity (and entirely in a direction at right angles to the entering stream axis) to clear the next following bucket. This resulting velocity will be the tangent of the discharge angle, multiplied by the bucket velocity.

"2. The air and friction surface must be maintained as small as possible by the use of a nozle which will give a perfectly circular and solid stream. The bucket surface and cutting edges must be of a shape which, with a minimum wetted surface, will allow the stream, without crowding at any point, to spread out in a thin fan-like discharge on each side. The surface must be such that the water will not adhere, and as smooth as possible."

In Figs. 229 and 230 is shown the variations in the design and disposition of the buckets from the trapezoidal Pelton type which are claimed to have advantages in efficiency.

Fig. 230 represents one of the latest improvements in the form of bucket for impact wheels, driven by a jet from a needle regulating nozle.

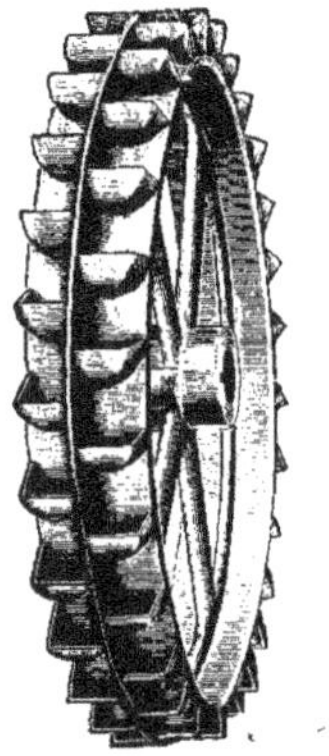

FIG. 229.—Leffel wheel step buckets.

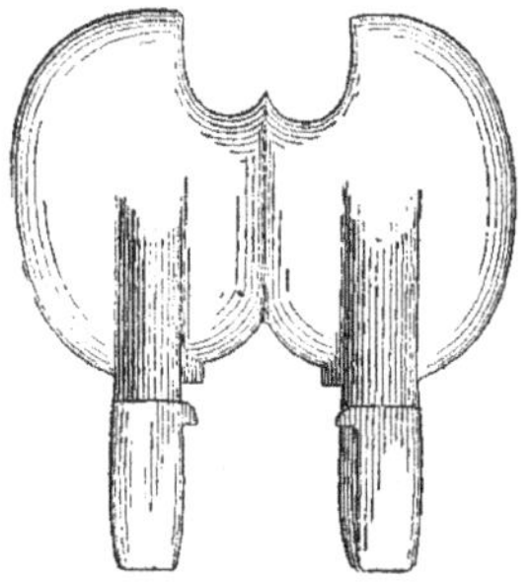

FIG. 230.—Ellipsoidal bucket, Doble water-wheel.

The Doble wheel is in operation under the highest heads available in the Pacific coast States with an efficiency above 90 per cent.

SMALL POWER-WHEELS

The increasing want of small power and its supply through the medium of water pressure made available by the great number of city and town water-works, artesian flowing wells, and the innumerable streams and brooks in the hilly parts of all countries—where the loss by the inefficiency of the common water-wheel bars its use, or makes it a negative quantity—has been met by inventive genius, stimulated by the requirement in the production of the very efficient forms of the small water-motors, as now found on the market, and of which but little is known of their efficiency and power by thousands who need the power, but lack the necessary information as to what can be done to meet the amount of water used per minute, and its power will be in ratio of the squares of the numerators, with equal denominators

Table XXVII.—Small Water-Motor Powers.

Head in feet.	Pressure, lbs.	Speed of periphery.	$\frac{3}{16}$-in. jet, gallons.	Horse-power.	$\frac{1}{4}$-in. jet, gallons.	Horse-power.	$\frac{5}{16}$-in. jet, gallons.	Horse-power.	$\frac{3}{8}$-in. jet, gallons.	Horse-power.	$\frac{7}{16}$-in. jet, gallons.	Horse-power.	$\frac{1}{2}$-in. jet, gallons.	Horse-power.	$\frac{9}{16}$-in. jet, gallons.	Horse-power.	$\frac{5}{8}$-in. jet, gallons.	Horse-power.	$\frac{3}{4}$-in. jet, gallons.	Horse-power.	$\frac{7}{8}$-in. jet, gallons.	Horse-power.	1-in. jet, gallons.	Horse-power.
23.1	10	1,155	3.1	.015	5.53	.027	8.58	.042	12.45	.062	16.95	.084	22.15	.11	28.03	.129	34.6	.17	50.0	.248	67.8	.336	88.5	.43
34.6	15	1,422	3.79	.028	6.76	.05	10.60	.079	15.20	.102	20.70	.154	27.03	.20	34.22	.255	42.2	.31	60.9	.454	82.7	.616	108.1	.80
46.2	20	1,632	4.45	.044	7.90	.078	12.50	.124	17.80	.176	24.20	.240	31.64	.31	40.04	.397	49.4	.49	71.4	.71	96.7	.96	126.5	1.25
57.7	25	1,833	4.92	.061	8.78	.110	13.78	.171	19.37	.230	26.85	.333	35.15	.43	44.48	.553	54.9	.68	78.9	.98	107.5	1.33	139.9	1.73
69.3	30	2,001	5.40	.080	9.63	.143	15.12	.225	21.70	.323	29.57	.448	38.56	.57	48.78	.727	60.3	.90	86.7	1.29	119.7	1.78	154.1	2.29
81.	35	2,166	5.80	.101	10.29	.180	16.25	.283	23.30	.406	31.71	.553	41.42	.72	52.42	.916	64.6	1.12	93.2	1.65	126.8	2.21	165.7	2.89
92.4	40	2,307	6.21	.123	11.01	.218	17.37	.345	24.91	.495	33.91	.674	44.29	.88	56.06	1.11	69.2	1.37	99.6	1.98	135.6	2.69	177.2	3.52
104.	45	2,454	6.58	.147	11.73	.262	18.41	.412	26.40	.591	35.93	.803	46.94	1.05	59.41	1.32	73.3	1.64	105.6	2.36	143.7	3.11	187.7	4.20
115.5	50	2,586	6.93	.172	12.37	.307	19.41	.482	27.83	.691	37.87	.942	49.48	1.28	62.61	1.55	78.3	1.94	111.3	2.76	151.5	3.76	197.8	4.92
127.	55	2,712	7.29	.199	13.00	.353	20.39	.557	29.23	.799	39.87	1.09	52.13	1.42	65.82	1.80	81.2	2.22	117.0	3.19	159.2	4.35	208.0	5.68
138.6	60	2,826	7.62	.227	13.53	.404	21.32	.628	30.58	.883	41.62	1.24	54.36	1.62	68.80	2.05	84.8	2.53	122.3	3.64	166.4	4.96	216.8	6.46
150.	65	2,949	7.91	.255	14.01	.455	22.15	.713	31.76	1.016	43.25	1.39	56.42	1.82	71.49	2.30	88.2	2.85	128.1	4.13	173.0	5.58	225.9	7.29
161.7	70	3,063	8.13	.282	14.67	.510	23.01	.800	33.00	1.148	44.92	1.56	58.67	2.04	74.26	2.58	91.7	3.19	132.0	4.59	179.1	6.25	234.7	8.16
173.2	75	3,168	8.51	.303	15.19	.566	23.85	.888	34.28	1.26	46.54	1.73	60.22	2.24	76.96	2.87	94.9	3.54	136.8	5.1	186.2	6.94	243.2	9.06
184.8	80	3,285	8.79	.338	15.67	.603	24.60	.947	35.36	1.36	48.00	1.84	62.70	2.41	78.36	3.01	98.0	3.77	141.0	5.43	192.0	7.39	250.8	9.65
196.3	85	3,390	9.11	.384	16.20	.684	25.49	1.10	36.56	1.53	49.76	2.10	65.00	2.74	82.77	3.47	101.5	4.29	146.7	6.19	199.0	8.40	260.0	10.98
207.9	90	3,480	9.35	.418	16.68	.746	26.17	1.17	37.53	1.68	51.09	2.28	66.70	2.98	84.46	3.77	104.3	4.65	150.1	6.71	204.3	9.14	266.9	11.94
219.4	95	3,564	9.60	.439	17.11	.807	26.85	1.26	38.50	1.82	52.30	2.47	68.56	3.23	86.64	4.09	106.9	5.05	153.4	7.24	209.6	9.90	273.8	12.93
231.	100	3,678	9.84	.489	17.55	.872	27.54	1.36	39.48	1.96	53.73	2.67	70.19	3.48	88.83	4.41	109.6	5.45	157.8	7.87	214.9	10.6	280.7	13.95

Jet columns = gallons per minute. For speed of any size wheel, divide speed of periphery in third column by the circumference of wheel in feet.

of the fractional sizes; as, for example, if a $\frac{1}{8}$-inch is to be used, and with the $\frac{3}{16}$-inch jet column in table XXVII, for a known quantity and power, the relative proportion will be $\frac{2^2}{16}$ and $\frac{3^2}{16}$ or as 4 to 9; hence to multiply any quantity and power in the fourth and fifth column of the table by 4 and divide by 9 will give the quantity and power for a $\frac{1}{8}$-inch jet under any of the stated heads or pressures, and in the same manner for a $\frac{1}{16}$-inch jet.

The speed of the periphery of any sized water-motor is given in column three, which for best effect is at one-half the spouting velocity of the water from the jet. In order to turn the stated speed into revolutions of any given size motor, it is only required to multiply its diameter in feet or decimals of a foot by 3.1416 and divide the numbers in the third column of the table by this product. For example, with a motor wheel 8 inches diameter its periphery is 0.66 foot $\times 3.1416 = 2.07$ feet, the divisor.

The quantity of water that will flow per minute through the best form of nozle in sizes from $\frac{3}{16}$ inch to 1 inch, under various pressures, with the corresponding horse-power at 85 per cent. efficiency, at which many of the best motors approximate, is carried out in the table, and from which the revolutions of any size wheel may be found by a simple computation.

The question of the cost of power by the use of small motors where water must be paid for by meter measurement is of frequent occurrence under the apprehension that quantity of water is all that is required; when in reality the hydrostatic head is the most important factor of economy of power in small hydraulic motors.

In the ordinary distribution of water to dwellings in cities and towns the water-taps in street mains are usually of uniform size, $\frac{3}{8}$ inch or $\frac{1}{2}$ inch in diameter, with the house pipes $\frac{1}{2}$ or $\frac{3}{4}$ inch if of iron, and $\frac{1}{2}$ or $\frac{5}{8}$ inch if of lead. The limit of pressure for dwellings in most of the cities and towns of the United States is under 50 pounds per square inch, so that motor jets attached to ordinary faucets in dwellings are limited to sizes under $\frac{1}{4}$ inch in diameter and their useful effect under $\frac{1}{4}$ of a horse-power.

It will be seen by inspection of the table that there is a large range of useful power for many purposes with much less than 50 pounds pressure and with jets of $\frac{3}{16}$ and $\frac{1}{4}$ inch in diameter.

As the power of a given flow of water is in direct proportion to the pressure, so the position of a water-motor should be placed at the lowest point that will allow of drainage for the most economical effect where water is to be paid for, which will be in proportion to the cost of water by meter per 1,000 gallons.

Where power is required greater than can be economically obtained from the ordinary piping of a dwelling, as for printing-presses, ventilating fans, small electric-light plants, blowing of organs, coffee-mills in stores, and the many requirements for more power than can be obtained from a $\frac{1}{4}$-inch nozle, a special service pipe is necessary of at least four times the diameter of the nozle for distances of 100 feet or less, and at least five times the diameter up to 200 feet in length from the street main. The operation of the sewing-machine has in this way become a source of pleasure instead of a tiresome labor.

TURBINE WATER-WHEELS

Turbine water-wheels rotate on a vertical or horizontal axis with balanced duplex runners on both types and also with outward and inward water flow and inward and downward flow with tail pipes.

The peculiar construction of this class of water-wheels allows of the economical use of any pressure head from 2 to 50 feet for ordinary wheels and up to 200 feet in specially constructed wheels, such as made for the Niagara power companies. The following figures show the various designs of turbines in use:

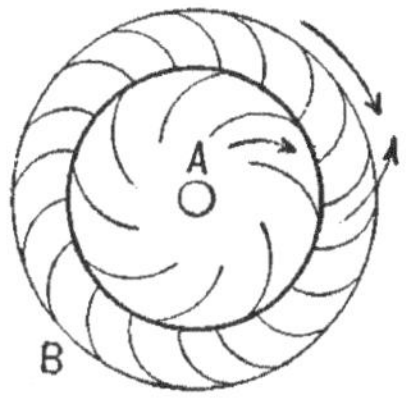

FIG. 231.—Outward flow. Fourneyron. A, the runner. B, fixed blades.

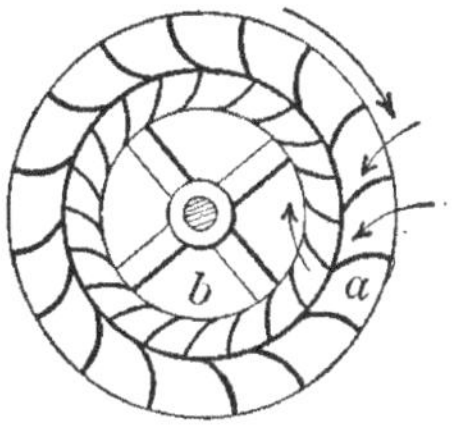

FIG. 232.—Inward flow. Warren and others. *a*, fixed blades. *b*, runner.

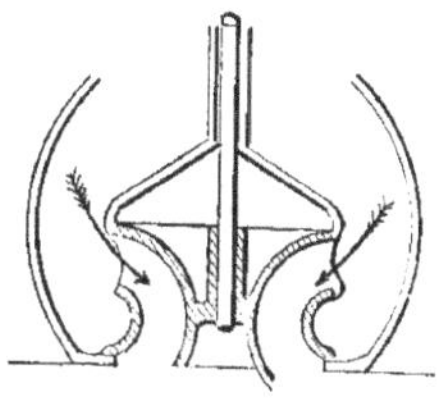

Fig. 233.—Inward and downward flow. Continuous curved blades.

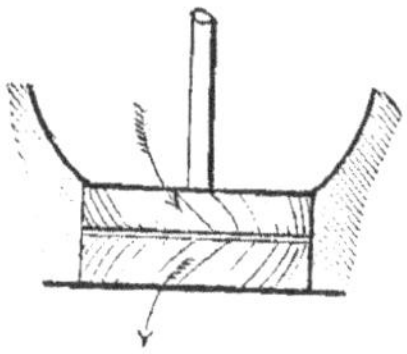

FIG. 234.—Jonval. Upper blades fixed; lower blades the runner.

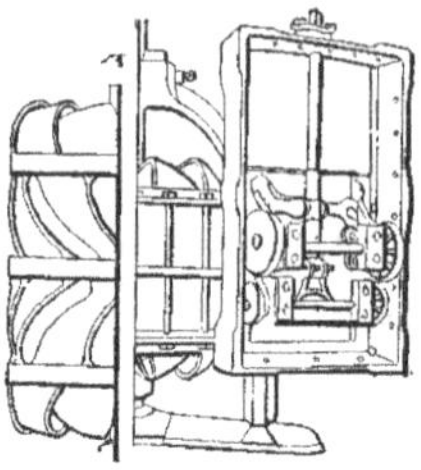

FIG. 235.—Munson. Double up and down discharge.

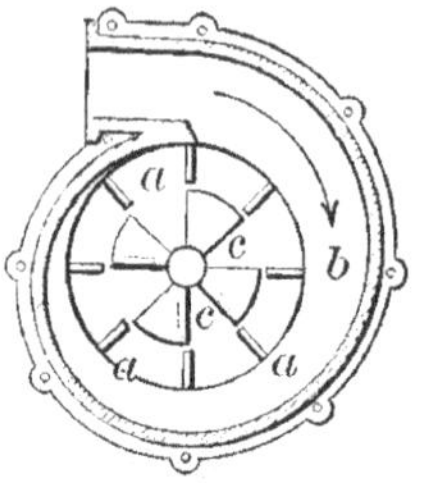

FIG. 236.—Volute with vertical shaft, inward and downward flow.

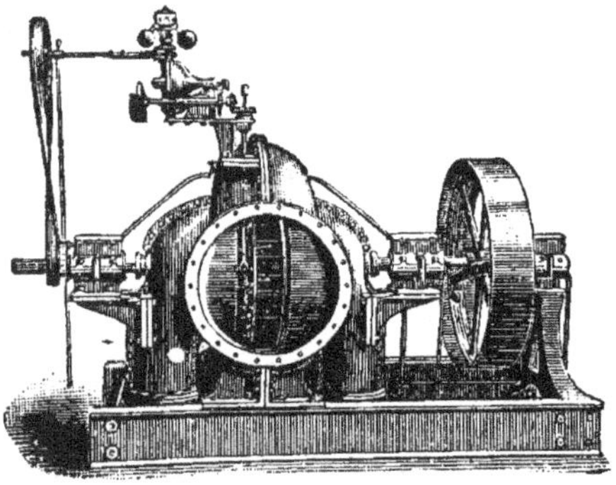

FIG. 237.—High-pressure turbine, "Leffel" model, with double draught-pipe and governor. End thrust on shaft is balanced by central inlet and double draught-pipes.

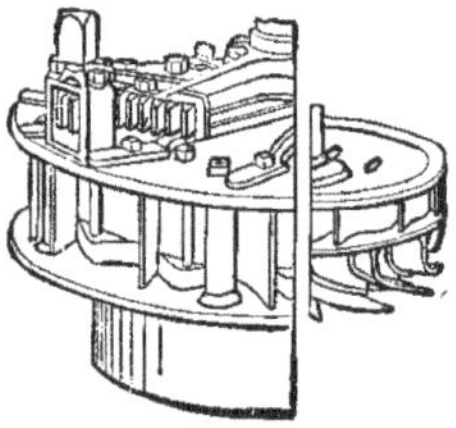

FIG. 238.—Double runner, "Leffel" model, inward and downward. Outside blades movable and geared for regulation.

Fig. 239 shows the longitudinal and cross-sections of a Swiss turbine of 1,000 horse-power designed for a high head and consists of a single nozle set on the inside of a curved bucket wheel. The nozle

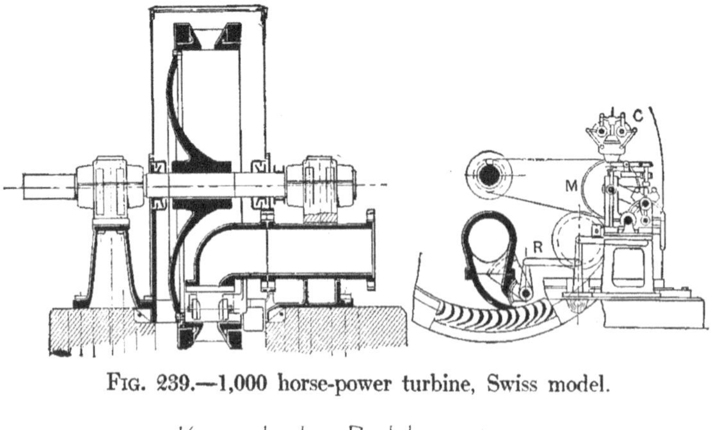

FIG. 239.—1,000 horse-power turbine, Swiss model.

is broad to match the buckets. The water flow is governed by the opening or closing of a sector slide valve controlled by a fly-ball governor, the action of which is shown by the letters C, M, R in the left-hand section of the figure.

There are many other models of turbines with good records for efficiency, which we cannot find space for illustrating in the limit of this work.

5,000 HORSE-POWER TURBINES OF THE NIAGARA FALLS POWER CO.

The turbines are of the modified Fourneyron type as improved by Messrs. Faesch & Piccard, the water being discharged horizontally and radially. Fig. 240 shows a turbine in section. Attached to the same vertical shaft are two wheels precisely alike, but inverted so far as their attachment to the carriers is concerned—that is to say, the upper wheel is pendent from its carrier outside of the guide-buckets, and the lower wheel rests on the carrier which is below the set of guide-buckets. The water enters the wheel-case midway between the two wheels at a point 136 feet below the water-level of the inlet canal, and passes through the penstock by an easy turn and without any interruption of velocity to its point of discharge.

Each of the two wheels composing the turbine unit are divided by partitions into three parts, forming in effect six perfect wheels, three at the top of the wheel-case and three at the bottom. The guide-wheels have 36 buckets, and the turbine wheels have 32. The form of the buckets was especially designed to produce the desired effects both with relation to power and velocity. The turbine wheels themselves are made of bronze, the rim and buckets being made in a single casting. The shaft that connects the two wheels is of steel, running in oil bearings in the wheel-case, no water being in any way accessible to these bearings.

The regulation of the wheels is effected by cylindrical gates with a clearance of $\frac{5}{32}$ of an inch between the gate and outside of each wheel, these gates being attached together and moving simultaneously with a difference of 1 inch in their position, so that one wheel is uncovered 1 inch before the other one begins to open, this difference being the width of the partitions that divide each wheel into three parts. These cylindrical gates, which are perfectly balanced and adapted to fully control the speed, are not capable of stopping the motion of the wheels by choking the discharge from them. Stoppage can be effected only by closing the sluice-gates in the canal leading the water to the penstock. When the cylindrical gates are closed en-

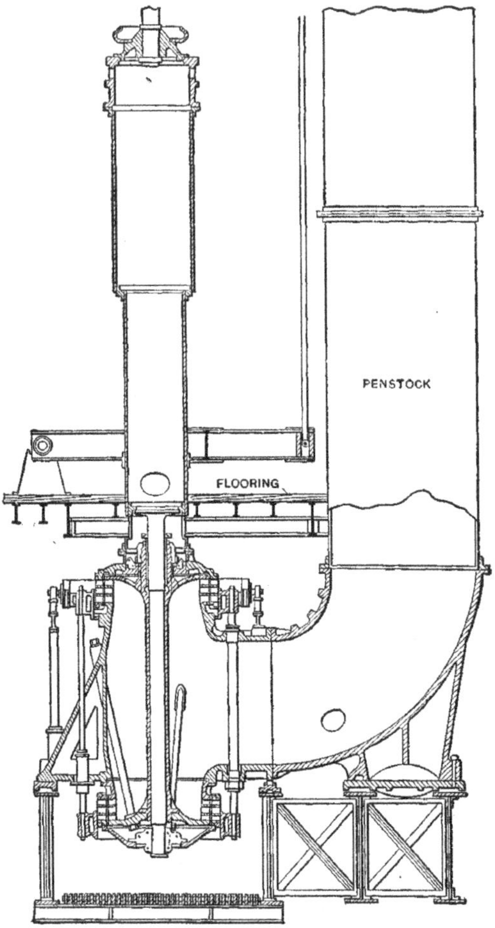

FIG. 240.—Section of 5,000 horse-power turbine.

tirely and no work is being done by the dynamos, the wheels will continue to run at about 40 revolutions per minute by reason of the gate leakage.

The vertical shaft which carries the power from the wheels to the dynamo is made of sections of tubing, 38 inches in diameter, varying in thickness from $\frac{3}{8}$ inch to $\frac{5}{16}$ inch, made without vertical seams, and united by steel couplings. Between the wheels and the dynamos there are but three bearings—two intermediate bearings, and one thrust-bearing, which is at the upper end immediately below the dynamo. The thrust-bearing has collars which control the position

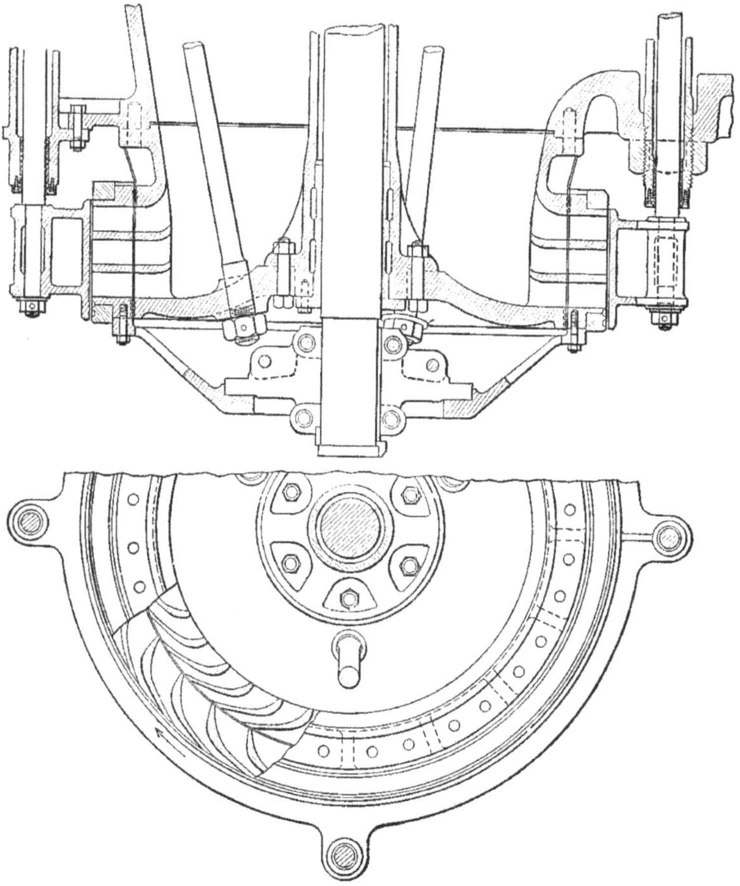

FIG. 241.—Enlarged section and plan of the 5,000 horse-power turbine.

of the shaft, whether the pressure be upward or downward. Where the intermediate bearings occur the shaft is reduced to 11 inches in diameter by means of solid steel shafts made fast to heavy steel couplings.

The vertical pipe or penstock conducting the water from the canal on the surface to the wheel at the bottom is 7 feet 6 inches in diameter, and when the wheel is running at full power the water will pass down the penstock with a velocity of about 10 feet per second. It is true that a considerably smaller penstock would answer, so far as the loss from friction is concerned; but with the larger penstock the energy stored in the falling column of water is less, and regulation of the speed of the turbine under changeable load is less difficult.

The weight of this vertical shaft, 136 feet in length, with the turbine at its bottom and the dynamo attached to the top, is carried by the hydraulic pressure in the turbine itself, thus avoiding the complications and trouble attendant on the use of any of the ordinary forms of footstep. The vertical section of the turbine shows the construction of the balancing piston; it is located just above the upper set of blades.

To make the loss from underload as small as possible the blades and guide passages in both the upper and lower turbines are divided into three sets by horizontal partitions, as shown in the vertical section. There are thus three rows of escape orifices, each orifice 4 inches high and about 3 inches wide. The valve by which these orifices are closed and the speed of the turbine is regulated is made in the form of a ring, surrounding the whole wheel and moved vertically by rods connected to the governing mechanism at the top of the shaft. The practical effect of this arrangement is that at either one-third or two-thirds load, the wheel is expected to work at the same efficiency as at full load. At any other load some one of the three sets of blades will be working more or less inefficiently, due to the loss of energy by the throttling of the escaping water.

In Fig. 242 is illustrated a turbine of 10,500 horse-power, which is probably the largest and most efficient turbine in operation in a single unit. It was designed and built by the I. P. Morris Company, of Philadelphia, Pa., for the Shawinigan Water & Power Company, Shawinigan Falls, P. Q., Canada. Its efficiency at full load is 86 per cent.

The illustration gives a good idea of the principal features of the design. It is of the horizontal shaft, "Francis" inward-flow type, with spiral (or "volute") casing. The water is discharged laterally from the centre of the wheel through two draught tubes, one on either side. The upper curved segment of one tube can be seen in the illustration.

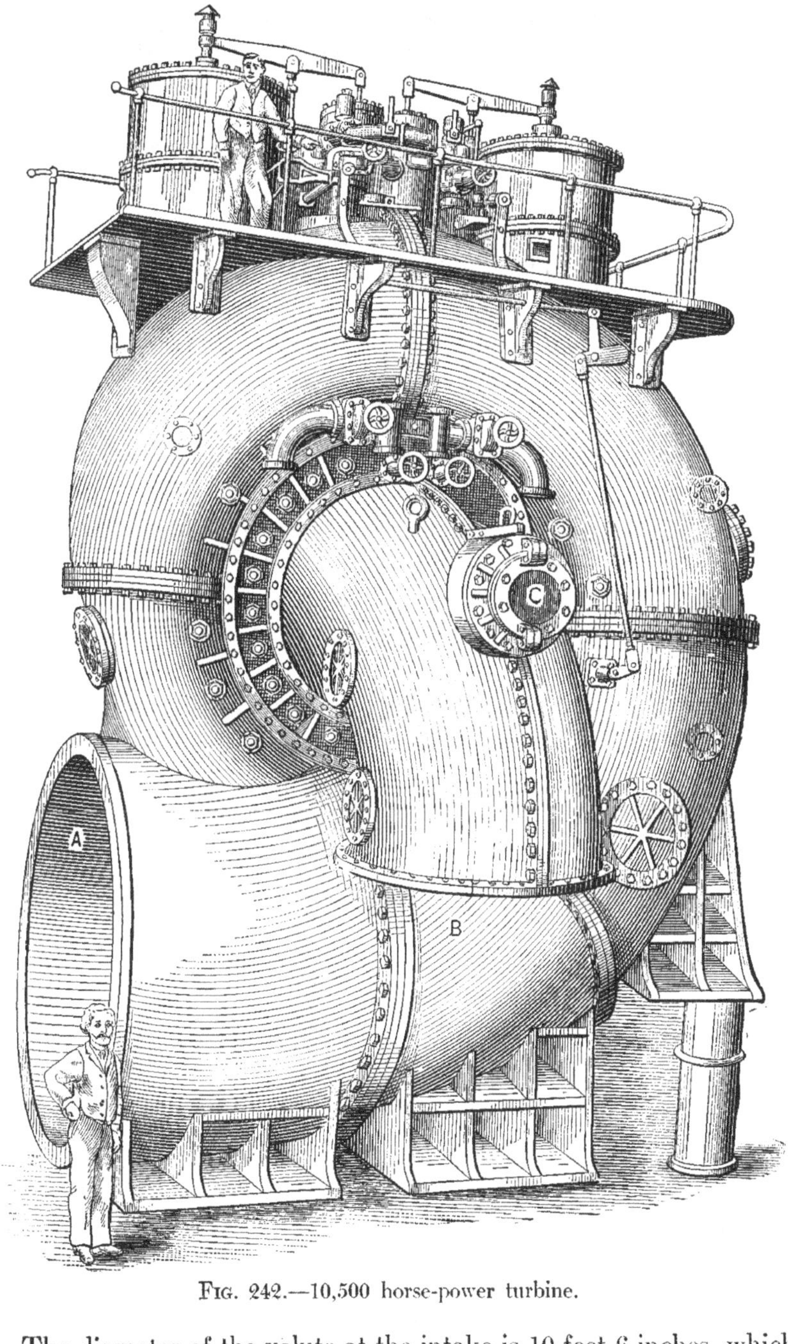

FIG. 242.—10,500 horse-power turbine.

The diameter of the volute at the intake is 10 feet 6 inches, which decreases gradually along the spiral in proportion to the amount of water flowing at different sectional areas. The height of the turbine is 30 feet, 22 feet wide, and weighs approximately 364,000 pounds.

The spiral casing permits the penstock to be received below the floor of the power-house, thus leaving room for oil switches under the switchboard gallery.

The guide vanes are operated by a vane ring, controlled by pistons from the cylinders. The area between the guides is regulated by the movement of this ring, and, therefore, the quantity of water entering the wheel per second is controlled.

The wheel is controlled by a Glocker-White mercurial hydraulic governor, and also a hydraulic hand-gear. The hydraulic cylinders at the top furnish power for moving the regulating apparatus.

The draught tubes from the wheel are of conical shape, designed so that the velocity of the water as it comes from the runner is decreased gradually until it is discharged into the tail-race. The velocity from the wheel and around the quarter turn is about 18 feet per second, while the velocity at the end of the draught tube as it enters the tail-race is about $3\frac{1}{2}$ feet per second, the draught tubes being gradually enlarged to 10 feet diameter at the ends in the tail-race. The head due to the velocity of the flowing water as it reaches the runners is about $4\frac{1}{2}$ feet. The draught tube was made a part of the turbine wheel and some of this $4\frac{1}{2}$ feet of head was regained by delivering the water in the tail-race at a velocity of $3\frac{1}{2}$ feet per second, with a corresponding velocity head of less than $\frac{1}{2}$ foot. The amount of head regained by the draught tubes would, theoretically, be the difference between the 4 feet and $\frac{1}{2}$ foot, or $3\frac{1}{2}$ feet. The efficiency of the draught tube is, however, only 50 to 80 per cent. so that the actual head regained is about 2 feet, and the efficiency is increased by the ratio of 2 feet to 135 feet, or $1\frac{1}{2}$ per cent.

The 10,500 horse-power is transmitted over long-distance electric lines 84 miles to Montreal, and there used for street-railways, electric-lighting and general power purposes. The current is "stepped-up" at Shawinigan Falls from the 2,200 volt, quarter-phase, to 50,000 volts, three-phase, and carried to Montreal over three cables, each composed of seven No. 7 aluminum wires. At Montreal it is "stepped-down," with a loss in transmission of 18 per cent.

CHAPTER XIV

PUMPS AND PUMPING MACHINERY

CENTRIFUGAL PUMPS

CENTRIFUGAL force finds one of its most useful effects in the work of the centrifugal pump.

Its measure of work is expressed by the formula $\frac{W\ V^2}{g\ R}$ in which W = the weight of a column of water 1 inch square and 1 foot deep or .434 of a pound. V^2 = velocity of the radius of gyration of solid disks or .7071 of the radius of the periphery of the wheel in feet per second. g = gravity or 32.16. R = radius of the periphery of the wheel.

Then for example, a pump of good design with a wheel 12 inches in diameter at 400 revolutions per minute, would have a peripheral velocity of $\frac{1256}{60} = 20.94$ feet per second and $20.94 \times .7071 = 14.8$, the velocity of the radius or centre of gyration. The formula expressed in figures will be $\frac{.434 \times 14.8^2}{32.16 \times .5} = \frac{95.06}{16.08} = 5.81$ pounds per square inch pressure, and the pressure $\times$ by 2.3 feet per pound = 13.36 feet total lift without overflow.

Below this head the pump will discharge inversely in proportion to the head and area of the pump and pipe connections. This corresponds with the tests of heads obtainable in practice, less the friction.

The action of compound and multiple centrifugal pumps multiplies the effect of a single pump in proportion to the number of units less the friction; they are constructed for any lift up to 1,000 feet.

There are many models of centrifugal pumps on the market, of which we illustrate enough to give a general idea of their various designs.

Fig. 243 shows Gwynne's centrifugal pump, which has six equidistant pallets inclined backwardly toward their outer extremities. Three of these extend from the axis, and the remainder only from the margin of the annular induction-space around the axis. The wheel

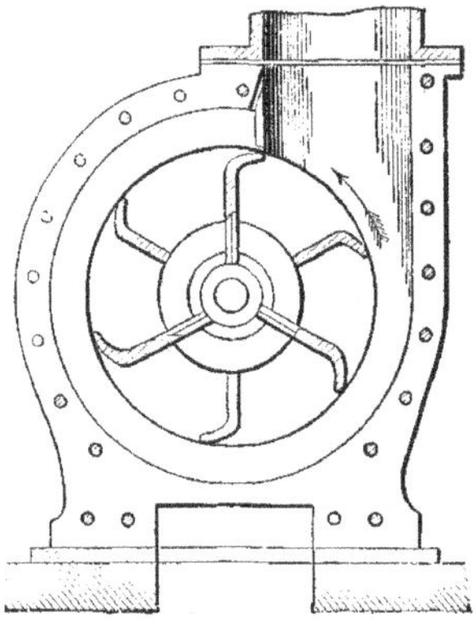

FIG. 243.—Gwynne's pump.

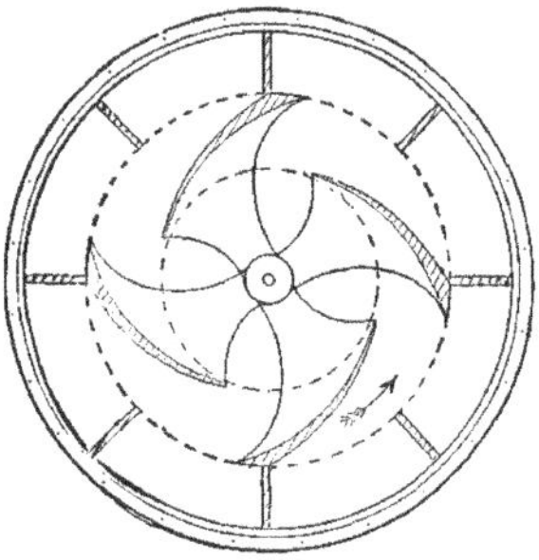

FIG. 244.—Golding pump.

rotates in a shell in the direction of the arrow, and delivers the water upward into the eduction-pipe L.

Fig. 244 shows the Golding volute pump. Four volute blades are attached to the shaft by arms. To the outer case are attached radial blades with their edges nearly touching the revolving volute blades. Suction at centre; discharge at sides of outside shells.

In Fig. 245 is shown a plan and section of the Wenzel pump consisting of four spiral wings on a conical drum which act as a gradual feeder to the main wings at the large end of the cone.

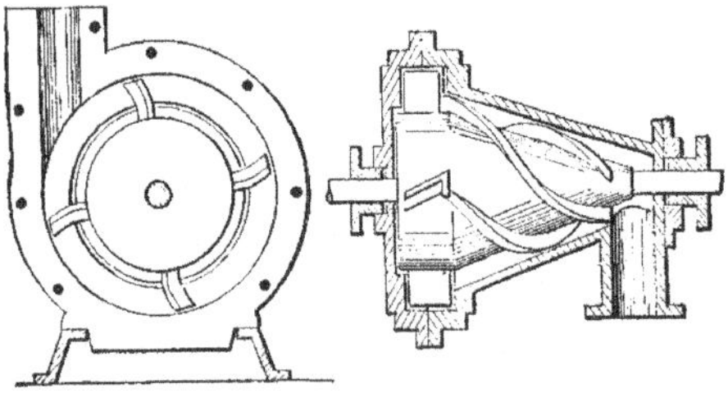

FIG. 245.—Wenzel conical pump.

The two sets of wings or blades are inclined at opposite angles to counteract end-thrust.

Andrew's centrifugal pump (Fig. 246) resembles a helix, which forms the base of a double cone placed with its axis in a horizontal position, the space between the inner and outer cones being the chamber of the pump, and occupied by a turbine-wheel shown in the detached view E. F is the stationary boss with spiral flanges *l*, which give the water a twist just as it enters upon the action of the wheel. *a* is the base of the pump, cast in one piece with the case *c*, to which is attached by flanges the conducting-case, forming a spiral discharge-passage, gradually enlarging to the outlet *f*. A series of grooves, which are fitted in a Babbitt metal box in the standard *h*, counteracts any tendency to end-thrust.

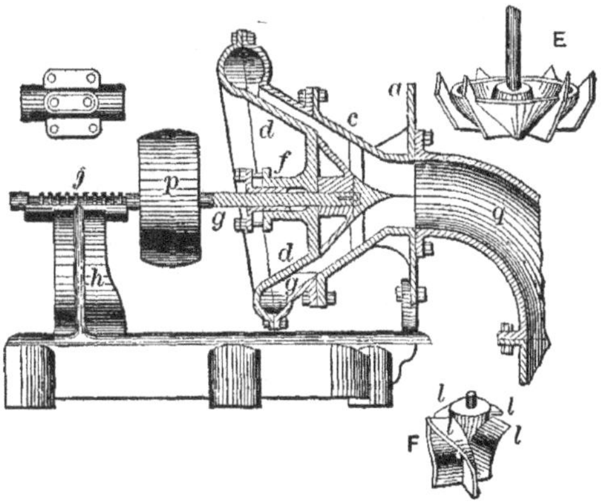

FIG. 246.—Helical centrifugal pump.

Fig. 247 shows a section of a disk pump of German type. The revolving disk receives the water on each side near the shaft in curved channels, and discharges through openings in the periphery of the disk opposite to a continued slot in the casing. A corrugated closure of the shell and disk near the shaft prevents back flow of the water escaping over the periphery of the disk, thereby adding to the efficiency of this class of pumps.

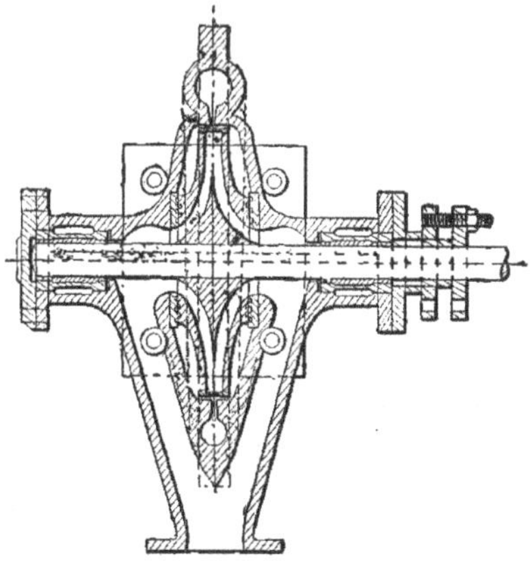
FIG. 247.—Single-disk pump.

In Fig. 248 is shown a section of the Worthington multistage turbine pump of three stages. In their design they are balanced for longitudinal thrust and their propelling power largely augmented by a series of diffusion vanes at the periphery of each impeller to check the tangential motion of the water and feed it radially to the next impeller under pressure.

In Fig. 249 is shown a section of a four-stage centrifugal pump, design of T. Reuter and made in Switzerland. The thrust balance is obtained by placing two impellers in each section together, so that their thrust pressure sides are opposite to each other, thus balancing the shaft-thrust. The

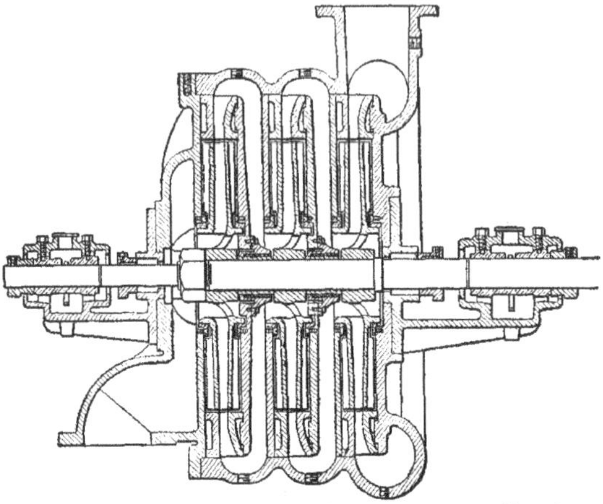

FIG. 248.—Worthington three-stage centrifugal.

water enters the centre of the first impulse wheel from the suction chamber A, is thrown out to the annular passage and returns to the centre of the second wheel through a passage at its back and so on to the second pair of wheels and the discharge chamber at D.

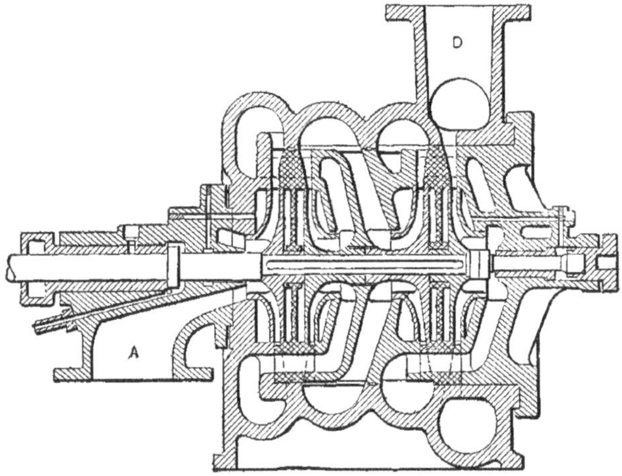

In the cross-section *d* shows the volute sections and *h* the cross passages in the partitions. At 900 revolutions per minute it sustains a forcing pressure of 240 pounds per square inch.

FIG. 249.—Four-stage centrifugal, Swiss model.

The single disk centrifugal pumps with vertical or horizontal planes of revolution are much in use for all purposes of simple water-lift in drainage, dredging, and sewage disposal. Having no valves or rubbing surface and a clear passage, mud and sand have a free way.

Fig. 250 gives a view of a centrifugal pump of the horizontal disk type for a submerged position in which it is self-charged and always ready for operation.

The efficiency of this class of pumps decreases with their smaller

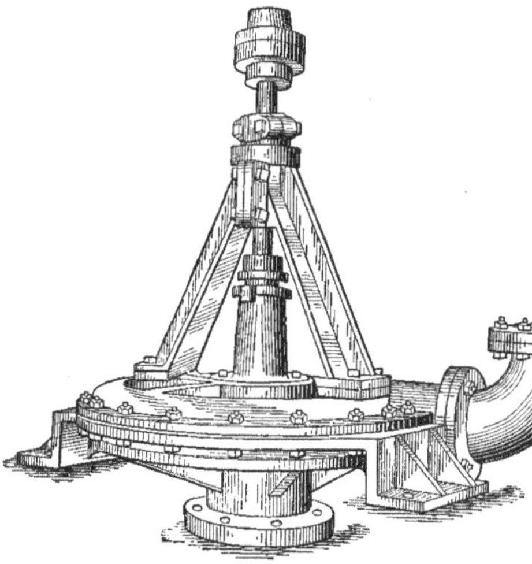

FIG. 250.—Submerged pump.

sizes as shown in the following table of speeds and capacities. Form of impeller arms not stated.

TABLE XXVIII.—SPEED, CAPACITY, AND EFFICIENCY OF SINGLE-DISK CENTRIFUGAL PUMPS.

Diameter delivery pipe.	Diameter suction pipe.	Capacity in gallons per minute 25-foot lift.	Horse-power required for each foot lift.	Side suction, revolutions per minute.	Double suction, revolutions per minute.	Efficiency, per cent.
1½	2	75	.05	1,450	750	.37
2	2½	120	.07	1,150	595	.43
3	3½	250	.14	820	500	.45
4	5	450	.25	725	425	.48
5	6	700	.34	600	380	.52
6	6	1,200	.64	525	345	.49
8	8	2,000	1.10	475	300	.46
10	10	3,000	1.62	450	265	.47
12	12	4,500	2.26	400	255	.50
15	15	7,000	3.50	325	240	.50
18	18	10,000	5.05	285	185	.50
20	20	14,000	7.06	275	140	.50
24	24	18,000	10.20	255	110	.45

The speed for a given capacity varies as the square root of the lift nearly.

The form of the arms makes quite a difference in the efficiency, as shown by the following table:

TABLE XXIX.—EFFICIENCY OF DIFFERENT ARMS AND LIFTS.

	Height of lift in feet.	Discharge gallons per minute.	Revolutions per minute.	Efficiency, per cent.
Radial arms	18	474	720	24
Straight inclined	18	736	690	43
Curved arms	8.2	2,100	828	59
Curved arms	9	1,664	620	65
Curved arms	18.8	1,164	792	65
Curved arms	19.4	1,236	788	68
Curved arms	27.6	681	876	46

PUMP LIFT OR SUCTION

Suction is a term commonly employed to denote vacuum or the absence of atmospheric pressure. When a pump raises water by suction the water is raised by virtue of the vacuum created in the pump cylinder and in the suction pipe. As a vacuum does not exist under natural conditions, but must be created by mechanical or other means, the natural tendency is to destroy the vacuum and to equalize the pressures inside and outside the pump and suction pipes. If the pipe were to be lifted free of the water, air would rush in and destroy the vacuum, but if the pipe be submerged and a vacuum then created above the water, the air pressing on the water forces the latter into the pipe and pump and destroys the vacuum. By creating a vacuum in the suction pipe a difference of pressure is caused, and the natural result is that the flow of water is from the higher to the lower pressure, and in the case of the pump, the lower pressure being within the suction pipe, the flow of water is naturally from the outside into the pipe.

The pressure on a unit of area is always the same at the surface of the water, and is equal to the pressure of the atmosphere whether the unit of area lies inside or outside the suction pipe.

It is well known that in practice it is impracticable to expel all the air from the suction pipe owing to imperfections in the pump, which allows more or less air to leak in. Sufficient air can be expelled to reduce the pressure to about $2\frac{1}{2}$ pounds, leaving about $12\frac{1}{2}$ pounds unbalanced pressure, which will raise a column of water in the suction pipe $27.7 \times 12.5 \div 12 = 28.8$ feet. A pump must be in practically perfect order to be able to reduce the pressure in the suction pipe to $2\frac{1}{2}$ pounds, and the majority of pumps when in good working order will not reduce the pressure to less than $3\frac{1}{2}$ or 4 pounds, thus leaving an unbalanced pressure of from 11 to $11\frac{1}{2}$ pounds, which corresponds to a height of from $25\frac{1}{2}$ to 26 feet. This, therefore, is the greatest lift at which pumps can be operated with a fair degree of economy.

Assuming 4 pounds to be the lowest pressure obtainable by the average pump in practice, it will be seen that, as the altitude above sea-level increases, the height to which water may be raised by the pressure of the atmosphere decreases because the pressure of the atmosphere which forces the water into the suction pipe is less and therefore it is unable to balance as high a column of water as at sea-level. As a matter of fact the pressure of the air in the suction pipe at moderate altitudes can be reduced below 4 pounds, because the difference of pressure between the air outside and inside the pipe would be less and consequently the loss by leakage of air would be correspondingly less.

These conditions apply to all pumps, save those of the centrifugal type, which require charging, except when submerged, that are located above their source of supply.

TABLE XXX.—ATMOSPHERIC PRESSURE AT DIFFERENT ALTITUDES WITH EQUIVALENT HEAD OF WATER AND THE VERTICAL SUCTION-LIFT OF PUMPS.

Altitude, miles above sea-level.	Pressure pounds per square inch.	Equivalent head of water, feet.	Practical suction lift of pumps, feet.
Sea-level	14.70	33.95	25
$\frac{1}{4}$	14.02	32.38	24
$\frac{1}{2}$	13.33	30.79	23
$\frac{3}{4}$	12.66	29.24	21
1	12.02	27.76	20
$1\frac{1}{4}$	11.42	26.38	19
$1\frac{1}{2}$	10.88	25.13	18
2	9.88	22.82	17

ROTARY PUMPS

Rotary pumps are employed for lifting and forcing water and other liquids. For low heads they are somewhat more efficient than the direct-acting pumps and the absence of close-fitting parts renders it possible to handle water containing a considerable quantity of impurities, such as mash, paper pulp, hot soap, and muddy water. This type of pump is compact and is generally self-contained, especially in the smaller sizes, and will deliver more water for a given weight and space occupied than the reciprocating types, while the simplicity of construction not only lessens the liability to derangement, but enables persons having a limited knowledge of machinery to handle them successfully. Rotary pumps are driven by means of belts from line shafting and by wheel gearing, and also by direct connection to any prime mover such as a steam- or gas-engine, or electric motor.

Rotary pumps may be divided into several classes according to the forms of and methods of working the pistons or impellers, as they

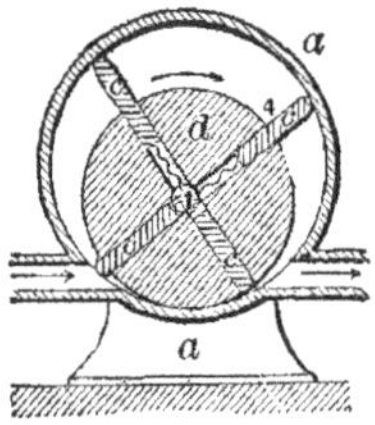

FIG. 251.

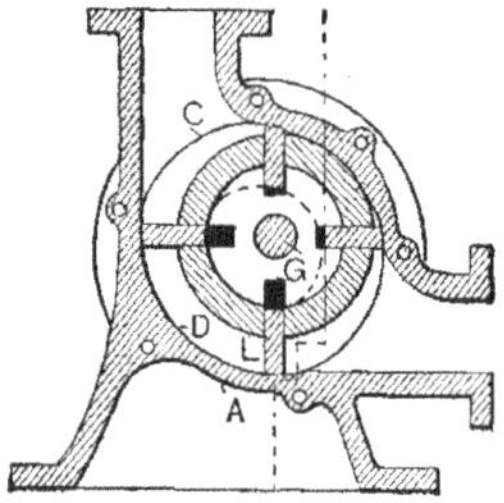

FIG. 252.

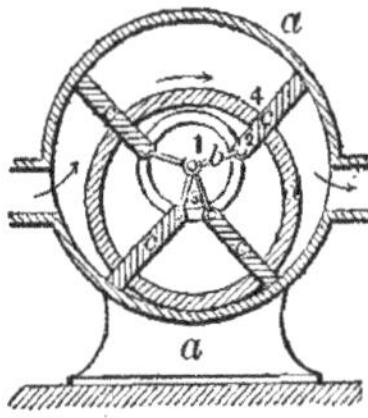

FIG. 253.

Wing Piston Rotary Pumps.

are usually called, that is, according to the construction and arrangement of the butments. The butment receives the force of the water when driven forward by the pistons or impellers and also prevents the water from being carried around the cylinder, thus compelling it to enter the delivery pipe. In the construction of the impellers or pistons, and of the butments, lies the principal differences in rotary pumps.

The rotary pumps with movable pistons, as shown in Figs. 251, 252, and 253, have been the theme of inventors during the past three centuries. Their wear from side-thrust on the blade pistons limits their usefulness to small pressures; but they are very compact and useful in the transfer of large volumes at low lift.

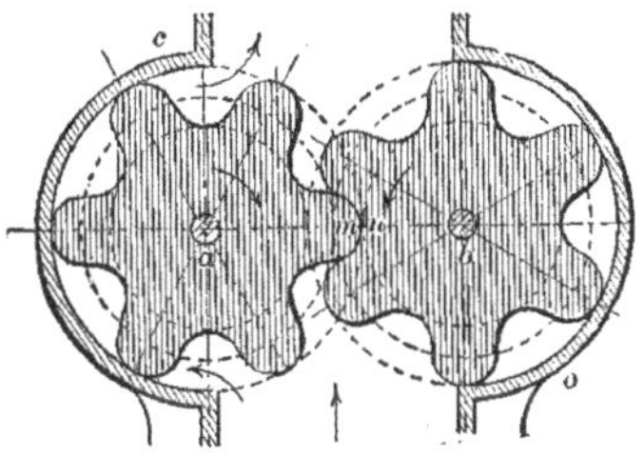

FIG. 254.—Pappenheim pump.

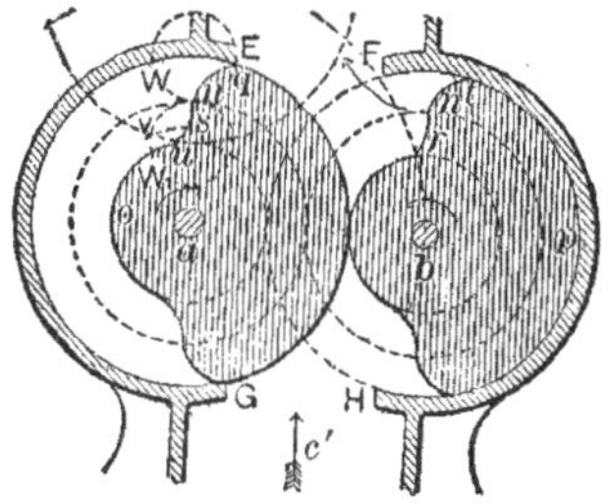

FIG. 255.—Repsold pump.

The double deep-toothed gear pump was an early rotary device with the design of obtaining simplicity of internal parts and of operating it at higher pressures than obtainable with the wing piston type. Fig. 254 shows a section of the Pappenheim pump and Fig. 255 shows a section of a modification of this principle in the Repsold model, consisting of two differential sector cylinders revolving in contiguous cylindrical shells. The greater and smaller sector surfaces match and alternately close the area between the centres of revolution.

An innovation of the principle in the design of the double deep-toothed gear pump is that of the Holley pump, Fig. 256, with two

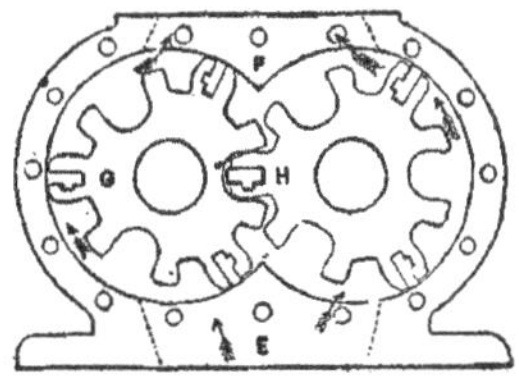

FIG. 256.—Holley pump.

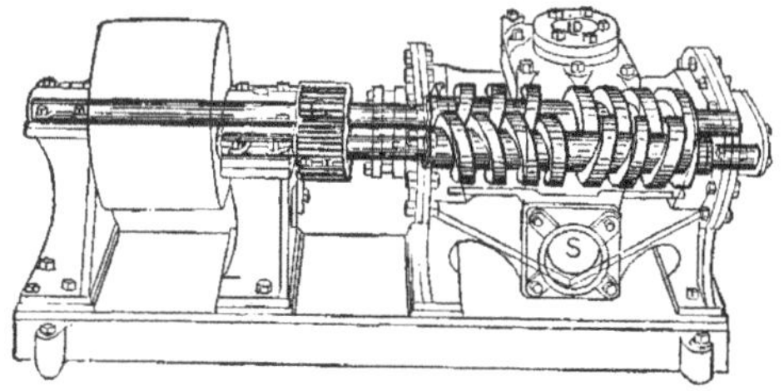

FIG. 257.—Screw pump.

or three long teeth in each disk, meshing with counterparts in the opposite disk, by which the volume per revolution was greatly increased.

The Quimby screw pump, Fig. 257, is a novelty in the line of rotary pumps, in which two screws revolve in mesh with each other and are enclosed in a close-fitted case. The suction is taken at each end from the chamber S and discharging at the middle D, thus eliminating shaft and pressure thrust. The screw threads are right and left hand on each shaft. The two-shaft stuffing-boxes are on the suction side and almost frictionless. The action of the screws on the liquid is continuous and without the effect of pulsation on the whole length of the pipe column.

CHAPTER XV

RECIPROCATING PUMPS

THE single and duplex cylinder piston pump is the accepted ideal model of the means for raising and for forcing water against any required pressure and for the highest duty of its impelling power.

Of the ordinary single- and double-acting pumps with single and duplex cylinders, there is little need for their description and illustration in this work as their domestic use has made their construction familiar to every one interested. However, there is one so different from the ordinary models, that we illustrate its construction for its novelty. It is the West double-acting differential pump combined with its air-chamber in one compact cylindrical shell. The lower section is of the same construction as the ordinary lifting pump. The upper section has a solid piston connected by rod to the lower bucket piston, and moving in an open cylinder projecting down from the cover, thus making the upper part of the pump an air-chamber.

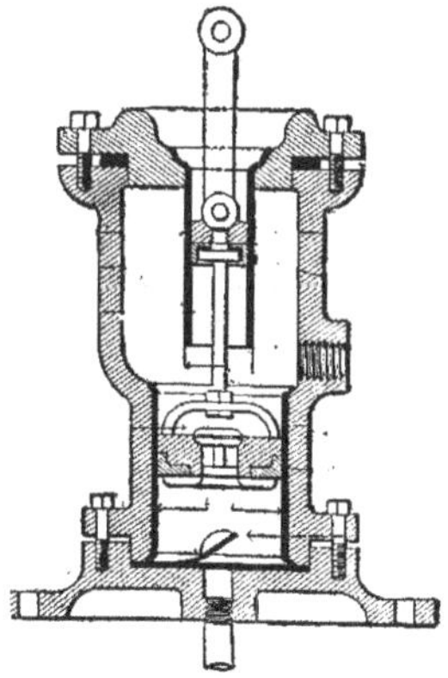

FIG. 258.—West's pump.

This combination makes the pump a favorite for domestic use.

The power pumps may be classed as of various types and models driven direct by steam, air, or crank and indirectly by belt or rope-transmission from a water-wheel or other motor, and may be designated as single power pumps, duplex power pumps, triplex and multiplex power pumps. When driven from a running shaft used for operating other machinery they develop their greatest economy.

The power required for operating pumps is computed theoretically as with other power formulas for raising water and other material from their natural level and of which the weight, height, and time are the principal factors. In the practical operation, the retarding elements of friction of the water in its devious passage through the pump and pipes; the slip or leakage of water past the piston and valves; the friction of the moving parts of the pump; the loss in the application of the power for driving the pump and with direct-acting steam-pumps, the losses in the cylinder of the steam end are functions affecting its efficiency. The efficiency of power-driven pumps usually ranges from 60 to 90 per cent.; while the direct-acting steam-pump of the sizes used for boiler feed and factory supply has an average efficiency of about 50 per cent. The efficiency has been found to vary with the height of lift from 45 per cent. at 15 feet lift to 87 per cent. at 87 feet, being the combined efficiency of the steam and water end of the pump. A piston speed of 100 feet per minute has been assumed by manufacturers as a standard of pump capacities; but well-regulated pumps may run at any speed from a mere motion to 125 feet or more per minute and for fire use still higher speeds are attained.

The theoretical capacity of delivery from a pump may be obtained from the following formulas in which:

Q = cubic feet per minute;

G = gallons per minute $= 7.48\ Q$;

$N = \frac{3.1416}{4}$ = ratio of the square of the diameter to the area of the piston;

d = diameter of piston in inches;

l = length of stroke in inches;

N = number of strokes per minute.

Then $Q = \frac{\pi}{4} \times \frac{d^2}{144} \times \frac{l\ N}{12} = .0004545\ Nd^2l$, and $G = \frac{\pi}{4} \times \frac{Nd^2l}{231} = .0034\ Nd^2l$; and for the diameter of the piston for a given quantity per minute in cubic feet $d = 46.9 \sqrt{\frac{Q}{N\,l}}$; 46.9 being derived from the square root of the inches in a cube foot multiplied by .7854.

Then for the piston speed S in feet per minute, make $d = 13.54\sqrt{\frac{Q}{S}}$. 13.54 being derived from the square root of the inches in a square foot divided by $.7854 = \sqrt{\frac{144}{.7854}} = 13.54$.

For the theoretical horse-power required to raise water to a given height; $H.P. = \frac{\text{Weight} \times \text{height}}{33{,}000}$ also $\frac{Q\,h \times 144 \times .433}{33{,}000}$ also $\frac{Q\,P}{33{,}000} = \frac{W\,h}{33{,}000}$ in which Q = quantity in cubic feet; P pressure in pounds per square inch; W = weight of the volume; h = height in feet, to which add the friction for the actual horse-power.

The actual power required for pumping water includes the friction of the pump, the water in the pipe lines and enough more to insure the free running of the pump, which must be added to the theoretical computation of the power.

The theoretical power required to force water to a given height depends principally on the vertical distance through which the water must be raised. The steam-cylinders of nearly all pumps are of larger diameter than the water-cylinders, the diameter of the steam-cylinders being usually from 25 to 50 per cent. greater. In pumps used for boiler feeding the ratio of diameter of steam- to water-cylinder is generally about 1.25. In pumps used exclusively for low-pressure work, that is, for moving large volumes of water under low pressure, the ratio is less, and for high-pressure work it is considerably greater.

The steam pressure required in any pump is found by dividing the pressure of water per square inch against the water-piston by the ratio of the area of the steam-cylinder to the area of the water-cylinder. The total pressure acting upon a piston is found by multiplying the area of the piston in square inches by the pressure in pounds per square inch, and the ratio of areas is found by dividing the area of the steam-piston by the area of the water-piston, both areas being taken in square inches.

It will be seen that the steam pressure is as much less than the water pressure as the area of the steam-piston is greater than the area of the water-piston.

It is also plain that the areas of the pistons are inversely proportional to the pressure, hence the following rules:

$$\frac{\text{Water pressure}}{\text{Area steam-cylinder} \div \text{area water-cylinder}} = \text{steam pressure.}$$

$$\frac{\text{Area water-cylinder} \times \text{water pressure}}{\text{Area steam-cylinder}} = \text{steam pressure.}$$

$$\frac{\text{Water pressure}}{\text{Steam pressure}} = \frac{\text{area steam-cylinder}}{\text{area water-cylinder}}$$

$$\frac{\text{Area water-cylinder} \times \text{water pressure}}{\text{Steam pressure}} = \text{area steam-cylinder.}$$

$$\frac{\text{Area steam-cylinder} \times \text{steam pressure}}{\text{Water pressure}} = \text{area water-cylinder.}$$

$$\frac{\text{Area steam-cylinder} \times \text{steam pressure}}{\text{Area water-cylinder}} = \text{water pressure.}$$

In the foregoing rules the pressures are to be taken in pounds per square inch, and the areas of cylinders in square inches.

The term power refers to resistance overcome through a given space in a given time. The resistance is generally expressed in pounds, the space or distance in feet, and the time, 1 minute. From this it is evident that it is not necessary to know what offers resistance, whether it is air, water, gas, the force of gravity or a spring. All that is required is to know what the resistance is, expressed in pounds. It is obvious that the same rule applies to all means of producing and utilizing power by mechanical means, consequently this is the method employed when finding the power required to raise water whether it is accomplished by suction or by forcing the water, or by a combination of these processes.

The resistance offered by a pump is that due to the weight of a column of water, which is found by multiplying the height of the column in feet by 0.433. This weight or resistance acts on each square inch of the water-piston, hence the total resistance on the piston is found by multiplying the resistance on 1 square inch by the number of square inches. The piston speed in feet per minute is the distance or space through which this resistance is overcome in that time, and resistance in pounds multiplied by distance in feet equals foot-pounds.

It will also be seen from an inspection of the rule for finding the power required that it is not necessary to know the area of the water-piston; all that is necessary is to know the resistance in pounds and the distance in feet through which it is overcome in 1 minute. Resistance to a pump piston is represented by the weight of water, so that by knowing the weight of water in pounds and the distance it is raised per minute, the power required can be found by simply multiplying together those two quantities. Suppose, for illustration, that 100 gallons of water, weighing 833 pounds, are to be raised 100 feet in 1 minute. The number of foot-pounds of work is $833 \times 100 = 83,300$, and, since this work is accomplished in 1 minute, the power required is 83,300 foot-pounds per minute. Since 33,000 foot-pounds per minute represent 1 horse-power, it will require $83,300 \div 33,000 = 2.5$ horse-power to raise 100 gallons of water a minute to a height of 100 feet.

This explains how the following rules for finding the horse-power of a pump are obtained:

(1) Horse-power =

$$\frac{\text{weight of water in pounds} \times \text{vertical distance in feet per minute}}{33,000}$$

(2) Horse-power =

$$\frac{\text{area water-cylinder} \times \text{water pressure} \times \text{piston speed}}{33,000}$$

and since area water-cylinder × water pressure = area steam-cylinder × steam pressure, we have:

(3) Horse-power =

$$\frac{\text{area steam-cylinder} \times \text{steam pressure} \times \text{piston speed}}{33,000}$$

The efficiency of a pump is measured by the work done in foot-pounds by the water end, divided by the work done in the steam end or by any other power and is generally taken at from 50 to 75 per cent. for small and ordinary service and with increased efficiency when working against the higher pressures.

When pumps take their supply from water-works mains their actual efficiency when driven by steam is decreased and is increased when driven by other power.

Of the three types of piston pumps in general use we illustrate their sectional details as shown in Figs. 259, 260, 261, and 262.

Among the many designs of pumps for special use other than before illustrated are the ordinary plunger with outside packing, much in use, although of very old model, for hydraulic high pressures, for mine-shaft sinking and deep-well service—wrecking pumps; ammonia pumps, and the pulsometer, which is driven by the direct pressure of steam. Pumps are made in multiple units and combined as the duplex, triplex, and more cylinders, the pistons of which are operated from a crank shaft; or, as with the duplex steam-pump, each piston motion is reversed by the movement of its opposite piston.

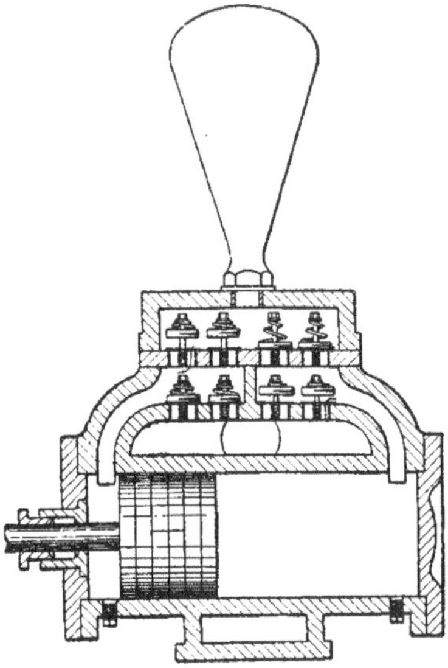

FIG. 259.—Piston pump.

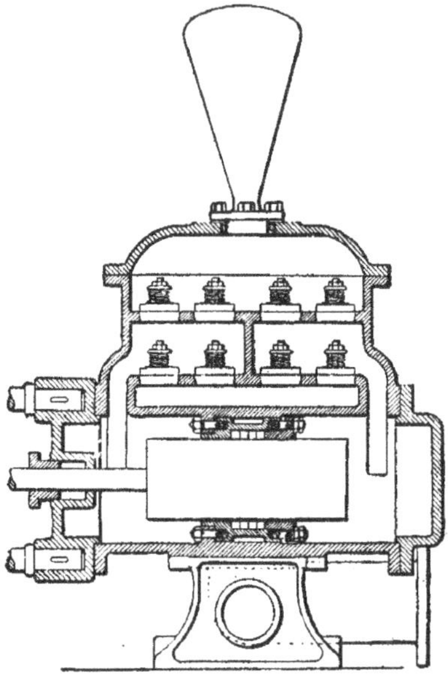

FIG. 260.—Inside plunger pump.

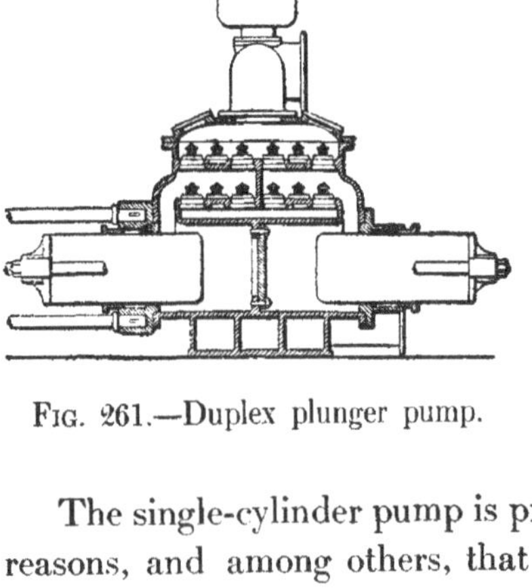

FIG. 261.—Duplex plunger pump.

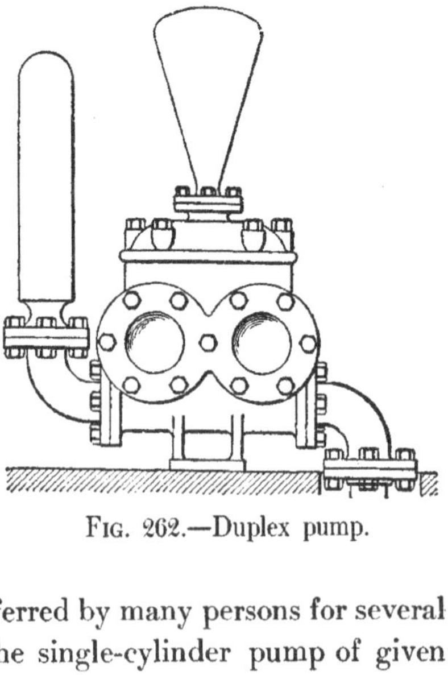

FIG. 262.—Duplex pump.

The single-cylinder pump is preferred by many persons for several reasons, and among others, that the single-cylinder pump of given capacity is said to deliver more water in proportion to the size and to steam used than the duplex. This is due to the fact that the volume of water delivered by each side of the duplex pump is dependent to a considerable extent upon the friction in each side. If the stuffing-boxes of one side, for instance, are tightened more than on the opposite side, the pistons on the tight side will move more slowly, and as the valve on that side is actuated by the more rapidly moving piston on the other side, the stroke on the tight side is apt to be considerably shortened, which reduces the volume of water delivered corresponding to a given number of strokes per minute. In the single-cylinder pump the piston never starts on its return stroke until it has traveled the full length of its stroke in either direction, since the admission of steam is controlled by the motion of the same reciprocating parts which it operates.

The better economy claimed for the single-cylinder pump is due to the smaller radiating surfaces for a given capacity, and to much smaller clearance space in the steam end. The latter is due to the fact that the single-cylinder pump has two main ports leading to the cylinder instead of four in each cylinder as in the duplex. In several makes of single-cylinder pump the number of working parts is less than in the duplex type, while the liability to derangement and the necessity for readjusting the valves is almost wholly eliminated.

The connection of a pump with its driver and the kind of motor is of importance and especially so with a steam-cylinder. In the ordinary steam-pump, the steam-cylinder capacity bears a certain relation to the water-cylinder capacity, according with the relative proportion of pressures. The valve gears are of many designs and their action is described and explained in the catalogues of their manufacturers; also in the work on "Modern Steam Engineering" by the author of this work.

The favorite design of valve for single-cylinder steam-pumps is the combination of the piston-valve as a driver of a slide valve as used on the Knowles, Blake, Cameron, and other makes. In Fig. 263 is shown a section of the Knowles single pump, with its valve-gear.

The main valve is of the B form with a flat seat; it has on top a stem which fits into a recess in the piston A. The piston has a slight rotation from the curved rocker R, which alternately covers and uncovers small ports, S, S, in each end of the cylinder that communicate with and throw the piston-valve and main valve by the cushion made by the overrun of the main steam-piston.

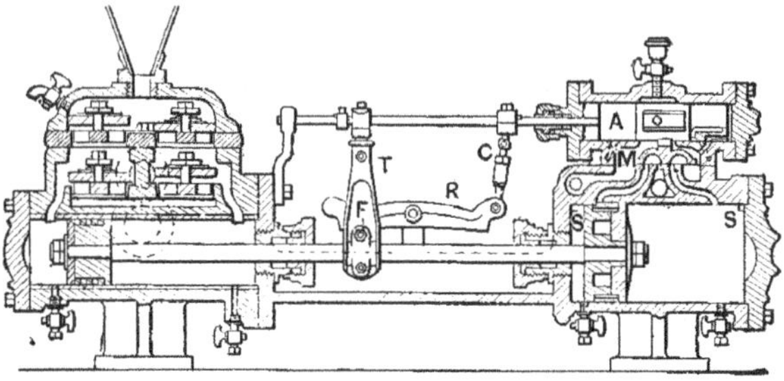

FIG. 263.—Knowles steam-pump.

The valve gear of the National steam-pump is shown in section in Fig. 264 in elevation and plan, consisting of a main valve and piston-driver as before described with the addition of a small D valve in a side chest for operating the piston-valve from the motion of a lever and sleeve on the main piston-rod.

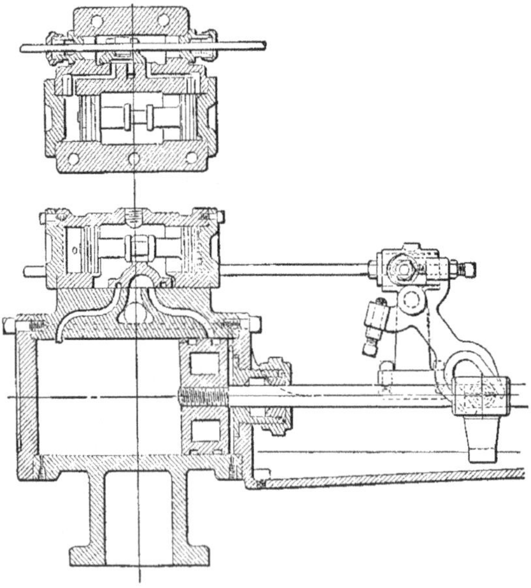

FIG. 264.—Main and auxiliary valves.

The valve piston or driver that operates the main valve is steam actuated, the steam being admitted to and released from the auxiliary cylinder by means of a small slide valve operated by the valve gear. The small slide valve is similar in every essential feature to the slide valve of an engine, admitting and releasing the steam in precisely the same manner. In this pump usual collars and tappets on the valve stem are dispensed with, the stem receiving motion by means of a roller carried by a slide block to which the valve stem is attached. The roller is given a lateral motion by set screws in the forked ends of the rocker-arm. By adjusting the position of the set screws the travel of the small slide valve is varied to suit the speed of the pump, thus preventing the main piston from striking the heads under varying speeds.

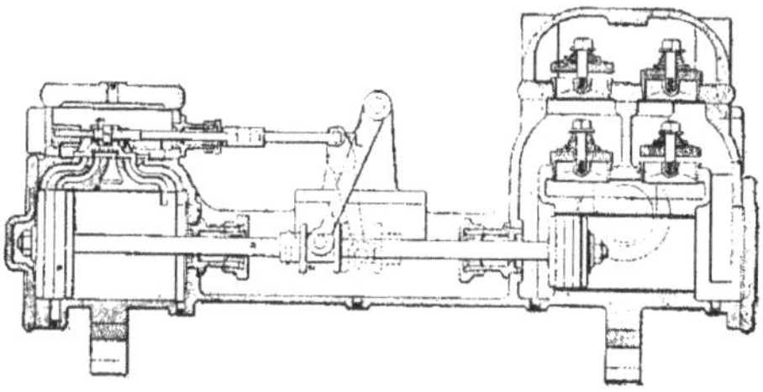

FIG. 265.—Dean duplex pump.

Many pumps of different makers have a small supplementary slide valve at the side of the main valve, operated by speeds on the main valve which is moved through part of its stroke by the main piston-rod and a lever and opening the ports to the piston-valve, thereby giving a full throw to the main valve. The duplex steam pumps have a double set of steam-ports which produce a cushion at each piston stroke by covering the inside ports alternately; a plain D slide valve making the closure by its movement. A rocker-arm linked to the piston-rod of each side of the pump

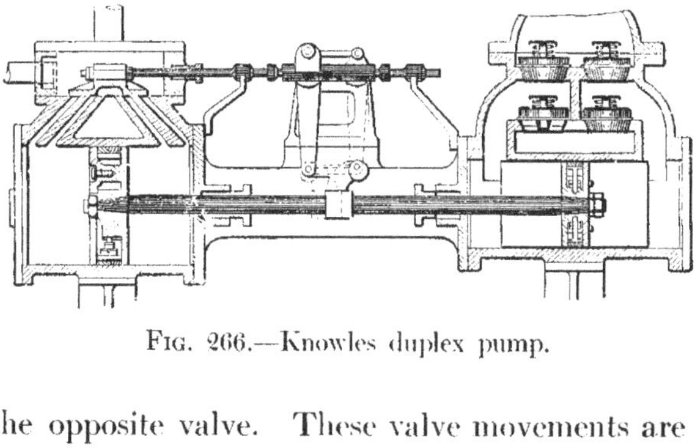

FIG. 266.—Knowles duplex pump.

operates the opposite valve. These valve movements are characteristic of the Dean, Knowles, Worthington, Blake, and other duplex pumps.

To describe and illustrate the large number of designs of steam-pumps on the market would fill a volume alone, and it is best to refer to the catalogues and descriptions furnished by their makers.

THE PULSOMETER

The pulsometer is a steam-pump, which dispenses with all movable parts except the valves. Fig. 267 represents a sectional view of a pulsometer, as manufactured by the Pulsometer Steam-Pump Company. It consists of two bottle-shaped chambers A A, joined together side by side, with tapering necks bent toward each other, and uniting in a common upright passage to which the steam-pipe is attached.

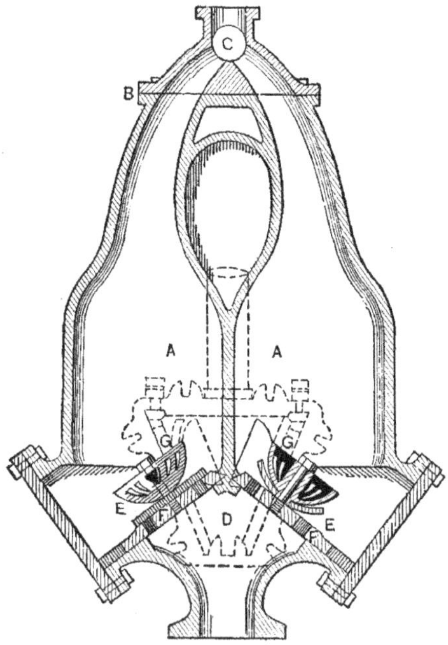

FIG. 267.—Pulsometer.

A small ball C is fitted so as to oscillate with a slight rolling motion between seats formed in the junction of the two chambers, which are alternately opened and closed for the admission of steam. Each chamber communicates by means of a separate suction-valve, E E, with the common vertical induction-passage D. I I are valve-guards to prevent the valves from opening too far. Each chamber communicates, by means of delivery-valves similar in construction to the suction-valves, with the common delivery-passage H. J is the air-chamber, cast between the necks of chambers A A, which connects only with the induction-passage D. A small brass air-check valve is screwed into the neck of each chamber A A, and one into the vacuum-chamber J, so that their stems hang downward. The check-valve in the neck of each chamber A A, allows a small quantity of air to enter above the water, to prevent the steam from agitating it on its first entrance, and to diminish the condensation of steam during the discharge stroke. The check-valve in the vacuum-chamber J serves to cushion the blow of the water consequent upon the filling of each chamber alternately.

The action of the pulsometer is as follows: When all chambers and pipes are empty, the air-check valves have to be closed, and the globe valve opened for an instant; then steam will enter one of the chambers, expel the air, and condense, forming a vacuum. This operation being repeated several times, both chambers will be filled with water through the induction-pipe. Each air-valve in the chambers must now be opened a little, to secure a regular and continuous action, which will be recognized by the steady pulsation and smooth working of the steam-ball without a rattle. Steam, being now permanently admitted, enters the chamber not closed by the ball, and forces out the water through the discharge-valves, until its surface is lowered below the discharge-orifice. At that instant the steam begins to escape into the discharge-pipe and condense; thus a partial vacuum is formed in the chamber. The water in the other chamber now presses the ball, which rolls over and closes the first chamber, where water enters through the induction-valves to fill the vacuum. The operation alternately changes from one chamber to the other.

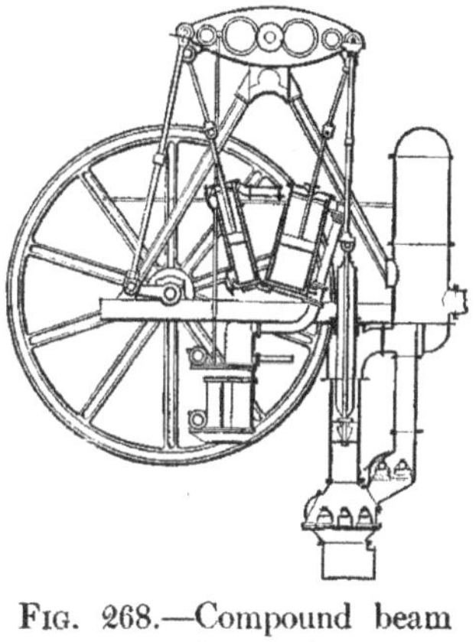

FIG. 268.—Compound beam pumping engine.

In Fig. 268 is illustrated an elevation of a type of compound beam pumping engine for water-works, not of recent date, but still doing good work at a duty of about 150,000,000 foot-pounds per 1,000 pounds of steam. The high- and low-pressure cylinders are inclined to make room for the direct connection of the pump and crank rods.

The duty records of steam pumping plants for water-works supply has gradually increased during the past few years by improvements in the design, mechanism, and by multicompounding with superheat, from 150,000,000 foot-pounds in 1890 to above 181,000,000 foot-pounds in 1906 per 1,000 pounds of steam.

The boiler horse-power required for each pump horse-power in modern steam pumping engines may be stated as follows on the basis of 10 square feet of heating surface per boiler horse-power:

TABLE XXXI.—DUTY OF STEAM IN BOILER HORSE-POWER.

Duty in foot-pounds per 1,000 pounds of steam.	Boiler horse-power per pump horse-power.	Duty in foot-pounds per 1,000 pounds of steam.	Boiler horse-power per pump horse-power.
40,000,000	1.63	140,000,000	.47
50,000,000	1.32	145,000,000	.46
60,000,000	1.10	150,000,000	.44
70,000,000	.94	155,000,000	.43
80,000,000	.83	160,000,000	.41
90,000,000	.74	165,000,000	.40
100,000,000	.66	170,000,000	.39
110,000,000	.60	175,000,000	.38
115,000,000	.57	180,000,000	.37
120,000,000	.55	185,000,000	.36
125,000,000	.52	190,000,000	.35
130,000,000	.51	195,000,000	.34
135,000,000	.49	200,000,000	.33

And to make this complete, it should be supplemented with the further statement that the limits of steam economy in the pumping engine are about reached, both theoretically and practically. Not to go too far back the duty records have been, and are, as follows:

Year.	Foot-pounds per 1,000 pounds steam.	Year.	Foot-pounds per 1,000 pounds steam.
1893	154,048,700	1900	178,497,000
1895	157,843,000	1900	179,419,600
1898	167,800,000	1906	181,068,605
1900	168,532,800		

And finally, the leading work has become crystallized into the vertical, triple-expansion machine, with outside packed plungers, and largely of the crank and fly-wheel type.

It is scarcely possible that the cost of pumping stations for water-works will be increased on account of a higher type of engine, because it is evident that the limit has been about reached with the new record of a little over 181,000,000 foot-pounds duty per 1,000 pounds of steam. Six years ago it nearly touched the 180,000,000 mark, and might have with some other refinement in the test; at any rate a gain of less than .6 of one per cent. in six years, with every nerve strained, is evidence of the top limit. The Mariotte curve is about the nearest approach to perfection in expressing the relation between the work done and the amount of steam used in doing it apparently possible for the modern steam-engine to accomplish; and, if the terminal pressure is taken as expressing the steam used and all of the steam is accounted for by the diagram, somewhere in the immediate neighborhood of 180,000,000 duty with 96 per cent. efficiency of the machine will be the resulting figure, with a reasonable amount of steam used in the jackets and reheaters charged against the account. If there were no necessity for the use of jacket steam the figure would approach 200,000,000 rather closely, and if superheating can save

FIG. 269.—Horizontal compound pumping engine.

the jacket steam and vitalize the working steam, the latter figure may in the near future be reached so far as the report on high duty accomplished in a calculation is concerned.

In Fig. 269 is illustrated a view of the Gaskill horizontal pumping engine at the Saratoga Springs water-works.

It is horizontal, of the rotative, nonreceiver, compound-beam type. The engine has four steam-cylinders, one high- and one low-pressure in each pair; the low-pressure (42″ diam., 36″ stroke) beneath the high-pressure (21″ diam., 36″ stroke). There is one pump to each pair of steam-cylinders, and each has a double-acting plunger 20″ diam., 36″ stroke. The fly-wheel revolves between, and has its pillow-blocks upon the pump cylinders. The cross-heads of high-pressure cylinders are connected by links to a beam, which is in turn connected, one end to the crank, and the other end to the piston-rod of the low-pressure cylinder and pumps.

The following data were obtained on trial:

Time	20 hours.
Average pressure by engine gauge	74.25 pounds.
Average pressure by, in force main	95.068 pounds.
Average vacuum	27.295″.
Average temperature feedwater to boilers	169.175° F.
Revolutions in 20 hours	21,449.
Revolutions per minute	17.8742′.
Piston speed per minute	107.2452.
Total coal burned	6,750 pounds.
Discharge per revolution	187.125 gallons.
Delivery at 18 revolutions per minute	4,850.280 gallons.
Net absolute duty	108,793,525 pounds.

A subsequent test gave a duty of 113,378,000 foot-pounds.

In Fig. 270 is shown a sectional elevation of the Worthington hor-

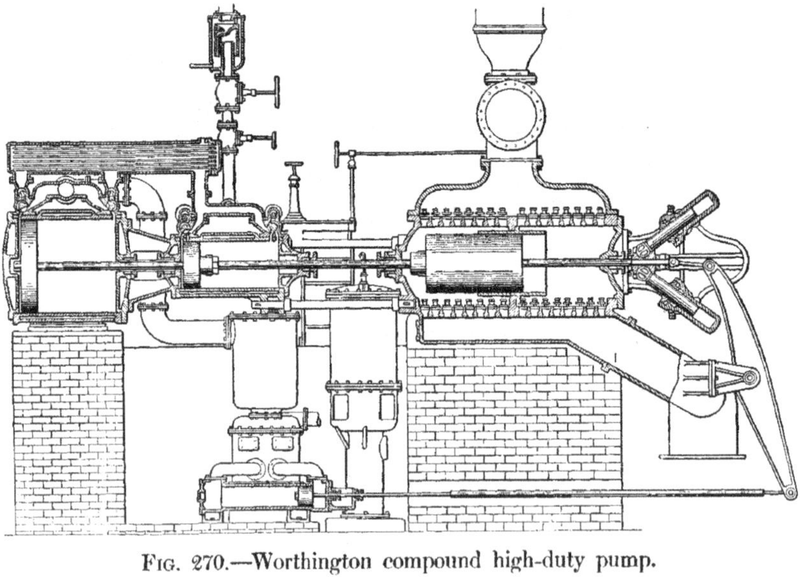

FIG. 270.—Worthington compound high-duty pump.

izontal high duty-pump, which is built in the duplex model with compound steam-cylinders and receiver reheater. The water-cylinder is designed on the inside plunger model with the addition of compensating cylinders on trunnions and plunger-pistons jointed to a continuous piston-rod from the water-piston through the back head of the water-cylinder.

These compensating cylinders, by their rocking motion and pressure, serve to equalize the pressure of expansion in the steam-cylinders

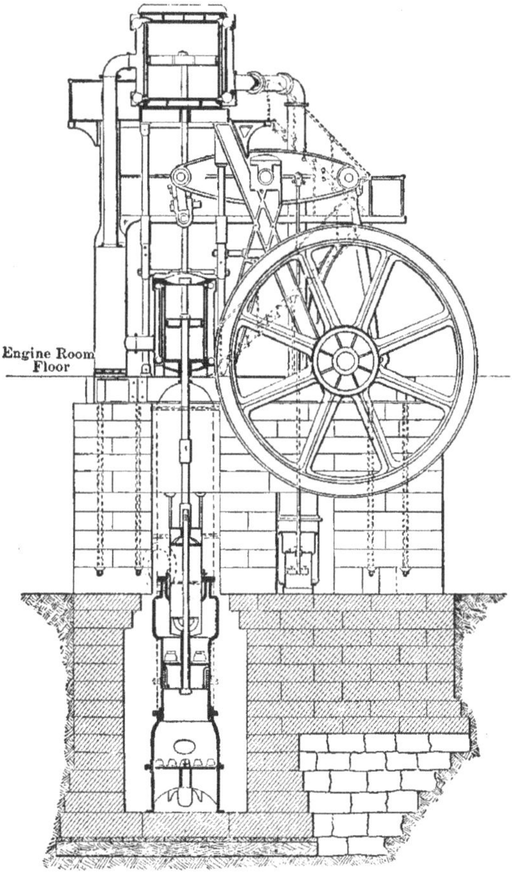

FIG. 271.—Allis compound pumping engine.

and thus enable the expansion system to be used in a direct-acting steam-pump.

The compression of air during the first half of the piston stroke and its expansive pressure during the last half from the varying angle of the rocking pistons as they alternately push against the full steam pressure during the first half and with it during the last half or expansion period of the stroke, produce an equalization of the pump-piston pressure throughout the stroke.

The pressures on the plungers within the compensating cylinders is produced by connecting these cylinders through their hollow trunnions with an accumulator the ram of which is free to move up and down as the plungers of the compensating cylinders move in and out.

Their pressure action is controlled by the pressure in the main delivery on the air-chamber which acts as a cushion; the air-supply is provided for by a small air-pump.

Fig. 271 illustrates a vertical section of a compound pumping engine constructed by E. P. Allis & Co. for the city of Milwaukee. It is of the walking-beam and fly-wheel type. The low-pressure cylinder is placed on the top of the wrought-iron framework, and directly central over the high-pressure cylinder, which is on a level with the engine-room floor, the pistons of the two cylinders being connected by two piston-rods. The rod for operating the bucket and plunger-pump is fastened to the high-pressure piston and extends through a stuffing-box in the bottom head to the bucket and plunger-pump placed in the pump-pit. By this means all the steam-cylinders are coupled solidly to the pump plunger. Both steam-cylinders are steam-jacketed and furnished with a device for regulating the point of cut-off and speed of the engine. The following are the principal items of interest from a test trial: Duration of trial, 48 hours; steam pressure in engine room, 74.81 pounds; vacuum by gauge, 26.25 inches; water-pressure gauge, 62.02 pounds; total head, including suction lift, 67.29 pounds; revolutions of engine per minute, 25.51; piston speed per minute, 255.10 feet; coal consumed, 32,395 pounds; duty in foot-pounds per 100 pounds of coal consumed, 104,820,431. The test was made under the ordinary every-day conditions, and the actual weight of coal consumed was charged up without deductions of any kind. This engine raised 12,000,000 gallons 150 feet high in 24 hours.

The record of a test of a high-duty triple-expansion pumping engine of the Allis-Chalmers vertical type, lately erected at the St. Louis water-works, is worthy of reference for its showing of thermal and mechanical efficiency.

TEST OF THE ST. LOUIS TRIPLE-EXPANSION PUMPING ENGINE

Duration of trial test	24 hours.
Diameter of steam-cylinders	34, 62, 94 inches.
Stroke of engine	72 inches.
Diameter of water-plungers	$33\frac{7}{8}$ inches.
Average revolutions per minute	16.539.
Piston speed per minute	198.44 feet.
Average steam pressure, engine	140.24 pounds.
Average first receiver pressure	26.36 pounds.
Average second receiver pressure	2.77 pounds.
Average vacuum pressure	13.21 pounds.
Average barometer	14.46 pounds.
Average head pumped against	238.23 feet.
Total water, cylinder condensation	220.129 pounds.
Average moisture in steam	0.13 per cent.
Indicated horse-power	865.23.
Delivered horse-power	842.69.
Friction	2.60.
Average moist steam per I.H.P. hour	10.60 pounds.
Average dry steam per I.H.P. hour	10.59 pounds.
Average thermal units per I.H.P. minute	201.39.
Total water pumped	20,070,690 gallons.
Mechanical efficiency	97.4 per cent.
Duty per 1,000,000 thermal units	158,851,000 foot-pounds.
Duty per 1,000 pounds steam	181,068,605 foot-pounds.
Thermal efficiency	21.06 per cent.

CHAPTER XVI

HYDRAULIC POWER TRANSMISSION AT HIGH PRESSURES

THE use of water under high pressure for operating machinery intermittently, such as cranes, haulage and elevator-lifts, was held in abeyance many years in long pipe-line service by the destructive effect of water-ram when several cranes or lifts started and stopped at the same moments. A single elevated reservoir, high towers, and air-chambers all proved unsatisfactory, and the solution of the hydraulic problem met its only success in the local application of the loaded accumulator; since which time the hydraulic-power transmission from a central pumping station through miles of piping for operating hydraulic appliances of all kinds has been a success and with greatly increasing usefulness in many countries.

In England, where the first instalments were made, the hydraulic-power system has extended throughout its largest cities, shipping ports, and to its colonies. For the lift type of canal locks, dry docks, and bridges, this system is in use in Great Britain, France, and Belgium.

London has by far the largest system in the world. The London Hydraulic Power Company has several stations each of about 1,200 horse-power, and has over 100 miles of mains laid in the streets. War vessels, merchant steamships, dredges, barges, crane punts, and other craft are in a great many cases provided with hydraulic plant to supply power for working turrets, guns, presses, lifts, cranes, capstans, winches, windlasses, steering-gear, grabs, etc., the water pressure being regulated by means of an accumulator loaded by steam pressure.

In some canals, such as that at Anderton, England, barges, with the trough in which they are floated, weighing altogether 240 tons, are raised to a height of 50 feet, to connect the river Weaver with the Trent and Mersey Canal. This invention, designed by Sir Leader Williams to dispense with a large number of locks, effected a very great saving of canal water, using only 1.7 per cent. of the quantity required if locks had been employed. Similar lifts designed by the late Sir Edwin Clark have since been constructed for canals in France at Les Fontinelles near St. Omer, and in Belgium at La Louvière on the Canal du Centre, near Mons, where the load on each ram is about 1,120 tons, the height through which the trough containing the canal boat is raised being 50 feet. Floating graving docks, notably at Bombay, London, Malta, and elsewhere, worked by hydraulic machinery, have been constructed by which vessels up to 6,500 tons and drawing 30 feet can be lifted clear out of the water for examination and repairs. Slip docks, where large vessels are drawn up on a carriage for a similar purpose, are provided at many ports in Great Britain. Swing-bridges over river and dock entrances, like those at the Tyne and the Clyde, and bridges of the bascule type, notably the Tower Bridge over the Thames, are operated by hydraulic power.

The pressure varies much in different locations with the intensity of the work; from 700 to 800 pounds per square inch and 1,000 or more pounds pressure are in use for special machines and presses.

Among the many advantages that may be claimed for water-pressure machinery are:

1. Steadiness, ease, precision, and comparative noiselessness of action.

2. Facility and promptness with which a machine may be started and stopped at any point of its travel and without waste of power.

3. Absence of risk from accident.

4. Simplicity of working, rendering the employment of skilled labor unnecessary.

5. Saving in charges for insurance as compared with machinery involving the use of fire or light.

6. Limitation of the expenditure of power to the time during which useful work is being performed.

7. The opportunity afforded of making special provision for extinguishing fires, by attaching fire-hydrants to the pipes laid to convey the water pressure.

The Hydraulic Power Company, London, England, furnishes power for above 2,500 hoists and other machines at a pressure of 750 pounds per square inch and the use of about 3,500 gallons per minute, or about 2,000 horse-power. The complete pumping plant is equal to 4,000 horse-power. The sizes of pipe-lines in use under high pressures do not exceed 8 inches in diameter with branches to meet the requirements of power.

The efficiency developed by hydraulic motors has been found to be from 50 to 60 per cent. of the power developed at the central station.

In the lift system by plungers on cranes the efficiency varies greatly with the load, from 10 per cent. with no load to 75 per cent. under full load; which relation is accounted for by the fact that it requires the same quantity of water per lift under each operation.

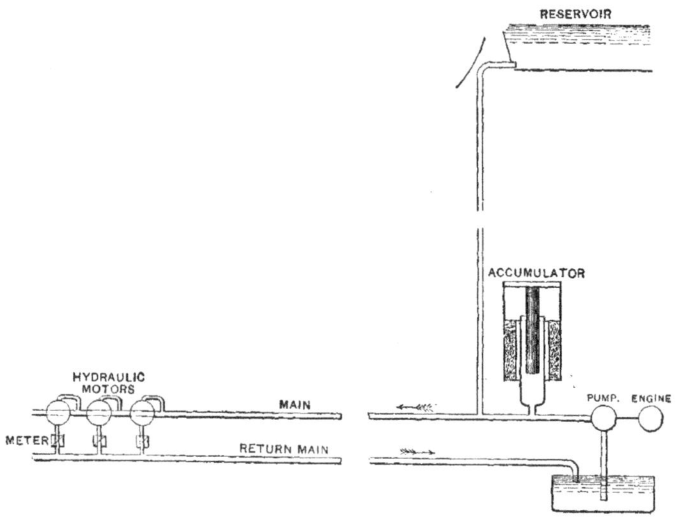

Fig. 272.—Hydraulic transmission.

In Fig. 272 is shown in a diagrammatic way the arrangement of an hydraulic transmission, which may be an open system receiving its pressure and supply from an elevated reservoir, or a closed system with high pressure from a pumping plant. In an open system the water from the return main may be pumped to the reservoir and in the closed system it may be pumped directly into the main with an accumulator to regulate the pressure shock.

In long transmission an accumulator should be installed near every motor and lift-cylinder.

Any of the various designs of rotary motors may be used if ports are made large enough to pass the water at its desired speed.

A reciprocating motor-engine comes under the same condition and is preferred for moderate speeds.

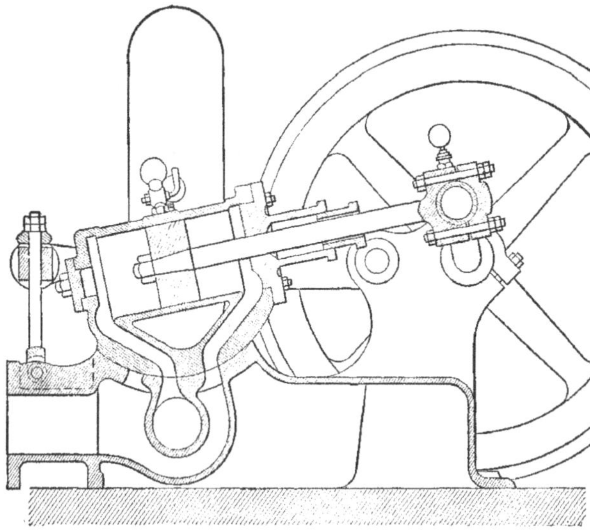

FIG. 273.—Hydraulic oscillating engine.

In Fig. 273 is shown an elevation in section of an oscillating engine of Swiss design so simple that no valve-gear is required. The cylinder trunnions are pivoted in side bars which in turn are pivoted to the main pillow-block with adjustment links to regulate the pressure on the port faces.

The efficiency of these engines is about 80 per cent. under full load.

For high efficiency under all loads there is probably no better motor than the impact or Pelton wheel as its power varies with the area of the nozle, which is made adjustable to meet the load.

In Fig. 274 is shown a section and view of a multiple lift or ram for a crane such as are much in use at the coaling and freight-loading docks in Great Britain. The height of lift equals the length of the ram lift, multiplied by the

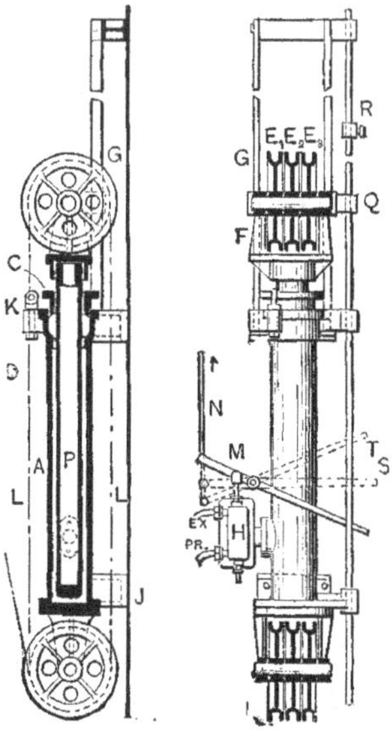

FIG. 274.—Hydraulic lift.

number of sheaves. The powerful lifts attached to the traversing cranes on the great coal docks raise a car and its load of coal (total 20 tons) to a height of 44 feet.

In Fig. 275 is shown one of the great traversing hydraulic cranes of the Wellington Harbor docks, New Zealand, with its lift com-

FIG. 277.—Hydraulic elevator, horizontal plunger.

pounded for the high-pressure system, lifting a package and swinging it over the hatch of a vessel.

These gantry cranes are illustrated at Fig. 275, which shows part of an installation of ten cranes constructed by the Hydraulic Engineering Company of Chester for the Wellington Harbor Board, New

Fig. 275.—Hydraulic goods-loading crane.

Fig. 276.—Fixed hydraulic crane.

Zealand. These are of double power for raising loads of 15 or 40 cwts., the height of lift being 84 feet, and they are fitted with hydraulic gear for luffing the load between the maximum rake of 37 feet and the minimum rake of 14 feet 6 inches.

The cranes sit upon gantries which span two lines of rails and admit of locomotives passing beneath, the gantries being also utilized

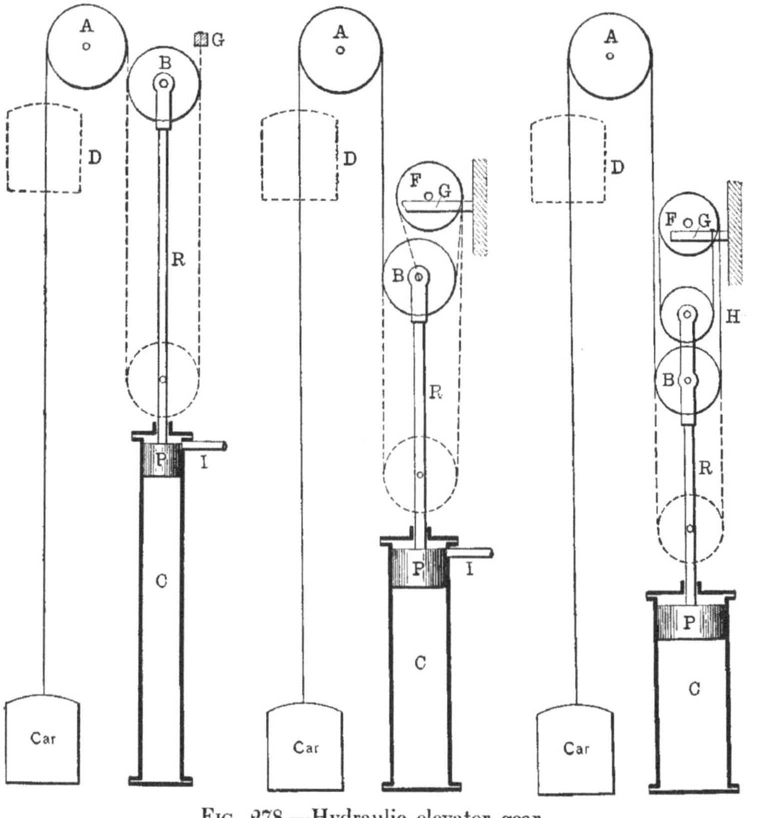

FIG. 278.—Hydraulic elevator gear.

as a platform or bridge to enable passengers to pass to or from the vessels and for crossing the railway lines.

Another model of the great hydraulic cranes is shown in Fig. 276, consisting of a base of masonry upon which rests the revolving crane and hydraulic lifts for lifting and swinging the cars over the hatches and dumping the coal by tipping the cradle and car.

Of the types of hydraulic lifts and elevators in use for passengers and freight, the single-plunger lift, after many years' trial at low lifts, has finally reached a lift height of 282 feet with a single plunger of $6\frac{1}{2}$ inches diameter, a total load of 1,617 pounds, and water pressure of 180 pounds per square inch, with a speed of over 400 feet per minute. Five of these lifts are running in the Trinity Building, New York City.

The multiple-geared short-plunger lift, as shown in Fig. 277, is much in use for passenger and freight service; the cylinder lying horizontally on the cellar floor gives easy access to every part.

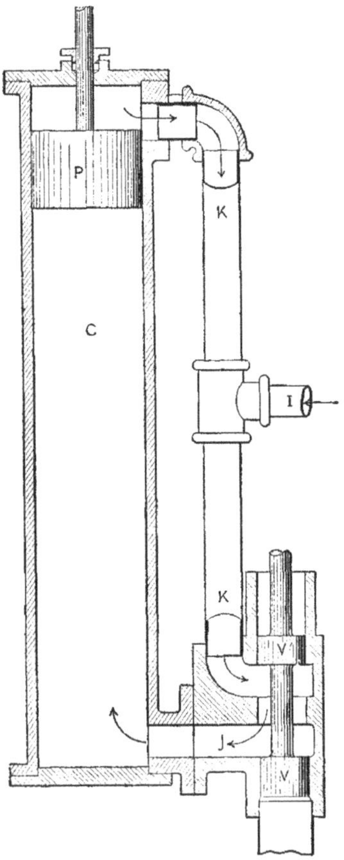

FIG. 279.—Transfer valve.

Various methods by multiple gear are in use, by shortening the cylinder and its piston travel in the pull-down system of hydraulic elevators, by which plan the water pressure is above the piston during the up trip of the car; the down trip being made by transferring the water from above to below the piston with the pressure equalized on both sides of the piston, the overbalance of the car being sufficient for changing the water during the down trip.

In Fig. 278 we illustrate the three kinds of gear used for shortening the cylinder and piston travel; the left-hand figure is a two-to-one lift, the central figure a three-to-one lift, and the right-hand figure a four-to-one lift.

In Fig. 279 is shown a section of the cylinder, transfer pipe, and valve of the pull-down system of operating elevators in which K, K, is the transfer pipe; I, the pressure inlet which is constant; V, V, the spool valve which, as shown, is transferring the water from above to below the piston by its upward motion produced by the descending overweighted car. The upward movement of the valve first locks the piston and a further movement exhausts the water through J, O, allowing the pressure to push the piston down.

CHAPTER XVII

HYDRAULIC MINING

FROM the earliest ages the washing of the auriferous sands and gravels of the earth's surface has been in practice for obtaining the gold that enriched the old potentates of wealth. Europe and Asia have had their eras of the golden fever, and in America has culminated its volume of product.

The earliest implement for washing the precious metal from its earthy envelope was probably the wooden tray, or pan since made of metal, and followed by the cradle, the tom or long tom, and the sluice.

FIG. 280.—Hydraulic gold washing.

In Fig. 280 is illustrated the primitive method of hydraulic mining as practised from the earliest times and now in general use in all countries.

The cradle retains the coarse gravel pebbles and large stones to be thrown out from the sieve hopper; the light sand washes away in the drain while the gold is caught on the riffles and in the cradle beneath the sieve hopper.

The cradle is generally of home-made design and manufacture as shown in section of a common type, Fig. 281. The hopper sieve may be of wire, No. 6 mesh, or of sheet iron punched full of holes, about $\frac{3}{16}$-inch diameter, which will let the fine sand and gravel pass, while the larger material may be readily thrown out and the nuggets seen and saved. The hopper is removable and the inclined apron may be riffled and supplied with mercury.

The riffle deposits are scraped out and further concentrated in a pan or separately amalgamated.

For facilitating the washing for gold in frosty weather and in places where the water-supply is limited or only snow or ice at hand,

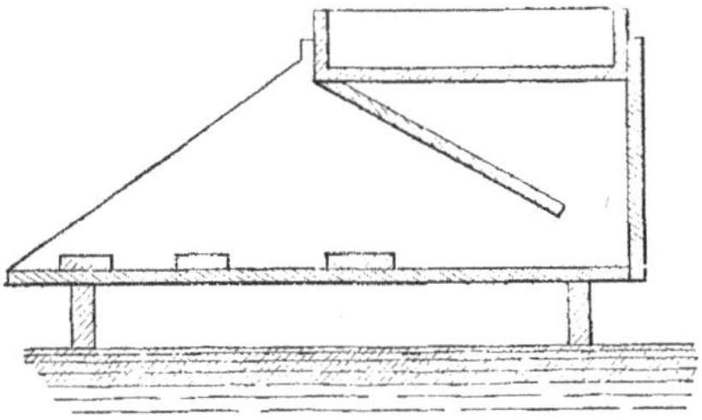

FIG. 281.—Hydraulic rocker.

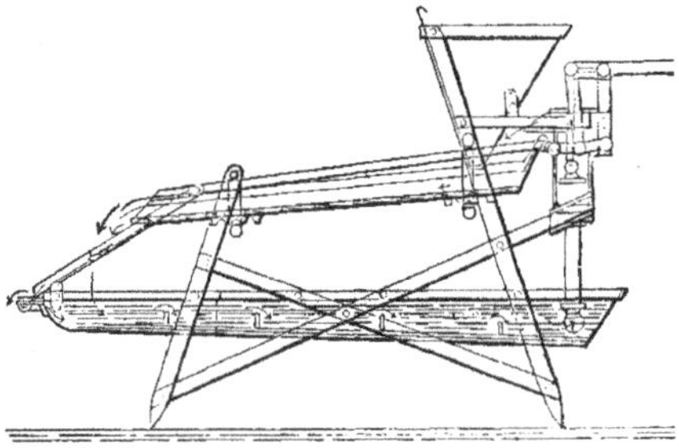

FIG. 282.—Klondike rocker.

the Klondike Mining Machine, Fig. 282, has furnished the means of washing the gold gravels in the winter months. In this machine the gold-bearing gravel is shovelled into a hopper and fed to the riffle pan, which is vibrated by a link from the pump handle. The pump supplies water from the settling pan to the rocking riffle, from which the gravel is separated by a fine sieve, the water falling into the settling pan beneath and kept from freezing by a fire underneath. This machine makes possible winter gold washing in Alaska.

The tom or long tom is usually a rough trough about 12 feet long, from 15 to 20 inches wide at the top, 30 inches at the lower end, averaging 8 inches in depth. It is supported on logs and set at an incline of about 1 inch per foot. The lower end is bevelled and closed by a perforated plate of iron nearly level, to separate the coarser gravel and allow the water and sand to pass over the riffle trough, which is also set on an incline.

In the use of the tom, continual stirring of the sand is required for its best work. By careful management the riffle trough may be charged with mercury and the flake gold amalgamated. There is a great variety in the forms and methods of manipulation of riffle

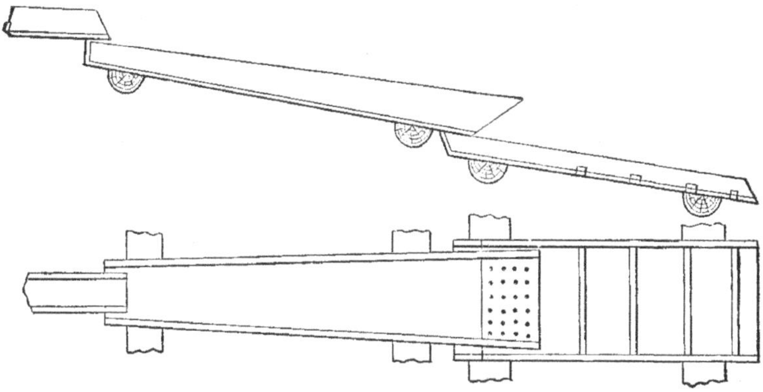

FIG. 283.—The tom or long tom.

troughs and other devices in use for amalgamating purposes in the various mining districts and complicated machines in the reduction mills, which we cannot give space to describe and illustrate. Hydraulic effect plays a part in all of their operations.

HYDRAULICKING

The washing down of auriferous gravel banks by means of streams of water under great pressure has developed into one of the most important sources of our gold-gathering industry.

Where natural falls or heads of water are not found near the proposed hydraulic washings, expensive ditches or canals, flumes, and pipe-lines have been constructed to furnish the great volume of water required for this method of mining. The pumping system of water-supply has also been made available in many places and localized for individual mining power.

In Fig. 284 is shown the simple appliance for excavating an

auriferous bank with a small stream and earth sluice with the nozle fixed in a simple frame.

A sketch of a hydraulic-balanced giant nozle is shown in Fig. 285. They are made with terminal outlets of from 2 to 8 inches

FIG. 284.—Simple hand-nozle stream.

diameter and for pressure streams of 50 to 200 pounds per square inch. They require a strong anchorage to hold them in place against the great back-thrust. The movable flange joint, B, B, held together by the through bolt K, and the trunnion joint at E, allows of an easy adjustment of the nozle direction.

These nozles are known in mining phrase as the "Monitor," "Hydraulic Chief," "Dictator," etc., and are made in various lengths from 6 to 15 feet and from 3- to 8-inch outlet. The amount of material that can be sluiced off per miner's inch of water depends much upon the location, grades, and dump, as well as upon the nature of the material, and varies from 1 to 6 cubic yards.

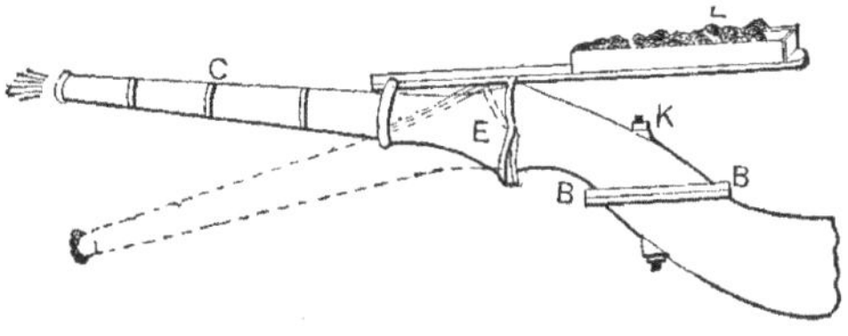

FIG. 285.—Giant hydraulic nozle.

A general view of an hydraulicking plant is shown in Fig. 286, illustrating the principal phases of this method of obtaining the metal from auriferous gravel banks.

The undercuts or amalgamated riffle-boxes are placed at convenient locations in the long sluices, where the work of saving the largest percentage of metal requires the greatest care in management.

One form of undercurrent that has given good satisfaction is shown in Fig. 287. The grade of the sluice A, leading from the dig-

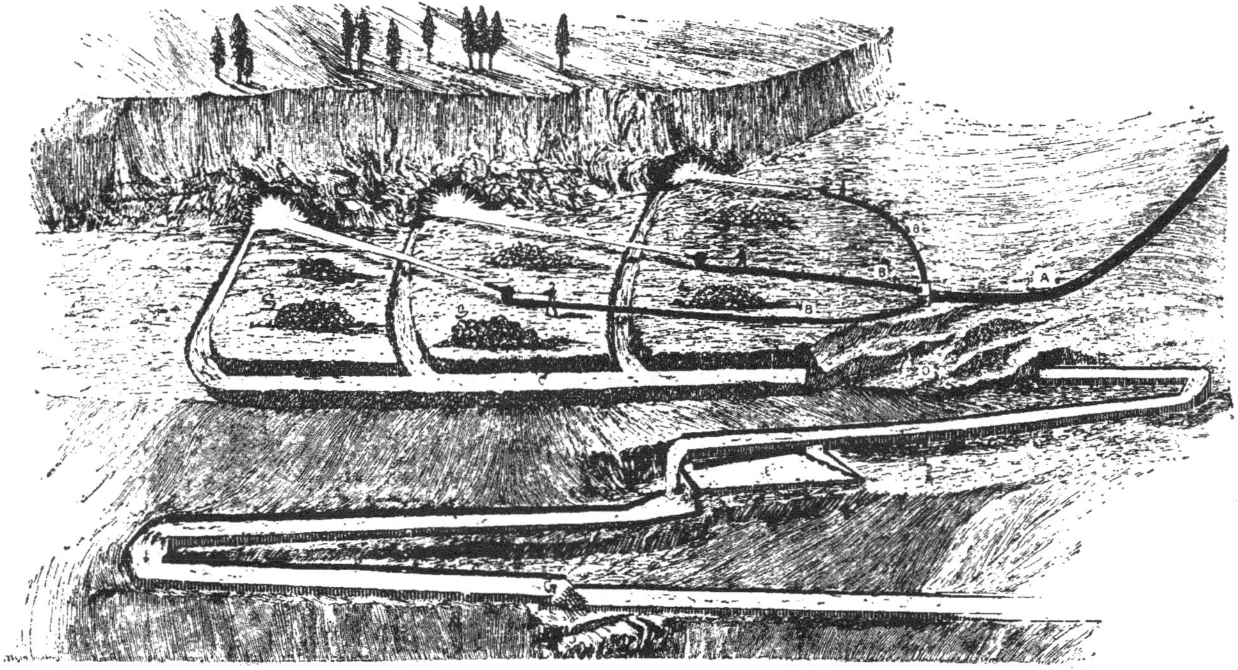

Fig. 286.—Hydraulic mining plant.

A, main pipe-line; B, B, B, distributing hose; C, main sluice; E, subcurrent; S, S, stone picked from primary sluices.

gings, may vary from one-third of an inch to one inch to the foot. From this point down the creek-bed the undercurrents are placed at proper intervals. In the bottom of the flume A, near the point of discharge, a "grizzly" of wooden bars covered with iron plates, or entirely of iron, is placed, with sufficient space left between the bars to precipitate all but the coarsest material into the box below. The bowlders, large pebbles, etc., go directly to waste over the grizzly. The eliminated material and water is cast upon and spread over a large platform, B, four or five times the width of the sluice above, and as long as circumstances will admit. While the grade is steeper, the velocity of the water is sensibly checked. Riffles, made in sections for convenience in removing and cleaning, are placed in the bottom.

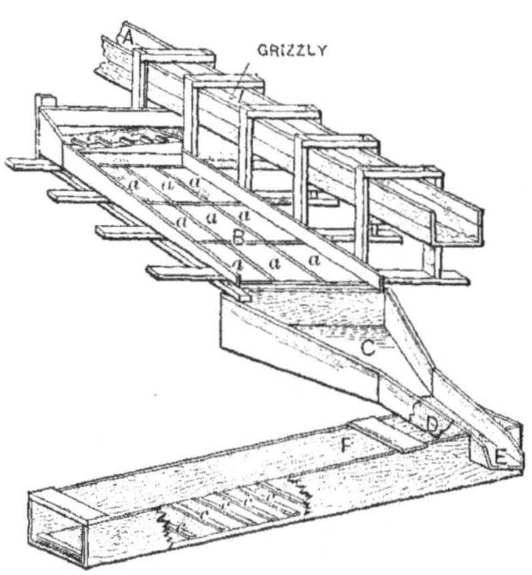

FIG. 287.—Undercurrent amalgamator.

A portion of the débris and presumably all the gold that has escaped lodgment or amalgamation, together with a portion of the water, are spread over B. The depth being materially decreased, some particles of gold that have been kept in motion by the velocity of the water and débris in the main flume come to rest, and the lighter earthy matters with some gold pass on. Sometimes, where the fall will admit, a second undercurrent is placed below the first, as shown at C, with a second and finer grizzly, as at D, the waste material passing off at E, and the sand, with any free gold or amalgam that may have escaped the upper undercurrent, passing into box F, which contains riffles C across the bottom, where it is caught and retained. Undercurrents are used quite extensively in the working of tailings.

Along river banks the facilities for hydraulicking are greatly increased by the use of timber floats or scows on which are mounted steam-pumps for supplying the giant nozles with water under any desired pressure, and thus enable the working of gold-bearing gravels along the banks of rivers at low cost.

It has been long known that the river bottoms in the gold regions are the harbingers of the precious metal; but the early attempts to rescue its wealth by suction and jet dredging were not profitable, and the successor, the chain bucket dredge, has come to stay and do the great work of lifting, washing, and separating the precious metal

FIG. 288.—Hydraulicking from a timber float, New Zealand.

from the mud, sand, and gravel of the river's bottom at a profit. In 1905 there were forty river dredges in operation in California alone, and were extending their useful work throughout the other gold-bearing States, Alaska, and the British colonies.

CHAPTER XVIII

CANALS, DITCHES, CONDUITS, AND PIPE-LINES

THE great ditches, canals, and pipe-lines in California, Colorado, Idaho, and Arizona; in India, Europe, Egypt, and the great barrage of the Nile, show the possibilities in like works in all the arid regions of the Western world. In the following tables are listed some of the leading irrigation and mining canals, not including lateral branches which may be many times the length of the main canal in their distributing systems.

In crossing ravines, passing along the abrupt faces of precipices, or connecting the ditch with the bulkhead, flumes are commonly used, although they are objected to on account of the danger from fire and cost of repairs. They are set on straight lines or very easy curves, and are of smaller area than the ditches with proportionally heavier grades. The grades sometimes reach even 35 feet per mile. Owing to the great irregularity of surface, many of these ditches have miles of fluming. The ordinary style of construction is shown in Fig. 289. The planking is commonly of heart sugar-pine, $1\frac{1}{2}$ to 2 inches thick and 12 to 18 inches wide. To effect a good seam, pine battens 3 inches wide by $1\frac{1}{2}$ inches thick are placed over the joints. Bents of square timber well mortised together, set from 4 to 6 feet apart, support and strengthen the flume. The posts spread out toward the bottom, so that the sills are somewhat longer than the caps, and they are generally sawed so as to extend about 2 feet beyond the foot of the posts. Where a flume is carried along the steep mountain-side, it is secured to the solid bed of rock as shown in Fig. 290. They are set in as close as possible to the bank, as a precaution against accidents from storms, winds, or snowslides.

As the importance of hydraulic mining began to be recognized, and the necessity of securing some means for conveying water across deep ravines and through very rough and broken districts of country not adapted to the erection of flumes, sheet-iron pipes were introduced

TABLE XXXII.—CANALS AND DITCHES.

NAME OF CANAL.	COUNTRY.	Length in miles.	Bottom width in feet.	Depth in feet.	Slope.	Discharge in cubic feet per second.
Upper Ganges	India	456	170	10	1 in 4,224	6,000
Lower Ganges	“	531	216	8	1 in 10,560	6,500
Western Jumna	“	433	......	......	......	2,372
Eastern Jumna	“	130	......	......	......	1,068
Varee Doab	“	466	120	5.5	......	2,500
Sutlej or Sirhind	“	503	190	6	1 in 4,800	3,500
Agra	“	137	70	10	1 in 10,560	1,100
Sone, Western	“	125	180	9	1 in 10,560	4,500
Sone, Eastern	“	170	180	9	1 in 10,560	4,500
Soonkasela	“	190	90	8	1 in 3,520	3,000
Ibrahimia	Egypt	170	113	Lois 6¼	1 in 16,600	......
Main Delta (Flood)	“	....	174	20	1 in 15,000	10,846
Main Delta (Summer)	“	....	174	10	1 in 12,000	3,943
Sirsawiah (Flood)	“	....	20	17	1 in 20,633	1,138
Nagar (Flood)	“	....	20	10	1 in 14,000	906
Sahel (Flood)	“	....	20	12	1 in 25,641	981
Subk (Flood)	“	....	13	6	1 in 20,000	114
Grand Canal of Ticino	Italy	31	......	......	1 in 1,860	1,851
Cavour	“	53	131	11	1 in 2,000	3,250
Ivrea	“	92	27.7	......	......	700
Cigliano	“	102	53	......	......	1,760
Rotto	“	84	......	......	......	600
Muzza	“	....	......	......	......	2,175
Martesana	“	....	......	......	......	738
Henares	Spain	28	8.23	4.9	1 in 3,067	177
Isabella II	“	50	......	......	......	89
The Royal Jucar	“	25	......	......	......	911
Marseilles	France	52	9.84	7.87	1 in 3,333	424
Ourcq	“	....	11.48	4.92	1 in 9,470	......
Crappone	“	33	26	6.5	......	500
Verdon	“	51	......	......	1 in 5,000	212
Alpines	“	....	......	......	......	480
St. Julien	“	18	......	......	1 in 3,333	165
Carpentaras	“	....	33	......	1 in 4,000	212
Del Norte	Colorado, U.S.A.	50	65	5.5	1 in 660	2,400
Citizens	“	45	40	5.5	1 in 1,760	1,000
Uncompahgre	“	32	24	......	1 in 1,560	725
Fort Morgan	“	28	30	3.5	1 in 3,300	340
Larimer	“	45	30	7.5	......	720
North Poudre	“	30	20	4.0	1 in 2,640	450
Empire	“	32	60	5.5	......	1,400
Grand River	“	....	35	5.0	1 in 2,880	......
High Line	“	70	40	7	1 in 3,000	1,184
Central District	California	60	60	6	1 in 10,000	720
Merced	“	8	70	10	1 in 5,280	3,400
San Joaquin and King's River	“	39	55	4	1 in 5,280	......
Seventy-Six	“	....	100	4	1 in 3,520	......
Calloway	“	32	80	3.5	1 in 6,600	700
Turlock	“	80	20	10	1 in 666	1,500
Idaho Mining and Irrigation Co.'s	Idaho	75	45	10	1 in 2,640	2,585
Idaho Canal Co.'s	“	43	40	4	1 in 3,520	......
Eagle Rock and Willow Creek	“	50	30	3	1 in 880	......
Phyllis	“	54	12	5	1 in 2,640	250
Arizona	Arizona	41	36	7.5	1 in 2,640	1,000

TABLE XXXIII.—MINING DITCHES IN CALIFORNIA.

NAME.	Length of ditch.	Top of ditch.	Bottom of ditch.	Depth of ditch.	Cost of ditch.	Average grade per mile.	Discharge in miners' inches.
	Miles.	Ft.	Ft.	Ft.		Ft.	
North Bloomfield.....	55	8.65	5	3.5	$422,000	14	3,200
Milton Company.....	100	6	4	3.5	259,000	145	3,000
Eureka Lake........	18	.....	.....	.....	430,000	....	2,800
San Juan............	45	.....	.....	.....	293,000	....	1,300
Excelsior............	33	8	5	4	....	9	1,700
Union...............	15	8	4	3.5	....	13	1,200
Boyer...............	15	8	4	3.5	....	13	1,200
Spring Valley*.......	52	6	4	3.5	....	....	2,000
Hendricks...........	46.5	6	4	2	136,000	9.6	....
La Grange..........	20	9	6	4	450,000	7.5	3,000

* Including 3½ miles of 30-inch iron pipe.

on account of their lightness and great tensile strength. They have been made of Nos. 16, 14, and 12 Birmingham gauge, from 11 to 40 inches in diameter, 20 feet long, and riveted in the horizontal seams, but put together in stovepipe fashion, without rivets or wire to hold

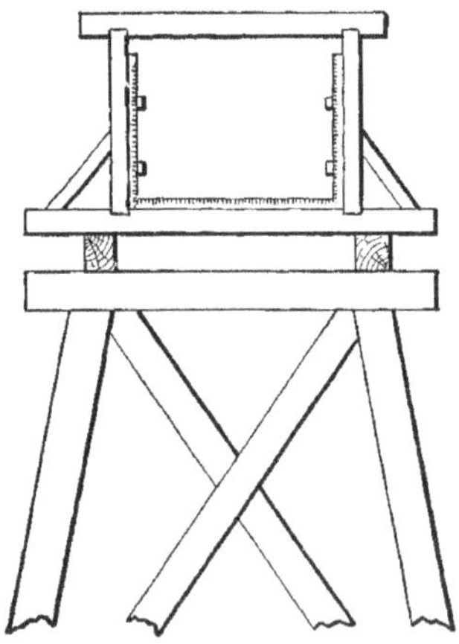

FIG. 289.—Valley flume.

FIG. 290.—Mountain-side flume.

the joints in place. As floating particles of matter readily render the joints comparatively water-tight even under a pressure of 200 pounds to the square inch, it is only under high heads and sudden shocks that lead joints such as shown in Fig. 292 are used. The

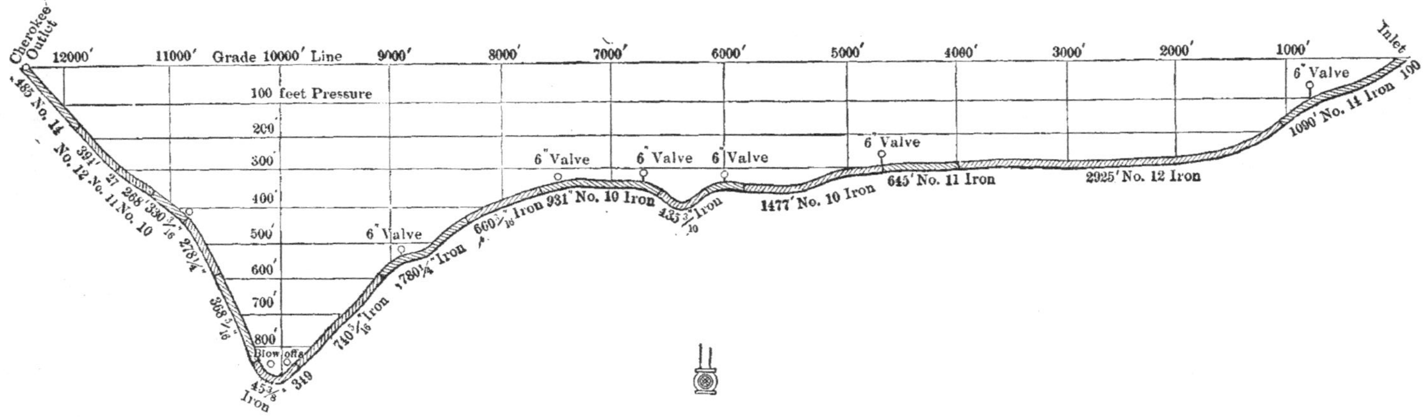

Fig. 291.—Profile of wrought-iron pipe for the Cherokee gravel mines, Butte Co., Cal.

This line of wrought-iron pipe, 30 inches diameter with circular seams single-riveted, longitudinal seams double-riveted, is 12,347 feet in length in an inverted siphon across a valley and gorge 900 feet below the source of supply. It discharges 52 cubic feet of water per second.

lead is forced between the iron sleeve and the pipe, while an internal flange is riveted to one length of pipe in such a way that the other fits over it. The spacing of the rivets has proved to be a matter of considerable importance. For example, a pipe 12 inches in diameter made of No. 18 iron was formerly riveted in the longitudinal seams every 1 to $1\frac{1}{4}$ inches, while the round seams have been left pretty open, with rivets set 3 inches apart. Now, in the better class of pipe adopted, the round seams are made with rivets three-quarters of an inch apart, and the longitudinal seams are double-riveted, with rivets 1 inch apart in the row, and about half an inch apart from one row to the other. If such pipes are dipped in asphaltum to protect them from the weather, they will last for many years. The thickness of the iron is usually proportionate to the head of water and the diameter of the pipe. Pipes made of the different sizes of iron here mentioned will stand the following strains per sectional inch:

No. of Iron.	Made to stand strain per sectional inch, pounds avoirdupois.
12	7,000 to 9,000
12 to 9	9,000 to 12,000
9 to $\frac{3}{16}$	12,000 to 14,000
$\frac{1}{4}$ to $\frac{3}{8}$	17,000 to 18,000

The head of the water in pounds avoirdupois, multiplied by the diameter of the pipe in inches and divided by the above coefficients, gives twice the thickness of the iron to be used. Allowance must be made for the security required; that is, if the breakage of the pipe

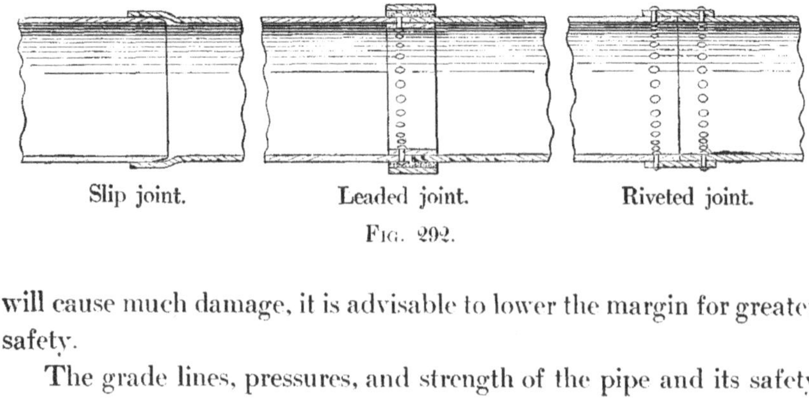

Slip joint. Leaded joint. Riveted joint.

FIG. 292.

will cause much damage, it is advisable to lower the margin for greater safety.

The grade lines, pressures, and strength of the pipe and its safety from water-ram are matters of importance in the installation of a long-distance transmission for power or for hydraulicking purposes.

In the early constructions the slip-joint or stovepipe joint was much in use and answered its purpose fairly well; later the pipe-joints were butted with a riveted sleeve or band inside on one end of each section and a sleeve or band on the outside with a driven lead packing; while the more substantial pipe-lines have their circular joints riveted and caulked throughout their length.

In Fig. 292 is shown the different methods of joining the pipe sections.

The air-valve is an essential feature of safety from water-ram in undulating long lines of pipe for liberating the air caught in the uptakes or swept into the pipe-line by a rapid intake. A section of a float air-valve is shown in Fig. 293. To the valve disk is fixed an inverted open copper float, which by its weight keeps the valve open and allows the air to escape until the water comes, when the air-balanced float is lifted by the water and closes the valve.

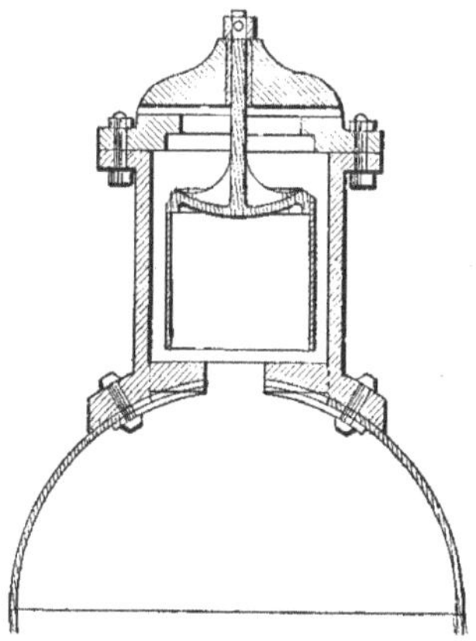

FIG. 293.—Air-valve.

GREAT MODERN WORKS

Of the great reservoirs of the world, old and new, the pride of Egypt, the artificial Lake Moeris leads in the vast volume of its contents and the great canal which fed it. Herodotus classed it as a greater wonder than the building of the Pyramids. Its capacity has been computed by modern engineers to have been equal to 330,000,000,000 cubic feet between high- and low-water mark. The greatest hydraulic work of modern times is the barrage of the Nile, the Assouan Dam.

At the spot where the dam was built, 600 miles above Cairo, the Nile is $1\frac{1}{4}$ miles wide during flood, but during the winter it is divided into five channels with intervening islands. It must not be supposed that it contracts to mere rills, for at its lowest it carries five or six times as much water as the Ohio River at mean annual flood, so that each of its channels is a fine river. The current runs at headlong rate down the Assouan Cataract, the speed in one channel being 15 miles an hour and the depth 30 feet. The fall is 3 feet in 200 feet. The difficulty of closing such a channel was enormous, even with the other channels left open as spillways. The barrage dam is a straight masonry wall, closing the passage of the Nile from shore to shore. Its length is 6,500 feet; the maximum height from foundation, 130 feet; the extreme difference of water-level above and below being 67 feet. The up-stream face is perpendicular, or nearly so, while that on the down-stream side is battered to reduce the width on the top to 22 feet. When full, the reservoir behind the dam will hold 40,000,000,000 cubic feet of water.

The first thought that arises in connection with the dam across such a river as the Nile is that the reservoir will silt up; and there have been engineers of reputation who have boldly prophesied such a result in this case. Of course, the danger was foreseen, and was provided for by arranging that the flow of the river shall be through openings in the wall. There are 140 sluice-gates 23 feet high by 6 feet 6 inches wide, and 40 gates 11 feet 6 inches by 6 feet 6 inches. Of these, 130 are on the Stoney principle, and can be moved by hand under a pressure of 450 tons. During the flood period, when the water is silt-laden, all the gates will be open, and the river will roar through the openings. After the flood, when the discharge has fallen to about 2,000 tons per second, the gates without rollers will be closed, and then some of those with rollers. Between December and March the reservoir will be gradually filled, the surplus running through the upper sluices. The reopening of the sluices will take place between May and July, according to the state of the Nile and the requirements of the crops.

It is of no advantage to have water merely flowing through the Nile bed in summer. Where it is wanted is in the irrigation canals that traverse the country up to the confines of the desert, and pass through the limestone range into the Fayoum. To enable the water to be discharged into the Great Ibrahimiyah Canal, a barrage has been built across the river at Assiut, to back up the water and divert it into the canals. This structure is similar in principle to that built at the head of the delta by French engineers many years ago. The total length is 2,750 feet, and it includes 111 arched openings of 16 feet

Fig. 294.—The Assouan Dam and Locks.

4 inches span, capable of being closed by steel sluice-gates 16 feet in height. The piers and arches are founded on a platform of masonry 87 feet wide and 10 feet thick, protected up and down by a continuous and impermeable line of cast-iron grooved and tongued sheet piling with cemented joints. This piling extends into the sand bed of the river to a depth of 23 feet below the upper surface of the floor, and this cuts off the water and prevents the undermining action which caused so much trouble and expense in the case of the old barrage. The height of the roadway above the floor is 41 feet, and the length of the piers up and down stream 51 feet. The river-bed is protected against erosion for a width of 67 feet up-stream by stone pitching, with clay puddle underneath to check infiltration, and down-stream for a similar width with stone pitching with an inverted filter-bed underneath, so that any springs which may arise from the head of water above the sluices shall not carry sand with them from underneath the pitching. The method of working was to enclose the site of the proposed season's work by temporary dams in November, then to pump out and keep the water down by powerful centrifugal pumps, crowd on the men, excavate, drive the cast-iron sheet piling, build the masonry platform, lay the aprons of puddle and pitching, and get the work some height above low Nile level before the end of June, so that the temporary dams should not need reconstruction after being swept away by the flood.

The busiest months were May and June, when, in 1900, the average daily number of men was 13,000. To keep the water down, seventeen 12-inch centrifugal pumps, throwing enough water for the supply of a city of two million inhabitants, had to be kept going, and in a single season as many as $1\frac{1}{2}$ million sand-bags were used in the temporary dams. A thousand springs burst up through the sand, each one of which required special treatment. It is these difficulties of construction which show us how far we have advanced beyond the engineering of the ancient Egyptians.

The Assouan Dam was constructed in four years. In 1898 some preliminary work was done, and surveys were carried out. The foundation-stone was laid by H. R. H. the Duke of Connaught on February 12, 1899. During that year excavation was completed over almost one-half of the total length of the dam, and embankments were formed on three out of the five channels. The low summer level of 1900 made it possible to excavate and lay the foundation masonry of all except the western channel, down which the whole discharge of the Nile was sent. After the flood of that year had passed, masonry work was continued, and preparations were made for damming the western channel, the only portion in which the foundation masonry was not laid by the end of 1900. Over 3,000 tons of masonry were sometimes completed in a day, and as much as 1,700,000 cubic feet of masonry were laid during the month just before the flood came down. The western channel was dammed in 1901, and the foundations got in before the flood. When it came this channel and the central channel were the only portions submerged. After the flood had diminished in October, the masonry was rapidly pushed forward. Since the beginning of 1902 the works have been practically completed; the dam was finished a year before contract time. The first cataract now no longer exists as a bar to navigation. A navigation canal has been constructed round it, about 6,500 feet in length, with a ladder of four locks, each 228 feet long and 32 feet wide. There are five lock-gates, 32 feet wide, and varying in height up to 60 feet. They are of a different type to ordinary folding lock-gates, being hung from the top on rollers, and moving like a sliding coach-house door. A channel 75 feet wide was cut through the narrow rapids north and south of the dam to improve the channel for boats.

This great work is a successful example of what can be done under difficulties in the barrage of great water-ways for the irrigation of the vast arid regions of the United States; of such work the entering wedge is the Truckee-Carson and Minidoka irrigation project of the United States Reclamation Service.

FIG. 295.—The Sluices of the Assouan Dam.

CHAPTER XIX

MARINE HYDRAULICS

WATER RESISTANCE, SKIN-FRICTION, STREAM LINES

THE resistance of sea water to a surface, say a thin plate when submerged and moving normally at right angles with its surface at 10 feet per second, has been found by experiment to be 112 pounds per square foot, Fig. 296, and also that when the plate moves

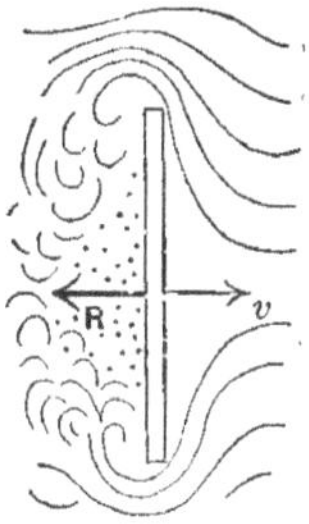

FIG. 296.—Resistance at right angle.

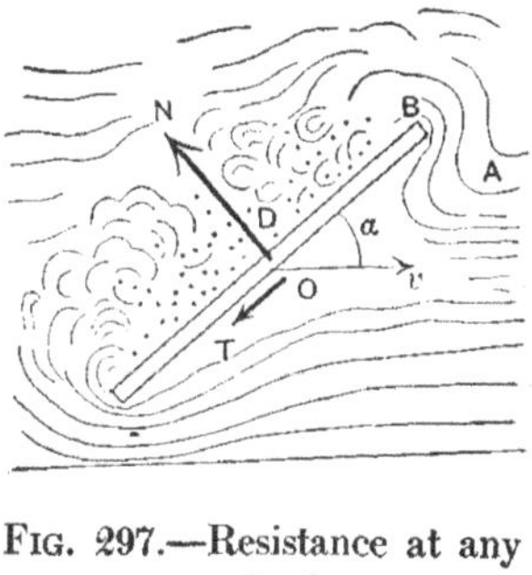

FIG. 297.—Resistance at any angle A.

obliquely, as in Fig. 297, the resistance was as 90°×sine of the angle a; to the normal resistance of 112 pounds per square foot; also, that the depth of submersion had no influence upon the resistance, and with sufficient velocity a vacuum is possible behind the plate. Then the formula for the resistance of a thin plate moving at right angles with its surface will be: $R = Caw\,\frac{V^2}{2g}$ in which C is a coefficient; a = area in square feet; w = weight of sea water per cubic foot (64 pounds). The coefficient as obtained from the experiments is: $\frac{2g \times 112}{64 \times V^2} = 1.127.$

SKIN FRICTION

The retardation of a body moving through sea water, by the friction of the water in contact with the surface of the body, has been the subject of much experiment, resulting in finding that the friction varies as the square of the velocity with a variable due to the roughness of the surface and to the density of the water; also that the friction is independent of the pressure and uniformly due to the amount of rubbing surface. It was also found that the coefficient decreased with the length of a continuous surface. The formula $CSw\frac{V^2}{2g}$ in which C is a coefficient based upon experiments at a velocity of 10 feet per second for surfaces 2 feet, 8 feet, 20 feet, and 50 feet long, with friction resistance of .41 – .32 – 28 and .25 of a pound per square foot, respectively, from which the following table was obtained for the different lengths and kinds of surfaces. S = square feet of surface.

By inverting the formula for a velocity of 10 feet per second we have $\frac{.41\times 64.4}{64\times 100}=C=.0041$ as in the table, and so on.

TABLE XXXIV.—COEFFICIENT OF SKIN FRICTION FOR VARIOUS LENGTHS AND ROUGHNESS OF SURFACE (FROUDE).

Character of Surface.	Value of C from equation $CSw\frac{V^2}{2g}$ When the board was 2 feet long.	8 feet long.	20 feet long.	50 feet long.
Varnish . . . . C =	0.0041	0.0032	0.0028	0.0025
Paraffine . . . . "	0.0038	0.0031	0.0027	. . . .
Tinfoil	0.0030	0.0028	0.0026	0.0025
Calico	0.0087	0.0063	0.0053	0.0047
Fine sand	0.0081	0.0058	0.0048	0.0040
Medium sand	0.0090	0.0062	0.0053	0.0049
Coarse sand	0.0110	0.0071	0.0059	. . . .

These numbers multiplied by 100 also give the mean frictional resistance in pounds per square foot of area of surface in each case (v = 10′ per second), considering the heaviness of sea water, 64 pounds per cubic foot, to cancel the 2g = 64.4 feet per second of equation of the preceding paragraph.

STREAM LINES, BUOYANCY, DISPLACEMENT, TONNAGE

The stream lines of a vessel that is to be propelled through water, in their form regulate its capacity for speed and load for an assigned power.

They should conform to the object for which the vessel is to be employed, such as for velocity or for carrying bulk; or for the relations of both as designed to meet the combined requirement.

When a vessel is immersed in water, it is buoyed up by a force equal to the weight of the water that it displaces and which is also the gross weight of the vessel in air, or previous to immersion.

The upward pressure or buoyancy of the water is assumed to be exerted at its centre of gravity or centre of pressure. In a vessel floating at rest, a line joining the centre of buoyancy and the centre of gravity of the floating body is vertical and is called the axis of equilibrium.

If by external force the vessel is careened, the axis of buoyancy still remains vertical; while the axis of the centre of gravity of the vessel becomes inclined and their point of meeting, the metacentre, decides the measure of stability of the vessel by its distance above the centre of gravity of the vessel and also above the centre of buoyancy of the water. The centre of buoyancy shifts in the same direction as the heeling of the vessel. (See Chapter II on flotation.)

The displacement of a vessel is the weight of the volume of water which it displaces, and is equal in tons of 2,240 pounds to the volume of the vessel beneath the water-line in cubic feet, divided by 35 for sea water.

The tonnage of a vessel is its cargo capacity or its volume in cubic feet between the no-load and the full-load water-lines, which divided by 35 equals tons of 2,240 pounds each.

The resistance of vessels or the net power required to propel them through the water depends upon the fineness of their water-lines fore and aft, gross displacement, and character of their wetted surfaces.

This is expressed by the general formula, $R = C\ S^2 D^{\frac{2}{3}}$ in which R = resistance in foot-pounds per minute; C = a coefficient of the form of the vessel, varying between .50 and .75. S^2 = square of the speed in feet per minute; $D^{\frac{2}{3}}$ = the cube root of the square of the displacement at the actual load line in tons of 2,240 pounds.

The velocity of progression of waves in channels or canals varies with their length and from experiments, that of a long free running wave is equal to the velocity that a body acquires in falling from a height equal to half the depth of the canal. Therefore, in a canal 8 feet deep, the velocity of a long wave is 16 feet per second, from which the following table was computed:

TABLE XXXV.—WAVE VELOCITIES AND LENGTHS IN CANALS (FROUDE).

Velocity of wave in knots per hour.	Wave-length, in feet.	Velocity of wave in knots per hour.	Wave-length, in feet.
6	19.513	17	156.646
7	26.559	18	175.618
8	34.690	19	195.672
9	43.904	20	216.812
10	54.203	22	262.343
11	65.585	24	312.209
12	78.052	26	366.412
13	91.602	28	424.952
14	106.238	30	487.827
15	121.956	35	663.987
16	138.760	40	867.248

A boat dragged along at any velocity less than the speed of a long wave in the canal would leave a train of waves behind it of such short

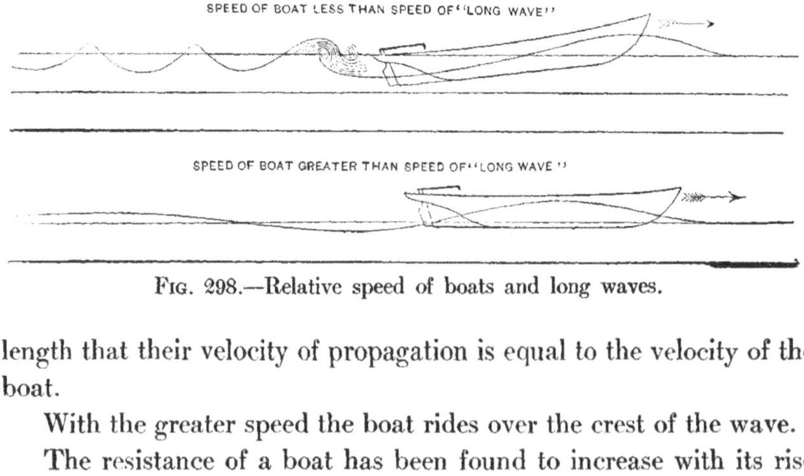

FIG. 298.—Relative speed of boats and long waves.

length that their velocity of propagation is equal to the velocity of the boat.

With the greater speed the boat rides over the crest of the wave.

The resistance of a boat has been found to increase with its rise behind a long-wave crest and decrease after passing the crest. In experiments on a canal with a long wave of eight miles per hour, the proportional force required for different speeds of a boat has been found as follows:

	Pounds.
At 6.2 miles an hour	250
" 7.6 " "	500
" 8.5 " "	400
" 9.0 " "	280

WAVES FROM BOAT MOTION

The waves describe the arcs of circles, parabolas, or hyperbolas, according as the line of motion is perpendicular, parallel, or oblique to the surface. Whatever the direction in which the body moves,

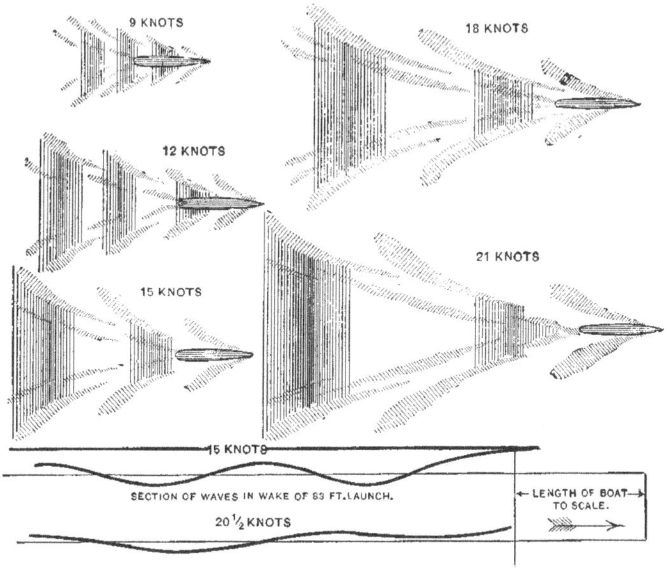

FIG. 299.—Wave forms at various speeds.

the wave lines at the surface show the arcs of curves formed by the intersections of a plane with a paraboloid. However, the movement of these waves shoreward always gives the impression of lateral displacement.

Fish swimming along beneath the surface of calm water create, at its surface, the wave lines mentioned. Should a body move slowly, its point of displacement will be near its section or the base of the paraboloid evolved, and hence its ratio of length to width is reduced and the generated parabolas correspondingly expanded. But if the body moves rapidly, the converse is true, the parabolic branches approaching the line of motion.

A ship moving at the rate of two or three knots per hour generates a bow wave which extends outward in a line almost at right angles with its course. When moving at a rate of twelve to fourteen knots per hour, the general direction of this wave line will form an angle of about 45° with its course. And if a higher rate of speed is attained, this angle is still further reduced.

These relative conditions are illustrated in Fig. 299 for various speeds from 9 to 21 knots per hour.

CHAPTER XX

TIDAL AND SEA-WAVE POWER

The power of the tides by their flow through the narrow-mouthed estuaries from the sea was a valuable contribution to the industries of our forefathers, and is still a valuable franchise wherever there is sufficient tidal flow and storage capacity in the enclosure of the estuary or bay. In past years there were many tide mills in operation in the United States and in Europe, mostly used for the milling of grain, and a few still in use doing good work.

The vast development during the past half century in water- and steam-power and their economy has overshadowed the tide-mills, so that there are but few places now where their intermittent power can be made profitable. The great advance in the efficiency of low-pressure head turbines has, however, given them as tide-wheels a better average speed as well as a longer range of action for each tide than that of the old current wheel, and this may continue the usefulness of tidal power. Wherever there is a 6-foot tide, a submerged turbine using 2,000 cubic feet of water per minute under an average head of 3 feet will give theoretically 11 horse-power. This will require a storage reservoir equal to 350 feet square and 6 feet deep.

There are many estuaries on the coast of the United States with far greater areas than the above and capable of yielding from 25 to 100 horse-power and furnishing water-power for mills during a part of each tidal run, or for the generation and storage of electricity during the four runs of the tide, flow and ebb, in twenty-four hours.

In this way tidal power may still become useful and profitable by the ability to concentrate the tidal work of twenty-four hours into the day's work of eight or ten hours for industrial and lighting purposes, and there are many river estuaries that can be closed and installed with turbines and electrical generators for transmitting current to distant localities for power and light.

The most remarkable variations in the tide are at Chepstow, England, where the rise of spring tides is about 60 feet; at Bristol it is 40 feet; in Mount St. Michael's Bay it is 46 feet; in the Bay of Fundy and on the coast of Nova Scotia it is about 60 feet; while in the Northern Atlantic it is on the average from 10 to 12 feet; at St. Helena only 3 feet; and on the shores of the islands of the Pacific it is barely perceptible.

Where the rise is so extreme, it is produced by the contraction of the sides of the river or estuary, as the Bristol Channel, for instance, or a convergence to one point of wide stretches of coast, as at the Bay of Fundy.

In some cases the phenomenon of the *bore* is produced, which is defined by Colonel Emery as being a peculiar undulation, which announces the arrival of the flood tide in many rivers. It consists of two, three, or sometimes four waves, very short, and succeeding one another rapidly, which bar the whole river, and ascend it to a great distance; they often break upon the crown, and upset everything they meet in their course, and are accompanied by a fearful noise. In the Severn, the bore is stated to be of almost daily occurrence, and sometimes even to attain a height of 9 feet; in the Dordogne it rises from 5 to 6 feet, and travels at the rate of about five miles in thirty-four minutes; in the Seine it does not exceed 3 feet; in the Thames it only exists in a rudimentary state; while in the Hoogly, at Calcutta, it rises about 5 feet, and is transmitted at the rate of about $17\frac{1}{2}$ miles per hour; and in the Menga the rise is said to be 12 feet.

The tide-wheel at East Greenwich on the Thames is, or was, a breast-wheel raised and lowered with the tide, so as to always have a submergence of 4 feet water. The buckets are divided into four steps, so as to prevent any jerking or irregular motion. The wheel turns both with the flowing and ebbing of the tide, having a sluice-gate and tail-gate on each side, one pair being opened when the other pair is closed.

WAVES OF THE OCEAN AND THEIR POWER

Many and widely different theories of wave-motion have been advanced from time to time, of which it is only necessary to state that the older theories are now definitely set aside as erroneous; and that the modern or trochoidal theory is generally accepted as very closely representing the actual phenomena which occur in nature.

While the motion of ocean waves in nature is usually quite complex in character, there are certain simple typical forms of such motion which have been satisfactorily studied, and the geometry and mechanics of which are well understood; and by making suitable combinations of these simple type-waves, the condition of more or less complex and irregular seas can be approximated to and their mechanics investigated with sufficient exactness for most practical purposes.

Prominent among the simple types of waves just referred to, and forming usually by far the most important element of actual ocean waves in nature, are those known as the deep-sea wave and the shallow-water wave. These are the forms of wave-motion which occur in nature in a long series of waves, in which each successive wave is an exact reproduction of the one just preceding it, so that the wave goes on repeating itself indefinitely. While these conditions are rarely complied with exactly in nature, yet they are often very nearly so; so much so that, from a study of these two types and of their combinations, we can draw conclusions which are practically applicable to nearly all the motions of the sea.

The possibility of "harnessing the waves" has long been an attractive one to inventors as the Patent Office records testify, and in a paper on this subject by A. W. Stahl, United States Navy, is given a discussion of the trochoidal theory of wave motion, which is the theory now generally accepted as a sufficiently close approximation to the actual motion of water waves. According to this theory each particle of water in a trochoidal wave moves in an elliptical orbit, whose major axis is horizontal, and the plane of which is vertical and perpendicular to the wave ridge or crest. The motion of the particle in the upper portion of its orbit is in the direction of advance or propagation of the wave; in the lower part of its orbit motion is in the opposite direction. The eccentricity of the ellipses depends on the ratio between the length of the wave and the depth of the water. When the depth is one-half the length of the wave or more, the ellipses cannot practically be distinguished from circles.

Of the motions of the wave particles which may be utilized for power purposes, we find the following:

1. Vertical rise and fall of particles at and near the surface.

2. Horizontal to-and-fro motion of particles at and near the surface.

3. Varying slope of surface of wave.
4. Impetus of waves rolling up the beach in the form of breakers.
5. Motion of distorted verticals.

Each of these wave-motions or their combinations have been assumed or used as a means of obtaining power, with more or less efficiency according to the perfection of design of the motive apparatus. The impact of sea waves upon the shore is so diversified by its slope and formation that no exact data can be made other than from practical tests at any location.

Calculations show that the transmitted energy of shallow-water waves decreases rapidly as the depth of the water increases. Stevenson found that the force exerted on the dynamometer by waves estimated to be 6 feet in height was 3,041 pounds per square foot, and that a ground swell 10 feet in height gave the same pressure. Waves 20 feet in height gave a pressure of 4,562 pounds per square foot, and similar waves during a strong gale gave 6,083 pounds. He also found that waves exert by far the greatest force at the level of high water, that the force of recoil is greater than the direct impact, and, with regard to a wall with a steep curved profile, that the vertical force was about 84 times greater than the horizontal. A true wave of oscillation, on reaching a wall having an inclination with the horizontal of from 45° to 90°, will not be broken, and will exert practically no dynamic pressure against it. On the other hand, a wall of this profile receives the greatest shock from translatory waves.

By means of dynamometers Lieutenant-Colonel H. M. Roberts, of the United States Corps of Engineers, obtained readings indicating that waves rising from 14 feet to 18 feet above still-water level, and having a velocity of from 30 miles to 40 miles per hour, exerted a pressure at still-water level of 400 pounds to 600 pounds per square foot; at 16 feet and at 8 feet below still-water level, less than 10 pounds per square foot; and at 8 feet above still-water level, 940 pounds per square foot.

The amount of force which is exerted by breaking waves is of importance in the matter of coast erosion. Captain Gaillard, of the United States Corps of Engineers, used specially constructed dynamometers for taking observations of this force in its direct application, and he found that breakers 2 feet in height and 46 feet in length gave a maximum pressure of 148 pounds per square foot; that those 3 feet in height and 75 feet in length gave 322 pounds per square foot; that those 4 feet in height and 82 feet in length gave 406 pounds per square foot; that those 5 feet in height and 120 feet in length gave 467 pounds per square foot; and that those 6 feet in height and 150 feet in length gave 667 pounds per square foot. All these were shore breakers, travelling nearly normal to the shore-line, over a hard and regular bottom. There are many instances on record of large blocks of masonry and concrete having been dislodged from breakwaters by the waves, and the computed pressures which must have been exerted on these occasions go to confirm the results obtained by dynamometers.

The total energy of one wave-length of a height H, from trough to crest, and of a length L, from crest to crest, and for one foot in breadth, is:

$\text{Energy} = 8\ LH^2\left(1 - 4.935\frac{H^2}{L^2}\right)$ in foot-pounds, and the time required for a crest to pass over a wave-length, is:

$T = \sqrt{\frac{L}{5.123}}$ in seconds, and the number of waves per minute $N = 60\sqrt{\frac{5.123}{L}}$.

The horse-power of continuous wave-motion as above $= \frac{E \times N}{33{,}000} = 0.0329\ H^2L\left(1 - 4.935\frac{H^2}{L^2}\right)$ from which the following table has been computed:

TABLE XXXVI.—TOTAL ENERGY OF DEEP-SEA WAVES IN TERMS OF HORSE-POWER PER FOOT OF BREADTH.

Ratio of length of waves to height of waves.	Length of waves in feet.							
	25	50	75	100	150	200	300	400
50	0.04	0.23	0.64	1.31	3.62	7.43	20.46	42.01
45	0.05	0.29	0.79	1.62	4.47	9.18	25.30	51.94
40	0.06	0.36	1.00	2.05	5.65	11.59	31.95	65.58
35	0.08	0.47	1.30	2.68	7.37	15.14	41.72	85.63
30	0.12	0.64	1.77	3.64	10.02	20.57	56.70	116.38
25	0.16	0.90	2.49	5.23	14.40	29.56	80.85	167.22
20	0.25	1.44	3.96	8.13	21.79	45.98	126.70	260.08
15	0.42	2.83	6.97	14.31	39.43	80.94	223.06	457.89
10	0.98	5.53	15.24	31.29	86.22	177.00	487.75	1,001.25
5	3.30	18.68	51.48	105.68	291.20	597.78	1,647.31	3,381.60

The table just shown gives the total energy, including both kinetic and potential, inherent in a regular series of waves. The figures are strictly correct for the trochoidal deep-sea wave with circular orbits only, though they give a close approximation for any nearly regular series of waves in deep water, and a fair approximation for waves in shallow water.

The practical utilization of wave-power depends upon the efficiency of the special design and friction of transmission for its effective work. Concussive pressure of from 600 with moderate waves to 6,000 pounds in great waves per square foot of surface have been observed.

Dr. Scoresby gives the following interesting facts with regard to the length and height of ocean waves. The mean height of waves in the Atlantic, driven by a westerly gale, is 18 feet. The greatest recorded height of a wave in the North Atlantic, from the trough to the crest, is 43 feet. In northwest gales, waves 40 feet in height have been measured off the Cape of Good Hope, while those off Cape Horn were 32 feet; Mediterranean, $14\frac{1}{2}$ feet; German Ocean, $13\frac{1}{2}$ feet; Bay

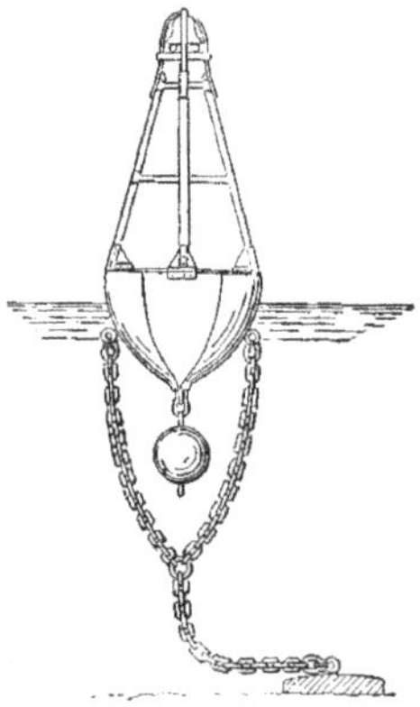

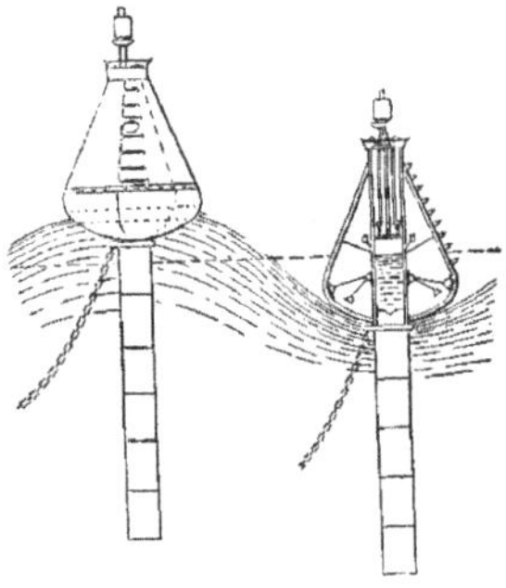

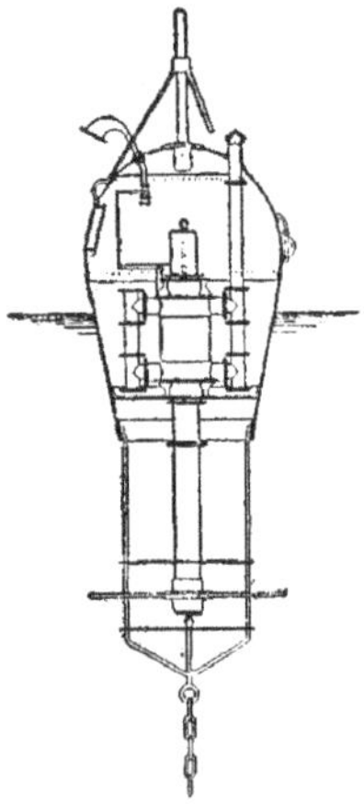

FIG. 300.—Bell-buoy. FIG. 301.—Whistling-buoy. FIG. 302.—Fog-horn buoy.

of Biscay, 36 feet. The velocity of ocean storm-waves in the North Atlantic is about 32 miles an hour, and that recorded by Captain Wilkes for the Pacific Ocean is $26\frac{1}{3}$ miles. In an Atlantic storm the breadth of the waves, measured from crest to crest, is about 600 feet.

One of the most useful effects of wave-power has been found in its adoption for operating signal buoys off the coast.

In Figs. 300, 301, and 302 are illustrated the bell-buoy, whistling-buoy and the fog-horn buoy.

The bell-buoy has a large bell mounted in a frame on a floating buoy. A radial grooved iron plate is made fast to the frame under the bell and close to it, on which is laid a free cannon-ball. As the buoy rolls on the sea, this ball rolls on the plate, striking some side of the bell.

In this design a very small roll of the sea makes a constant ringing of the bell.

In the whistling-buoy a hanging tube below the float is open at the bottom. In the vertical motion of the float and tube by the waves,

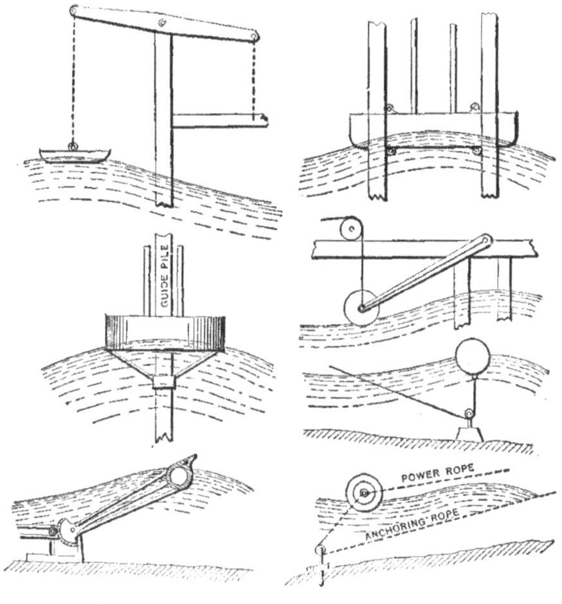

FIG. 303.—Vertical motion wave-motors.

the water in the tube reacts as a piston, drawing in air at the top of the buoy and compressing it to blow the whistle.

The fog-horn buoy consists of a float anchored at the edge of banks with an air-pump operated by the waves. The action of the sea is utilized in such a manner as to blow desired blasts through a fog-horn by means of the compression and release of air into and from a suitable air-tight chamber forming a portion of the buoy, this chamber being charged by means of a pump actuated by the movement of the sea.

Following we illustrate the many designs that have been proposed or used for obtaining power from the action of the waves. In Fig. 303 is illustrated seven methods of applying devices for obtaining power from the vertical motion of waves which may be used for pumping or other work.

In Fig. 304 the horizontal motion of waves acting against swinging blades pivoted above the crest; a float with rope attachments and

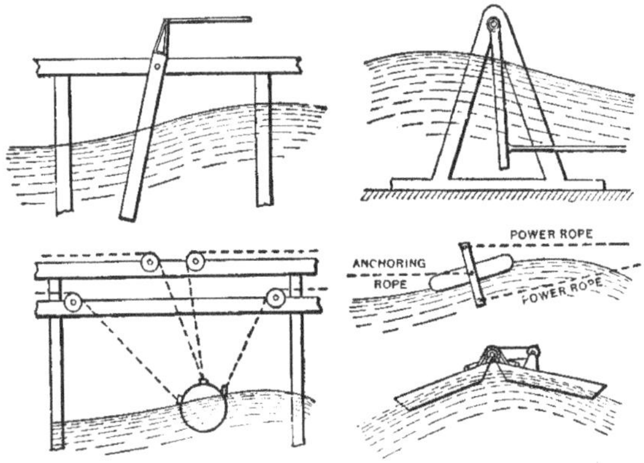

FIG. 304.—Horizontal and rolling wave-motors.

both vertical and horizontal motion; an anchored float with arms rocking by the rolling of the float over the wave-crest; and the ratchet drive of two boats hinged together as they roll over a crest and through the trough of the seas.

A more complicated movement is shown in Fig. 305 in which the

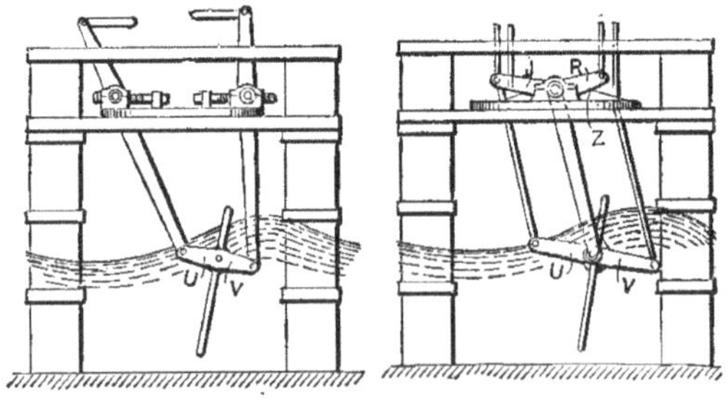

FIG. 305.—Differential motion wave-motors.

differential movement of the surface and bottom sections of a wave rock the propelling blade together with its fore-and-aft motion, representing a single double-acting transmission.

The storage of water in an elevated reservoir offers many advantages, and seems on the whole to be the best adapted for use in connection with a wave-motor. When this plan is adopted it becomes unnecessary to regulate the power at the motor, as the latter may simply drive a system of pumps for delivering the water into the reservoir, the pumps always working at a practically constant pressure, and the number of such pumps in action at any particular time being regulated by the magnitude of the waves, and the consequent amount of energy in the same.

On the shores of the Great Lakes, the wave-motor would have its best work in a water-supply for towns and for fire service from storage reservoirs. On the sea-coasts, the fire service and the flushing of sewers and streets would be the most economical service from wave-motors.

INDEX

BIOGAS
POWER FOR TOMORROW
D. S. HALACY JR.
HYDROGEN
Movable Insulation
William K. Langdon
SUNSHINE
TO DOLLARS

Made in the USA
Monee, IL
04 November 2026

67384893R00193